Making Things Move

DIY Mechanisms for Inventors, Hobbyists, and Artists

Dustyn Roberts

New York Chicago San Francisco
Lisbon London Madrid Mexico City
Milan New Delhi San Juan
Seoul Singapore Sydney Toronto

The McGraw-Hill Companies

Cataloging-in-Publication Data is on file with the Library of Congress

McGraw-Hill books are available at special quantity discounts to use as premiums and sales promotions, or for use in corporate training programs. To contact a representative, please e-mail us at bulksalesmcgraw-hill.com.

Making Things Move: DIY Mechanisms for Inventors, Hobbyists, and Artists

8 9 0 DOC DOC 1 0 9 8 7 6 5 4

ISBN 978-0-07-174167-5
MHID 0-07-174167-4

Sponsoring Editor
Roger Stewart

Editorial Supervisor
Patty Mon

Project Editor
Howie Severson, Fortuitous Publishing Services

Acquisitions Coordinator
Joya Anthony

Copy Editor
Marilyn Smith

Proofreader
Paul Tyler

Indexer
Jack Lewis

Production Supervisor
James Kussow

Composition
Glyph International

Illustrator
Sean Comeaux

Illustration
Glyph International

Art Director, Cover
Jeff Weeks

Cover Designer
Jeff Weeks

Contents

About the Author

Dustyn Roberts is a traditionally trained engineer with nontraditional ideas about how engineering can be taught. She started her career at Honeybee Robotics as an engineer on the Sample Manipulation System project for NASA's Mars Science Laboratory mission, scheduled for launch in 2011. While at Honeybee, she also designed a robotic drill; led field operations of a robotic truck in an Australian mine; supported proposal efforts for DARPA, NIH, NASA, and DOD; and led a project with Goddard Space Flight Center to create a portable sample manipulation system for lunar operations. After consulting with two artists during their residency at Eyebeam Art + Technology Center in New York City, she founded Dustyn Robots (www.dustynrobots.com) and continues to engage in consulting work, ranging from gait analysis to designing guided parachute systems. In 2007, she developed a course for New York University's (NYU's) Interactive Telecommunications Program (ITP) called Mechanisms and Things That Move, which led to the book you are now holding in your hands.

Dustyn holds a BS in Mechanical and Biomedical Engineering from Carnegie Mellon University, with minors in Robotics and Business, and an MS in Biomechanics and Movement Science from the University of Delaware, and is currently pursuing a PhD in Mechanical Engineering at NYU-Poly. She has attracted media attention by Time Out New York, PSFK, IEEE Spectrum, and other local organizations. She currently lives in New York City with her partner, Lorena, and cat, Simba.

Acknowledgments

First, I'd like to thank all my family and friends for putting up with far too many "I can't—I have to write" excuses. To my dad, for being an engineer and encouraging my whims, even when they didn't make good business sense. To my mom, for her confidence in my abilities, even when she had no idea what I was talking about.

Thank you to NYU's Interactive Telecommunications Program (ITP), specifically Red Burns and Tom Igoe, for hiring an engineer to teach artists. Tom quickly became more than just the area head for my class. He offered support and encouragement from day one and has become a mentor. When I started teaching, I was an engineer, but now I'm a maker, too. Thank you for challenging me to make my field accessible and to empower others through making. I have no doubt learned more than I have taught. And thank you ITP for attracting students who are a pleasure to teach. Every student I've had the opportunity to interact with has shaped this book.

Thanks to Eyebeam Art + Technology Center for supporting this work through their artist in residency program and for attracting great interns. This book would have taken much longer and been less fun to work on without my team of interns, who worked for little more than free lunch and the promise of certain fame and fortune. To Sean Comeaux for all the illustrations and for making me find new ways to explain things. To Sam Galison and Stina Marie Hasse Jorgensen for their enthusiasm and their amazing work on the projects, photography, and video editing for the website.

I'm sure neither of them will forget Chapter 6 or the Not Lazy Susan any time soon. Thanks to the other residents, fellows, and staff for making it an inspiring place to work.

To everyone who helped edit remotely or made it to my Book & Bribe parties (and Tom for seeding the idea), where I stealthily convinced friends and colleagues to read through early drafts by serving food and drinks: Matt Bninski, Lee Carlson, Joanna Cohen, Stephen Delaporte, Russ de la Torre, Heather Dewey-Hagborg, Rob Faludi, Eric Forman, Michelle Kempner, Jenn King, Adam Lassy, Ben Leduc-Mills, Adi Marom, Gale Paulsen, Jennifer Pazdon, Lauren Schmidt, Greg Shakar, Ted Southern, Becky Stern, Mike Sudano, Corrie Van Sice, Dana Vinson, Irene Yachbes, and any others I may have forgotten.

To the team at McGraw-Hill for being patient and answering every last question of mine. Thanks to my book agent, Neil Salkind, who has nurtured this first-time author with enthusiasm from our very first email contact.

To Kickstarter.com and all our backers for helping Ben Leduc-Mills and me fund the SADbot project. And thanks to Ben for having the idea and roping me in—you'll make a great computer scientist.

And finally, to my partner Lorena, for her unconditional love and support. I will never be a good enough writer to describe in words how much you mean to me.

Introduction

What This Book Is

In a conversation I had with Bre Pettis, one of the creators of the CupCake CNC at MakerBot Industries (www.makerbot.com), I asked if any of the creators were mechanical engineers by training. He replied "No, if we were, it would have been impossible." The CupCake CNC is a miniature 3D printer that uses computer models to create real 3D objects about the size of a cupcake out of melted plastic. The MakerBot team members were able to build it from available materials with the tools they had on hand. A trained engineer would have known how difficult this project would be and might not have attempted it without the proper resources or funding, but the MakerBot team members didn't have the experience to know what they were getting themselves into. They just kept their goal in mind and figured out a way. This book is written for anyone who wants to build things that move but has little or no formal engineering training. In fact, as Bre said, not having engineering training may help you.

In this book, you will learn how to successfully build moving mechanisms through nontechnical explanations, examples, and do-it-yourself (DIY) projects. Maybe you're a sculptor who wants a piece of art to come alive, a computer scientist who wants to explore mechanics, or a product designer who wants to add function to complement the form of your product. Maybe you've built projects in the past, but they fell apart easily.

Or maybe you didn't grow up making things move but want to learn. The students in the class I teach at New York University's (NYU's) Interactive Telecommunications Program (ITP) in the Tisch School of the Arts have been all of these things, and they gave me the inspiration to write this book.

The class is called Mechanisms and Things That Move, and was created to fill a gap in the program between what students were already learning how to do (basic electronics, interaction design, and networked objects) and what they wanted to make (baby strollers that autonomously climb stairs, wooden mechanical toys, and stationary bikes that power televisions). The objective is to start with their seemingly impossible project concepts, inject some basic engineering know-how, and end up surprisingly close to the original concept. You can see these projects and more on the class site at http://itp.nyu.edu/mechanisms. I realized in the first year of teaching this class that the practical experience I had gained from engineering design work could be applicable to a completely different audience of nonengineers. I was told by one student, "Your class gave me a whole new world" and by another, "It's unbelievably satisfying to design and build something that works." This book is designed to bring this level of satisfaction to all the people who want to learn about mechanisms but don't know where to start.

There is little purpose to building circuits for an electromechanical project if the mechanism to be controlled is too weak to handle the task. You can protect projects from costly overdesign with a basic knowledge of mechanics and materials. To address these ideas, I'll cover a breadth of topics, ranging from how to attach couplers and shafts to motors to converting between rotary and linear motion. You'll be guided through each chapter with photographs, drawings, schematics, and images of 3D models of the components and systems involved in each project. All the illustrations were drawn by an actual illustrator (and nonengineer) in order to minimize the intimidation factor of difficult-sounding concepts and graphs. The resulting interpretation of the concepts is in a playful style designed to be eye catching and friendly.

I emphasize using off-the-shelf components whenever possible, and most projects will also use readily available metals, plastics, wood, and cardboard, as well as accessible fabrication techniques. Simple projects are placed throughout the book to engage you in applying the material in the chapter at hand. At the end of the book, you'll find more complex projects that incorporate material from multiple chapters.

I guarantee that you will gain a general understanding of mechanisms and save time, money, and frustration by avoiding mechanical design mistakes that lead to failure. Anyone can become a mechanism maker—even if you've never set foot in a machine shop.

What This Book Isn't

This book is not an engineering textbook. It assumes no prerequisite knowledge of electronics or robotics, and you do not need to know what a microcontroller is or how to program one to get the most out of this book. I don't assume you've grown up with a metal shop in your garage, know what a lathe is, or can estimate motor torque by looking at a rotating shaft.

Each chapter could be expanded to a book of its own, and there are many other places to look for detailed technical explanations. This book is about getting things made, and it includes the necessary information for you to do just that. The small amount of theory and background presented will help you understand how mechanisms work, so you can concoct and manipulate your own creations. If these sections get too heavy for you, or you already know the background, skip right to the hands-on stuff.

How to Use This Book

As the White Rabbit was told by the King in *Alice's Adventures in Wonderland,* "Begin at the beginning, and go on till you come to the end: then stop." If you really have no background in making things, this is probably the best way to approach the book. You would only get frustrated when you read about estimating torque in Chapter 6 if you had not read Chapter 4's discussion of torque and don't know what it is. Do the small projects to start getting your hands dirty and used to making things. The chapters are organized in a way that builds up knowledge of all the parts that go into building things that move, so when you get to the end of the book, you will have all the tools in your tool belt and be ready to conquer the final projects in Chapter 10.

Each project in the book has two sections: shopping list and recipe. I've heard that baking is more of a science, and cooking is more of an art. Making things move is a bit like baking in the beginning. You want to make sure you measure every ingredient just right, follow every step, and do everything by the book. But once you get used to making things move, it becomes more like cooking. After you get the basic recipes down, you can start adding your own ingredients and experimenting.

You can also use this book as a reference manual, especially if you have theoretical knowledge of how things work but want a practical guide to making things move. This latter scenario is where I was after my undergraduate education in engineering. I could figure out the torque or force I needed to make things move, but couldn't tell you how to choose a motor or attach something to its shaft. They don't teach that kind of stuff in school (at least not where I went), so you need to learn it through experience. I hope this book will help you start higher on the practical learning curve than I did.

Your Ideas Are Your Biggest Assets

Although very little prior knowledge of mechanisms is assumed in this book, anything you *do* know will help you, and I do mean anything.

The most important thing you bring to the table is an idea. Some of the most amazing projects I've seen have come from people with no prior experience in hands-on projects, and certainly no engineering degree. If you're a passionate musician who has an idea for a guitar that plays itself, you are more likely to end up with a great project than if you're an engineer who thinks you know how a guitar works but have never picked one up. This book will give you the tools to make your passions into projects and your concepts into realities. The tools are here, along with examples of how to use them, but the ideas on how to apply them come from you.

I don't claim to be an artist. My right brain is not nearly as developed as most of the students and designers I've had the pleasure of working with. However, I do claim to know how to talk nonengineers through the process of creating things that move. You could use this book as a light read to kill time on a Saturday night, but what I'm banking on is that this book will give you the tools and techniques you need to take that concept for a human-powered smoothie blender out of your head and into reality.

The book includes plenty of projects that you can build, but the applications of the concepts and skills are limited only by your imagination. Mechanisms can seem a little scary at first, but once you break down a complicated project into its elements, you'll learn that it's not so daunting after all. This book will enable you. And the more you learn, the more inspiration you will have for future projects.

What You Need to Know

Although prior engineering and fabrication expertise is not required, you do need to know a few things to get the most out of this book. One of the most important is knowledge of how to use the Internet. There are at least three reasons for this.

- When it comes to mechanisms and all things related, we are standing on the shoulders of giants. From the Instructables website (www.instructables.com) to Leonardo Da Vinci's first mechanical sketches, a lot of inspiration can be found online to help form ideas for projects and learn from similar ones. The goal of this book is to get projects done, not to learn everything there is to know about a topic before getting started. Are you trying to convert rotary motion to linear motion? Guess what—you're not the first person to do that. Take advantage of the basics explained in this book, and the dozens of websites devoted to examples of converting rotary to linear motion, to inspire the mechanism you need to realize your idea. Borrow the idea, and then customize it to make it your own, and always give credit where credit is due. As Aiden Lawrence Onn and Gary Alexander say in their book *Cabaret Mechanical Movement: Understanding Movement and Making Automata* (London: G&B Litho Limited, 1998), "If you want to make things move, be sure to spend some time studying how other things move."
- Making things requires parts and tools. You will most likely need to order some of these things online. Although you can do a lot with cardboard boxes and straws, you may not have a local big box store that sells DC gearhead motors for your Not Lazy Susan (Project 10-1 in Chapter 10). Luckily, you can order parts and tools online, no matter where you are. You can also get better deals on most things—from hand drills to alligator clips—than you can at your local hardware store. Resources are listed for each project, but a few I will refer to often are McMaster-Carr (www.mcmaster.com), SparkFun (www.sparkfun.com), and All Electronics (www.allelectronics.com).
- This book has a companion website: www.makingthingsmove.com. Color photographs and videos that cannot be included in the book will be posted here. You will also find a blog and other resources. By purchasing this book, you have become part of a maker subculture that is bigger than you may know. The makingthingsmove.com website will help you connect with those who share similar interests. Links to digital files to download, make, and buy

will be posted there, or you can search for "dustynrobots" on Thingiverse (www.thingiverse.com), Ponoko (www.ponko.com), or Shapeways (www.shapeways.com) for a full listing of everything I have posted.

Along with knowing how to use the Internet, I also expect you to have a working vocabulary of geometry, trigonometry, and basic algebra skills. If you can solve for the ? in the equation 2 × ? = 6, and know what sine, cosine, and tangent mean, relax—that's about as complicated as we'll get in this book. You need to know what words like diameter, circumference, tangent, and perpendicular mean. If any of your knowledge in this area is a little rusty, do a quick search online to review.

What You Need to Have

Each project in the book has a shopping list of parts and tools, so you can pick and choose what you need. However, if you want to get a head start, here are some common tools that will serve you well (see Figure 1):

FIGURE 1 Basic tools and supplies to get you started

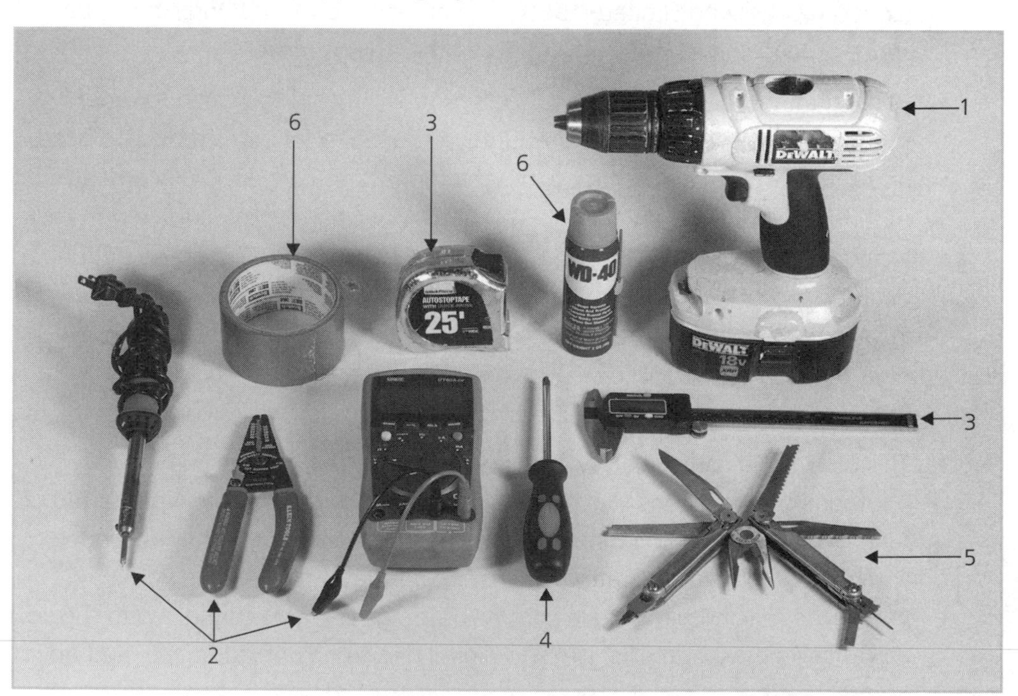

1. **Hand drill** You will use this for drilling holes in wood and thin metal for screws and dowels during project construction. I prefer the cordless, rechargeable kind like the Dewalt model pictured, but any drill will do. Make sure it can hold small drill bits (down to 1/16 in diameter). A Dremel rotary tool will also do the trick for most small jobs, and can be used for cutting and sanding small parts as well.

2. **Multimeter** You will use this any time you're working with electricity to check if your battery is dead and if your circuit is hooked up correctly. Make sure the multimeter you get measures voltage, resistance, amperage, and continuity. Do yourself a favor and get a model that is autoranging. This means that you don't need to estimate the thing you're measuring before you measure it to choose the correct setting. Autoranging will be a little more expensive, but it will save you time and frustration if you're not well versed in electronics. Auto-off is a nice battery-saving feature. The one pictured in Figure 1 is SparkFun's TOL-08657. It's autoranging and can measure higher current than cheaper models, so it will come in handy when working with motors. A basic soldering iron (RadioShack 64-280 pictured) and wire stripper (SparkFun TOL-08696) will help when you start working with circuits.

3. **Measuring tools** A tape measure for large things, a metal ruler for small things and to use as a cutting edge, and a caliper for even smaller things. I recommend a digital caliper for ease of use (SparkFun TOL-00067).

4. **Screwdrivers** Phillips and flat head styles. Having a few different sizes on hand is a good idea. Jameco Electronics (www.jameco.com) sells a handy two-sided miniature tool for about $2 (part number 127271). A larger, multipurpose option is the Craftsman 4-in-1 (model 41161). Cheaper ones will be made of soft metal, and the tips will get bent out of shape easily, so go for a step above the bargain-basement models.

5. **Multitool** More commonly referred to as a Swiss Army Knife or Leatherman, multitool is the general name. It is handy to have around and may save you from buying a lot of separate tools to do little jobs. Multitools come in all different shapes, sizes, and prices, but I recommend getting one that has at

least screwdriver tips, scissors, a file, a knife, and a saw blade. I've had a Leatherman Blast for years, and at a cost of around $45, it has earned its spot in my toolbox many times over. Check here for Leatherman brand models: www.leatherman.com/multi-tools. The Maker Shed (www.makershed.com) sells a few laser-etched models, aptly named "warranty voider" and "bomb defuser." For Swiss Army brand tools, check www.swissarmy.com/multitools under the Do-It-Yourself category. For particularly frustrating projects, look for a multitool with a corkscrew and/or bottle opener.

6. **Duct tape and WD-40** "If it moves and it shouldn't, use duct tape; if doesn't move and it should, use WD-40." I'm not sure where I first heard this, but it may have been on a page-a-day calendar my boss had on his desk at my first engineering internship called "365 Days of Duct Tape." Most readers will be familiar with the standard wide silver duct tape you can use for just about anything. WD-40 is also handy to use on everything from squeaky hinges to lubricating gears and other moving parts.

The most important thing to have is not a tool. It is a commitment to safety. Don't drill a hole to mount your motor without wearing safety glasses, and don't drill into a piece of wood right on top of your kitchen table. You are likely to end up with sawdust in your eye and a hole to cover up with a strategically placed placemat. Use gloves when handling sharp things or rough edges that might cut you or cause splinters. I will point out safety concerns in each specific project, but get in the habit of thinking through an action before you do it to identify safety hazards and eliminate them. Although cuts and scrapes heal, it's very hard to grow back your sense of hearing after too many hours listening to a loud drill, or to regain your sense of sight after the Dremel cutting wheel flies off in an inconvenient direction. At the very least, have a pair of safety glasses and earplugs around and use them. Safety precautions should always be the first step of any project you do.

Introduction to Mechanisms and Machines

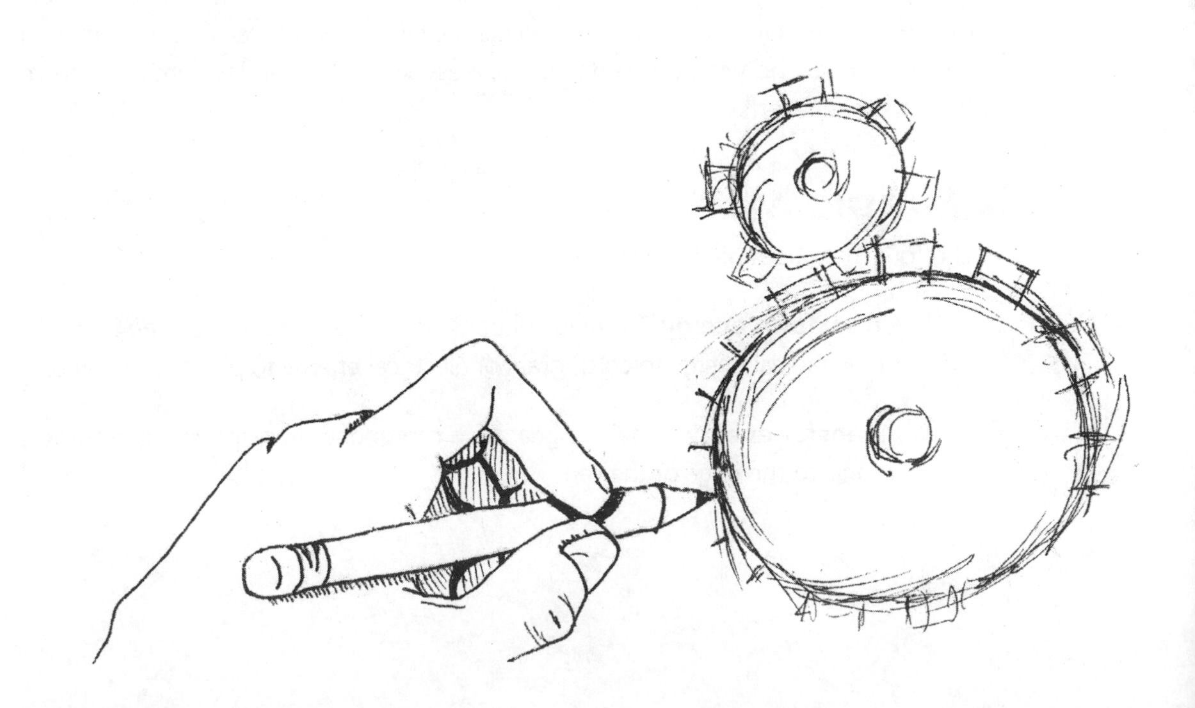

Mechanical systems come in many shapes and forms, and they have various definitions. Before we can start making machines, we need to know what we're talking about:

- A *mechanism* is an assembly of moving parts.
- A *machine* is any device that helps you do work, from a hammer to a bicycle. A hammer is a machine because it makes your arm longer, so you can do more *work*.

In this book, we use the mechanical definition of work:

$$Work = Force \times Distance$$

Force (F) equals *mass* (m) times *acceleration* (a), and is written as $F = ma$ (also known as Newton's second law).

For example, imagine that you're stomping on a bunch of grapes to make wine. The force the grapes feel when you stand still is equal to your weight, but the force the grapes feel when you stomp is your weight plus the acceleration your muscles give to your foot. The grapes would feel less force, however, if you were stomping them on the moon, which has just one-sixth of the Earth's gravity. *Mass* refers to the amount of stuff you're made of, which doesn't change. Gravity and acceleration depend on where you are and what you're doing. So, mass is the stuff, and weight is the force that the mass exerts.

Six Simple Machines

The four main uses of machines are to:

1. **Transform energy** A windmill transforms energy from the wind into mechanical energy to crush grain or electrical energy to power our homes.
2. **Transfer energy** The two gears in a can opener transfer energy from your hand to the edge of the can.

3. **Multiply and/or change direction of force** A system of pulleys can lift a heavy box up while you pull down with less effort than it would take to lift the box without help.

4. **Multiply speed** The gears on a bicycle allow the rider to trade extra force for increased speed, or sit back and pedal easily, at the expense of going slower.

It turns out that all complicated machines are made of combinations of just six classic simple machines: the lever, pulley, wheel and axle, inclined plane, screw, and gear. These machines are easy to spot all around us once you know what to look for.

1. Levers

You can consider a *lever* a single-mechanism machine. It's a mechanism, by our definition, because it has moving parts. It's a machine because it helps you do work.

A lever is a rigid object used with a pivot point or fulcrum to multiply the mechanical force on an object. There are actually three different classes of levers. Each kind of lever has three components arranged in different ways:

1. Fulcrum (pivot point)
2. Input (effort or force)
3. Output (load or resistance)

First Class Levers

In a first class, or simple, lever, the fulcrum is between the input and output. This is the classic seesaw most people think of when they hear the word *lever*, as shown in Figure 1-1.

Things can balance on a seesaw in three ways:

1. The two things can weigh exactly the same amount, and be spaced exactly the same distance from the fulcrum (the way it looks in Figure 1-1).
2. You can push down on one side with the same amount of force as the weight on the other side. Your parents may have done this with you on seesaws when you were a kid.

FIGURE 1-1 The classic playground seesaw is an example of a first class lever.

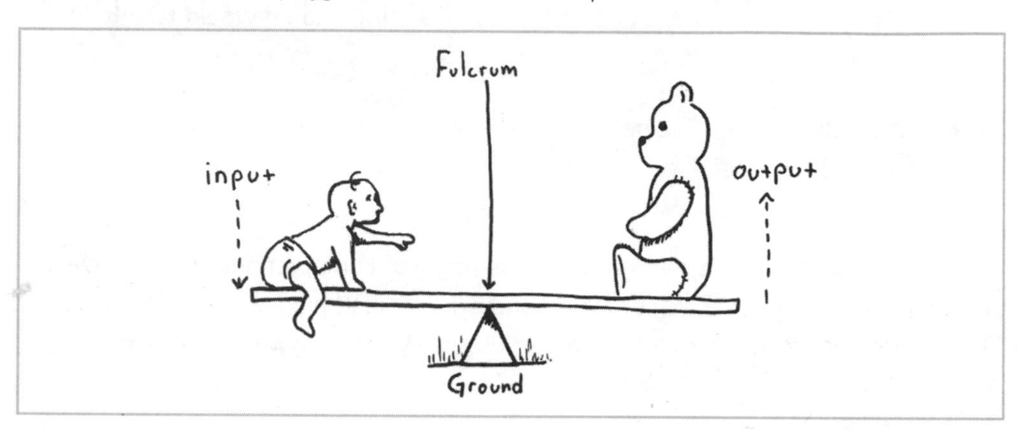

3. The two things can have different weights, and the lighter one must be farther from the fulcrum in order to balance. If you've ever been on a seesaw with someone heavier than you, you've probably done this without thinking about it. If you were the lighter one, you backed up as far as you could to the edge of the seesaw, and your heavier friend probably scooted in toward the pivot point.

In order to apply these balance rules to machines, let's replace the word *thing* with *force*. But first, meet Fido and Fluffy.

Fido is a big dog. Fluffy is a small cat. Because their names both start with *F*, I'll use F_1 for Fido and F_2 for Fluffy when I abbreviate them. Fido is heavier, so his arrow (F_1) on the left side of Figure 1-2 is bigger. He is sitting at a certain distance (d_1) from the fulcrum. Similarly, Fluffy (F_2) is at a distance d_2 from the fulcrum on the right side. In order to balance the seesaw, F_1 times d_1 must equal F_2 times d_2:

$$F_1 \times d_1 = F_2 \times d_2$$

You can see from Figure 1-2 and the equation that if $F_1 = F_2$, and $d_1 = d_2$, then the seesaw will look like Figure 1-1 and balance. But if Fido (F_1) is a 50 pound (lb) dog, and Fluffy (F_2) is a 10 lb cat, then they must adjust their distances to the fulcrum in order to balance. Let's say that Fido is 3 feet (ft) away from the fulcrum (d_1 = 3 ft). How far away from the fulcrum does Fluffy need to be to balance? Now our equation looks like this:

$$50 \text{ lbs} \times 3 \text{ ft} = 10 \text{ lbs} \times d_2$$

FIGURE 1-2 Balanced first class lever with different forces

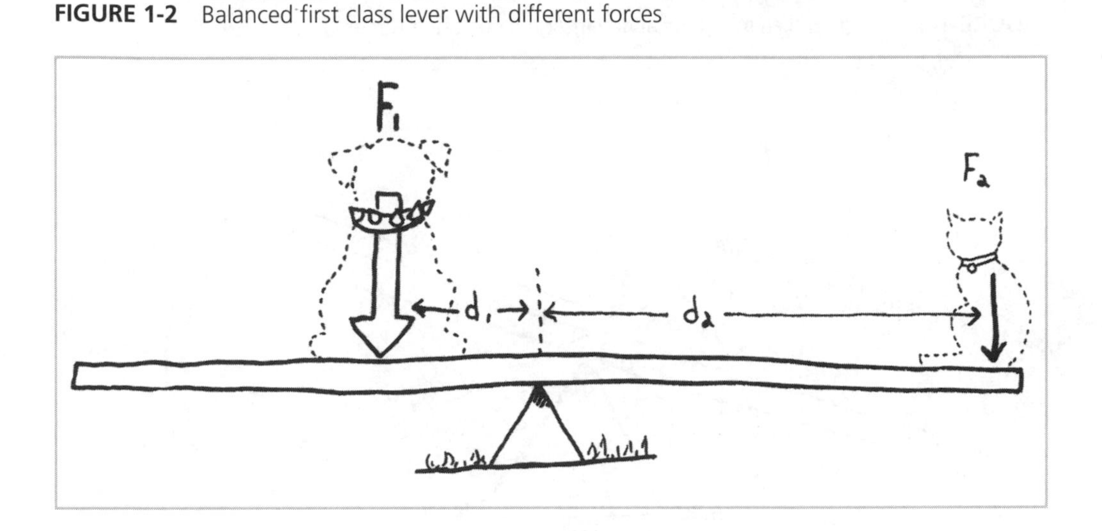

In order to balance the equation (and the seesaw), d_2 must be 15 ft. Although Fido and Fluffy helped us illustrate this point, the forces F_1 and F_2 can be anything—boxes, birds, buildings . . . you name it.

So, the lighter cat can balance a dog five times her weight if she just scoots back farther. You'll also notice that if Fido and Fluffy start seesawing, or pivoting on the fulcrum, Fluffy will go up higher because she is farther from the pivot point. I'll call the distance from the ground to Fluffy's highest point the *travel* (see Figure 1-3).

So the lightweight cat can lift the heavy dog, but she must travel farther to do it. This is how levers give us *mechanical advantage*: A smaller force traveling through a longer distance can balance a heavier force traveling a shorter distance. We could also say the lighter cat is using a 5:1 mechanical advantage to lift the heavy dog by being five times farther from the fulcrum. In our example, the travel of the light cat Fluffy (F_2) is five times that of the heavy dog Fido (F_1).

There are many places you can see levers at work every day. A hammer claw acts as a first class lever when pulling a nail out of a board (see Figure 1-4). You pull at the far end of the hammer handle with a light force, so a big force pulls the nail out with the hammer claw that is just a short distance from the hammer head. The hammer head creates a pivot point that acts as the fulcrum.

FIGURE 1-3 Levers utilize mechanical advantage to balance forces.

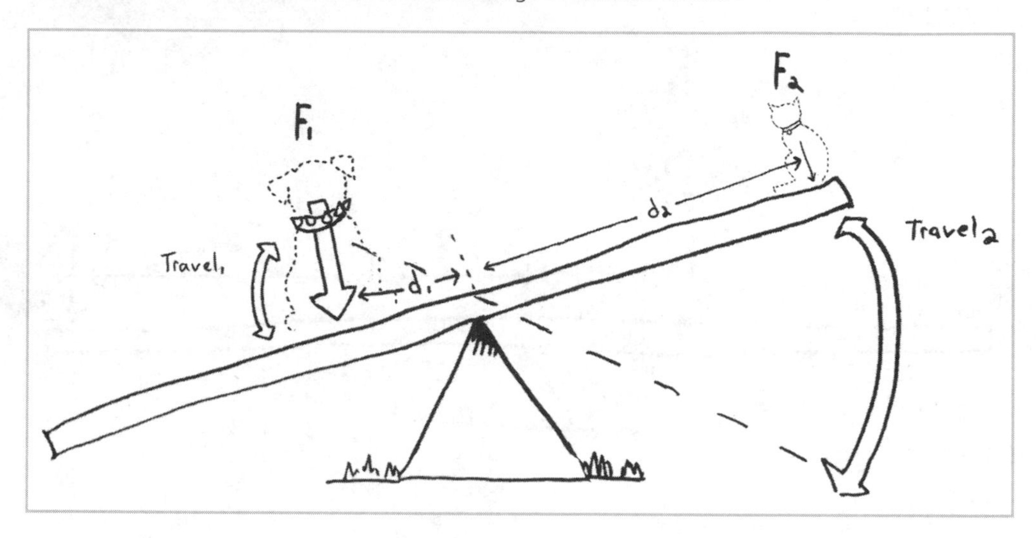

Here are some other examples of levers:

- A crowbar is a first class lever in the same way as a hammer claw.
- Oars on a boat work as first class levers.
- If you've ever used a screwdriver to pry the lid off a paint can, you were using the screwdriver as a first class lever.
- A pair of scissors is like two first class levers back to back. Scissors designed to cut paper don't have much of a built-in mechanical advantage, but think of the long handles of garden shears or bolt cutters. The long handles make the cutting force much higher—that's mechanical advantage at work!

FIGURE 1-4 A hammer being used as a first class lever

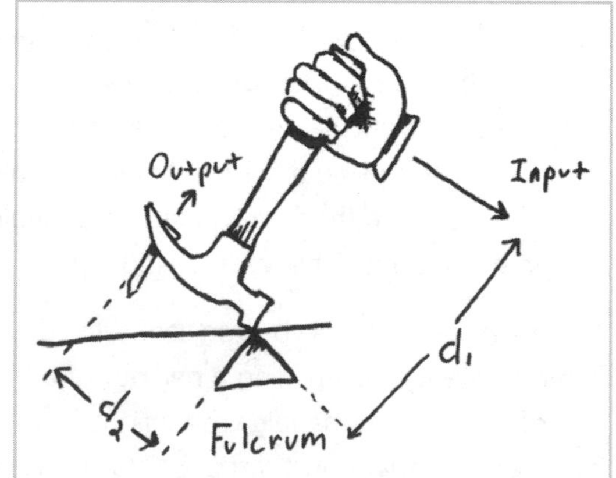

Can you think of some other first class levers?

Second Class Levers

In a second class lever, the output is located between the input and the fulcrum. The classic example of this is the wheelbarrow. As you can see in Figure 1-5, the stuff in the wheelbarrow is the output or load, and we use the handles as the input.

FIGURE 1-5 The wheelbarrow as a second class lever

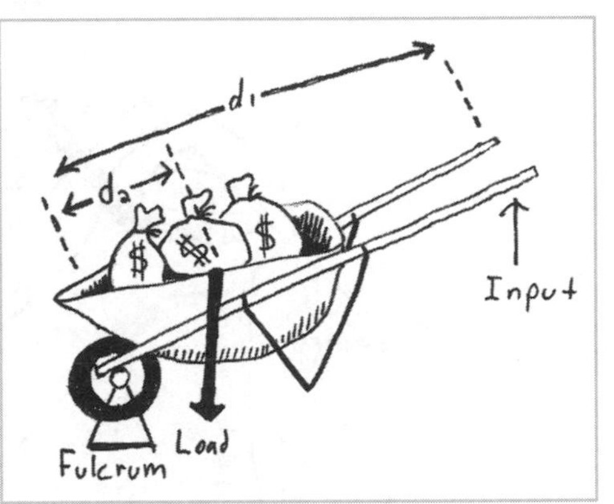

We can use the same equation as for first class levers to figure out the balance of forces. Let's say we have a 50 lb load (F_2) of bags of gold in the wheelbarrow, and the distance from where the bags of gold are to the wheel is 1 ft (d_2). If the handles are 5 ft long from the grip to the wheel (d_1), how hard do we need to pull up to lift the bags of gold? Let's put what we know into our equation:

$$F_1 \times 5 \text{ ft} = 50 \text{ lbs} \times 1 \text{ ft}$$

So in order to lift the bags of gold, we must pull up on the handles with at least 10 lbs of force (F_1). See that? We can move 50 lbs of bags of gold with only 10 lbs of pull force, for another 5:1 mechanical advantage—the same as we saw with Fido and Fluffy on the seesaw.

Another household item that uses a second class lever is a bottle opener. In Figure 1-6, you can see the input, fulcrum, and output identified. The handle of the bottle opener goes through a lot of travel to get the cap of the bottle off, but the force at the lip of the bottle cap is relatively high. A nutcracker is another example of a second class lever. Can you think of any other second class levers?

First and second class levers are *force multipliers*, which means they have good mechanical advantage. The trade-off in both cases is that the input, or effort, must move a greater distance than the output, or load.

FIGURE 1-6 A bottle opener as a second class lever

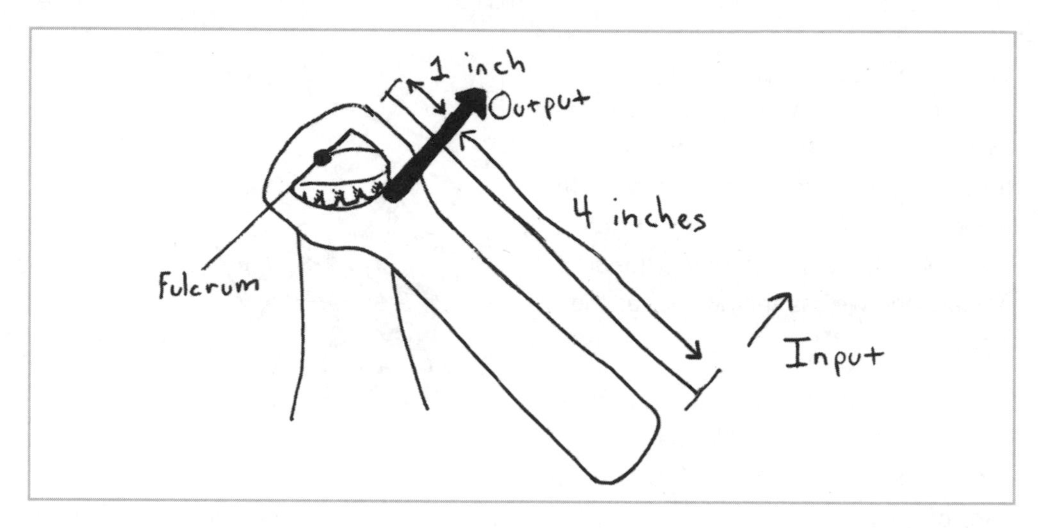

Third Class Levers

In a third class lever, the input is applied between the fulcrum and the output, as shown in Figure 1-7. This is known as a *force reducer*.

Why would you want a machine that reduces force? Most of the time, it's used when this arrangement is the only option available to lift or move something, due to space or other constraints. Although a higher force is needed at the input, the advantage of a third class lever is that the output end moves faster and farther than the input.

FIGURE 1-7 Using a ladder as a third class lever

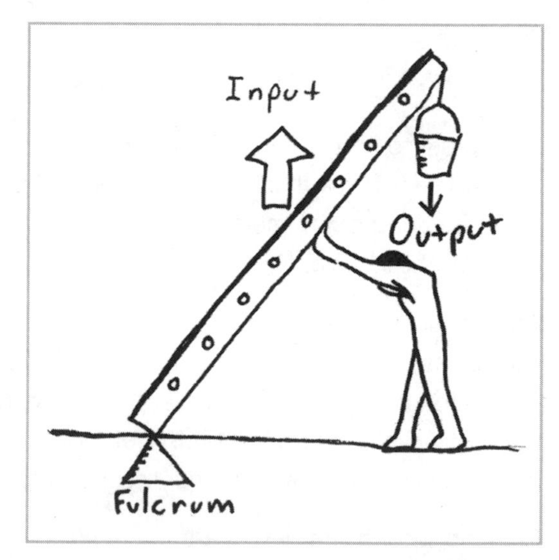

Your arm is a good example of a third class lever. As you can see in Figure 1-8, your bicep muscle is attached between your upper arm near your shoulder and forearm just past your elbow. Your bicep must work hard to lift even a small weight in your hand, but the weight can travel through a long distance since it's far from the pivot point at your elbow. A triangular arm that allowed your bicep to attach near your wrist would be more efficient, but it would have a very limited range of motion. Fishing rods and tweezers also work as third class levers.

FIGURE 1-8 Your arm as a third class lever

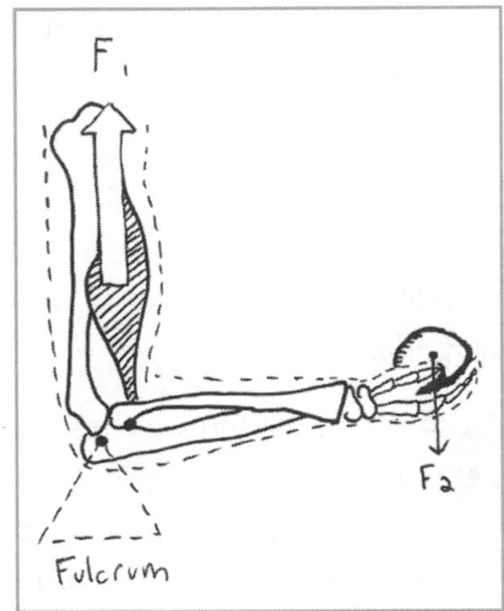

You can also combine levers into linkages, which we'll talk more about in Chapter 8. For now, take a look at a project from some former students of mine, shown in Figure 1-9. The two weights are being balanced by a first, second, and third class lever all at once. The fulcrums of each are circled. Can you tell which one is which? (Go to http://itp.nyu.edu/~laf333/itp_blog/2007/03/lever_madness.html to confirm your answer.)

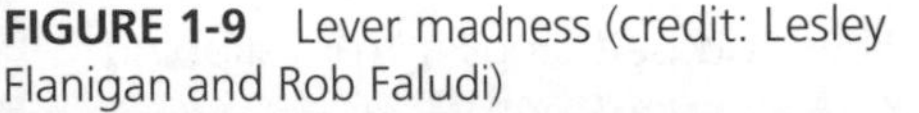
FIGURE 1-9 Lever madness (credit: Lesley Flanigan and Rob Faludi)

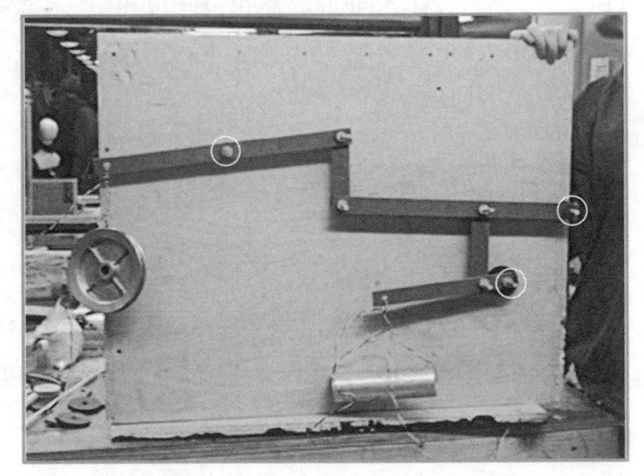

2. Pulleys

A *pulley*, also known as a *sheave*, *block* (as in block and tackle), or *drum*, is basically a wheel with a groove along the edge for a rope or belt. It's another simple machine we can use to gain mechanical advantage in a system. The two types of pulley systems are closed and open.

Closed Systems

I will call a pulley system on a fixed-length rope or belt that's constantly tight a *closed system*. A common example of this is the timing belt in a car, as shown in Figure 1-10. Timing belts use pulleys with little teeth on them that mesh with matching teeth on the belt. This helps the motor drive the belt without slipping, called *positive drive*, because the belt and the teeth on the pulley mesh together.

FIGURE 1-10 Timing belt on the engine of a car as a closed pulley system

You can find a similar system inside cameras that use 35mm film. The holes on the edge of the film actually match up with little teeth on the pulley wheel the film wraps around.

Closed pulley systems can also use smooth belts and pulleys that are spaced so the belt is tight enough not to slip on the pulleys. This is called *friction drive*, because the belt is made to fit tight around the pulleys so the friction between the pulleys and belt stops it from slipping. LEGO systems use pulleys with belts that are color-coded depending on length, as shown in Figure 1-11.

Closed pulley systems are used to translate rotational motion between axes. There is a mechanical advantage only if the *driven*, or *input*, pulley is smaller than the *output pulley*, as shown in Figure 1-11.

Any pulleys in between the input and output are called *idlers*, because they don't do anything other than redirect the belt. Sometimes the idlers are spring-loaded, or mounted such that they are adjustable, so the tension on the belt can be controlled.

The mechanical advantage of closed system pulleys is easier to calculate than with levers. It's just the ratio of the pulley diameters. If a 1 inch (in) diameter pulley is stuck on a motor and drives a 3 in diameter pulley, the mechanical advantage is 3:1. This means that the system can turn something that's three times harder to turn than the motor could by itself.

FIGURE 1-11 LEGO motor using a friction drive pulley system. The large pulley is 1 7/8 inches and the small one is 3/8 inch, which creates a 5:1 mechanical advantage.

Open Systems

Open systems are what most people think of when pulleys come to mind, but will be less useful to you when making projects like the ones in this book. In an open system, one end of the rope or belt is open or loose. A good example of this is a flag hoist. A flag hoist is just a pulley attached to the top of a long flag pole with a rope going around it, so you can stand on the ground and pull down on the rope to raise the flag. One pulley *fixed* in place like this does not magnify force or give you a mechanical advantage. The rope moves the same distance that the flag does when pulled. However, it does allow you to change the direction of movement.

On the other hand, one *unfixed* pulley does magnify force. Unfortunately, as with levers, we don't get something for nothing. The ability to decrease the effort we put in comes at the expense of needing to pull the rope or belt on the pulley a longer distance. As shown in Figure 1-12, an unfixed, movable pulley (also called a *runner*) gives us a 2:1 mechanical advantage. Because each length of the rope carries half the weight, the weight is twice as easy to pull up as it would be to lift the weight alone. The trade-off is that you must pull the rope twice as far as the distance you want the weight to move, since your effort is cut in half.

This last configuration is never very convenient. In order to be able to lift something standing on the ground, most people would prefer to pull down instead of up. By adding another pulley to the system, we maintain the 2:1 mechanical advantage but change the pull force direction to be more convenient. The arrangement in Figure 1-13 is called a *gun tackle* and does exactly that.[1]

The next logical step in this progression is to get a mechanical advantage of 3:1. There are at least two ways to do this. One is called a *luff tackle*. This uses a compound pulley (two independent pulleys in the same housing). Notice in the left image of Figure 1-14 that the weight is suspended by three parts of rope that extend from the movable single pulley at the bottom. Each part of the rope carries its share of the

FIGURE 1-12 One unfixed pulley, or runner, gives a mechanical advantage.

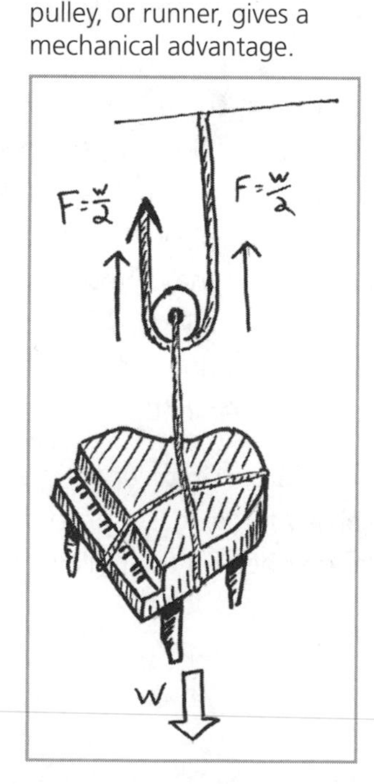

FIGURE 1-13 A gun tackle arrangement gives a 2:1 mechanical advantage, and a convenient pull direction.

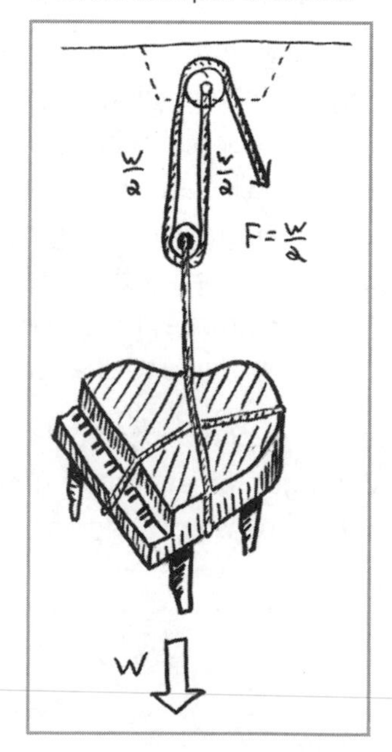

FIGURE 1-14 Pulley arrangements that give a 3:1 mechanical advantage

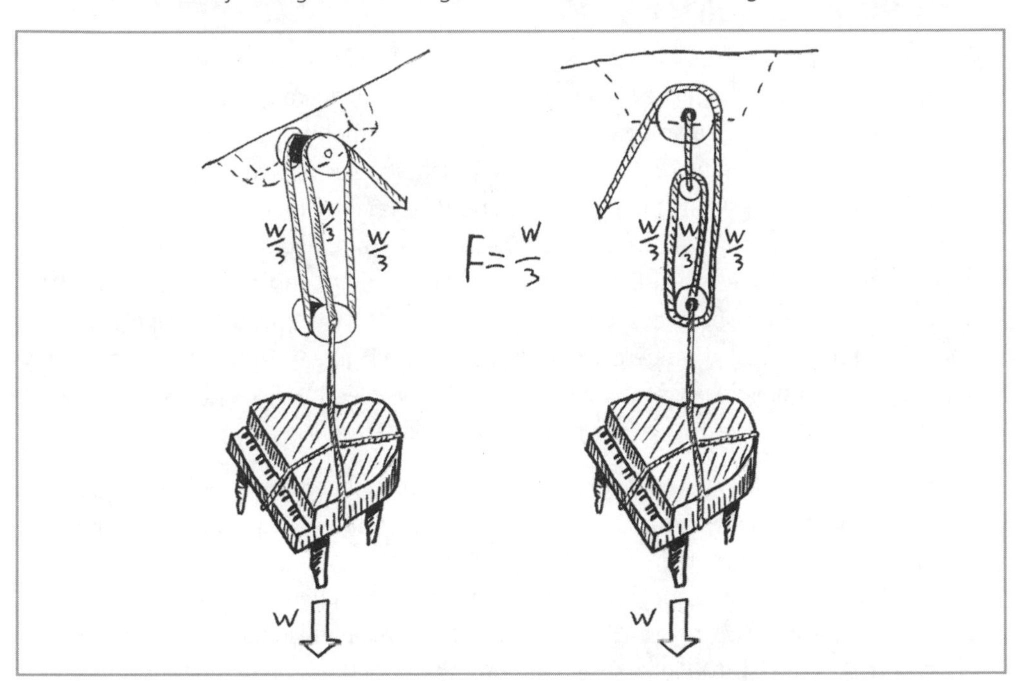

weight being suspended. So in this case, each part of the rope carries one-third of the weight, and that is the mechanical advantage we feel when pulling on the rope: It's three times easier to lift the weight using this arrangement than it would be to lift the weight on our own. That's a 3:1 mechanical advantage.

> ***TIP*** ***If you count the number of parts of rope going to and from the movable pulley that suspends the weight, you can figure out the mechanical advantage. If there are three pieces of rope going to and leaving one movable pulley, the mechanical advantage is 3:1.***

Another way to get the same 3:1 mechanical advantage is by using three simple pulleys, rather than one simple and one compound pulley. You can see this arrangement in the right image of Figure 1-14. The more pulleys you add to the system, the more mechanical advantage you can get.

Pulley systems can get pretty complex and allow you to do things like lift a piano to guide it into a second-story window with significantly reduced effort (though you might be pulling for a very long time).

Speed and Velocity

Speed is how fast something is moving. It's measured in distance over time. Velocity is the same thing, just in a specific direction. Common units are miles per hour (mph) or feet per second (ft/s). If you tell someone to drive 60 mph north, you are actually expressing a velocity. *Rotational velocity* (also called *angular velocity*) is exactly what it sounds like: the speed of something spinning. This is commonly expressed in revolutions per second (rps) or revolutions per minute (rpm) and distinguished from straight-line velocity (v) by using the symbol ω (the Greek letter omega). *Tangential velocity* describes the speed of a point on the edge of the circle, which at one split second in time is moving tangentially to the circle. See Figure 1-15 to visualize this. In the bicycle example, think of rotational velocity as the speed the rear wheel spins by itself, and tangential velocity as the speed of the bike along the ground.

As an example, let's say you ride a bicycle with a cog attached to the rear axle that has an 8 in diameter, and your tire is 32 in across. Circumference is equal to π (or 3.14) multiplied by diameter, so the circumferences of the sprocket and wheel are about 25 in and 100 in, respectively. This means that if you pedal at the rate of 1 rps, a tooth on the sprocket travels 25 in per second, while a corresponding spot on the wheel travels through 100 in. So the point on the wheel has a tangential velocity four times higher than the sprocket, even though they have the same rotational velocity of 1 rps. If the wheel shrunk down to the size of the sprocket, you would need to pedal really fast to get anywhere (and look pretty funny doing it). So instead, use the 1:4 *mechanical disadvantage* to help you cover more ground.

FIGURE 1-15 The rear sprocket on a bicycle wheel magnifies the speed of the wheel.

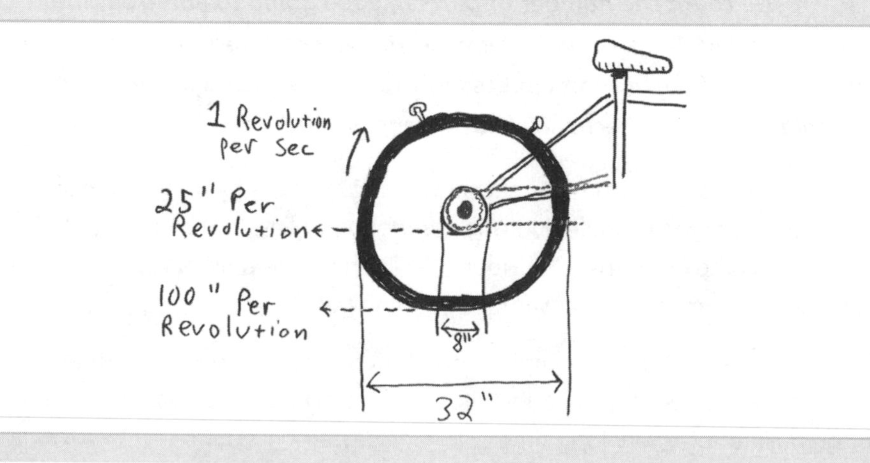

3. Wheel and Axle

You have probably never thought of the steering wheel in a car as a machine, but that's exactly what it is. The large diameter of the steering wheel is fixed to an axle, which acts on the steering system to turn the wheels. Let's say the steering wheel has a diameter of 15 in, and the axle it is fixed to has a diameter of 1 in. The ratio of input to output size here is 15:1, and that's our mechanical advantage. (For more on how steering systems work, check this link: www.howstuffworks.com/steering.htm.) Similarly, a screwdriver with a thick grip handle is much easier to use than one with a handle the size of a pencil.

You can use a wheel and axle to magnify force, as in the steering wheel example, or to magnify speed, as in the wheels of a bike. A bicycle's rear cog is fixed to the rear axle, so when you pedal, the chain turns the rear cog that turns the rear wheel. This is the opposite setup as in a steering wheel. In a steering wheel, you turn a big thing (steering wheel) to make it easier to turn a small thing (steering wheel axle). In a bicycle, you turn a small thing (rear cog) in order to turn a big thing (rear wheel). You don't gain mechanical advantage in this setup, but you do gain speed.

4. Inclined Planes and Wedges

If you've ever done the move yourself from one home to another, you might have used a ramp coming off the back of the moving truck to help you roll boxes on and off the truck bed. This ramp, or *inclined plane*, is a simple machine.

Let's say you have a 100 lb box of books you need to load into the truck. If you lift it yourself, you obviously need to lift the whole 100 lbs to get the box into the truck. However, if you use a 9 ft long ramp that meets the truck at 3 ft off the ground, you can set the books on a dolly and roll them up the ramp. Since you are rolling 9 ft to go up 3 ft, instead of just lifting the box 3 ft straight up, the ramp gives you a 3:1 mechanical advantage. So with the ramp, you can get the books into the truck with only one-third of the force of lifting it directly. The mechanical advantage of a ramp is the total distance of the effort exerted divided by the vertical distance the load is raised.

You've also probably used an inclined plane to prop open a door. A few horizontal kicks to a triangular wooden stopper drive it under the door, and the vertical force created by the inclined plane keeps the door propped up and open.

A *wedge* is like two inclined planes set base to base. Wedges can be found on knives, axes, and chisels. If you drive an axe into a piece of wood, as shown in Figure 1-16, the mechanical advantage is the length of the blade divided by the width of the base. In this case, you see a 6:1 mechanical advantage. That means that if you swing the axe and it has a downward force of 100 lbs when you hit the wood, the splitting force that the wood feels coming off the axe is 600 lbs on each side.

FIGURE 1-16 An axe uses a mechanical advantage to split wood.

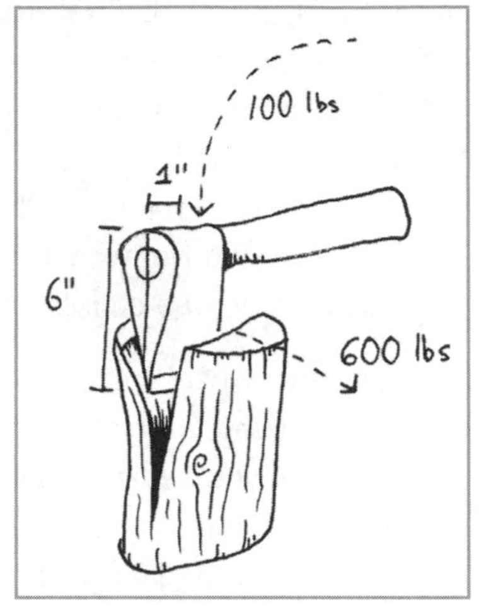

5. Screws

A *screw* is really just a modification of an inclined plane. There are two main types of screws:

1. Screws used for fastening parts together.
   Fastening screws use their mechanical advantage to squish two or more pieces of material together.

2. Screws used for lifting or linear motion (called *power screws*). Power screws have a slightly different geometry thread to allow them to lift or push an object that slides along the threads, like in the screw jack in Figure 1-17.

TRY THIS ***Cut a piece of 8 1/2 × 11–in paper in half along the 11-in side, and then cut one of the remaining pieces diagonally from corner to corner. Next, line up the shorter side of the triangle with a pencil and start wrapping the triangle around the pencil. Notice the spiral shape? This shows how a screw is a modification of an inclined plane—the triangle.***

As with any simple machine, the mechanical advantage is the ratio of what you put in to what you get out. One example of a power screw is a screw jack that you might use to prop up your car before changing a tire. Let's say the screw jack has a handle length of 12 in, as shown in Figure 1-17. The pitch of the screw is the distance between threads, and is the distance the screw will move up or down when turned

FIGURE 1-17 Screw jack used to lift a heavy load

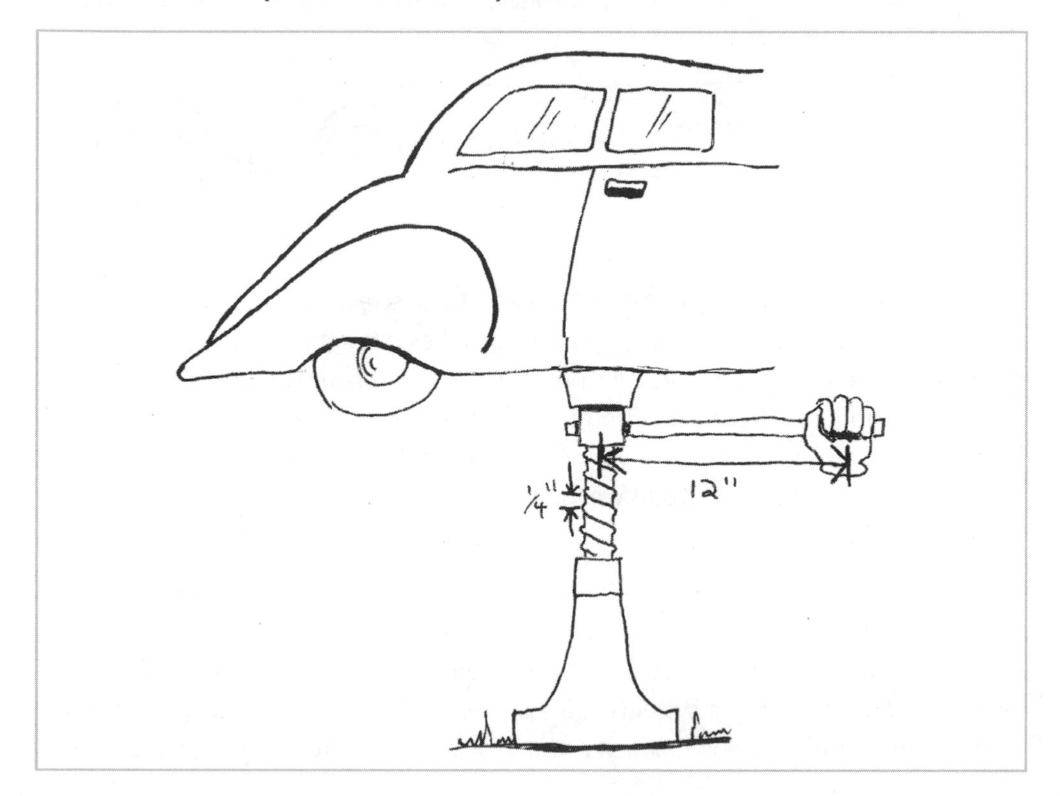

one full revolution. In our example, let's use 1/4 in. To use the screw jack, we need to turn the 12-in handle through a full circle for the jack to raise up 1/4 in. The end of the handle traces out a circle with a radius of 12 in., and the circumference equals 2π multiplied by the radius ($C = 2 \times \pi \times R$). So, our mechanical advantage is the input ($2 \times \pi \times 12$ in) divided by the output (1/4 in), which is about 300!

Power screws like in our screw jack example can achieve very high mechanical advantages in a compact space, so they are great for lifting jobs when rigging up a pulley system wouldn't be practical. A lot of this mechanical advantage is lost to friction, and we'll talk more about that in Chapter 4.

Another place you may have seen power screws at work is in turnbuckles. These are used to tension ropes and cables that are already secured. As indicated in Figure 1-18,

FIGURE 1-18 A turnbuckle can be used to tighten or loosen the tension in a cable.

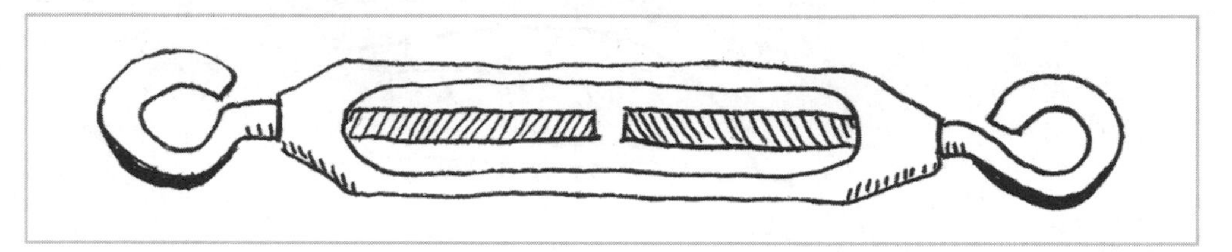

the turnbuckle has left- and right-hand threads. Most screws that you've encountered have a standard right-hand thread, which means they get tighter as you turn them clockwise, or to the right. Left-hand threads get tighter when you turn the screw to the left, or counterclockwise. By using one of each, the turnbuckle can either draw in both sides at once to tighten or loosen both sides simultaneously. This same idea can be used in leveling mechanisms as well. You can also find power screws in C-clamps and vises.

You'll also find power screws in positioning systems where precise location, rather than mechanical advantage, is the main concern. These types of systems use motors to turn a power screw that positions a table or other mechanism horizontally or vertically. You can see these systems in 3D printers and precision lab equipment. (For some good examples of power screws, visit www.velmex.com/motor_examples.html.)

6. Gears

Gears are used to magnify or reduce force, change the direction or axis of rotation, or increase or decrease speed. Two or more gears in line between the input and output are known as a *drive train*. Drive trains that are enclosed in housings are called *gearboxes* or *gearheads*. The teeth of the gears are always meshing while they are being turned, so a gear drive train is an example of a positive drive.

Gear Types

There are many different types of gears and ways to use them. We'll cover the details in Chapter 7. Here, we'll take a look at the five basic types of gears: spur, rack-and-pinion, bevel, worm, and planetary.[2]

Spur Gears The most commonly used gear is called a *spur gear*. Spur gears transmit motion between parallel shafts, as shown in Figure 1-19. Individual spur gears are primarily described by three variables:

FIGURE 1-19 Spur gears in a drive train

1. Number of teeth (*N*)
2. Pitch diameter (*D*)
3. Diametral pitch (*P*)

The last two variables sound alike, which can be confusing, because they represent very different things. The *pitch diameter* of a spur gear is the circle on which two gears effectively mesh, about halfway through the tooth. The pitch diameters of two gears will be tangent when the centers are spaced correctly. This means that half the pitch diameter of the first gear plus half the pitch diameter of the second gear will equal the correct *center distance*. This spacing is critical for creating smooth running gears.

The *diametral pitch* of a gear refers to the number of teeth per inch of the circumference of the pitch diameter. Think of it as tooth density—the higher the number, the more teeth per inch along the edge of the gear. Common diametral pitches for hobby-size projects are 24, 32, and 48.

NOTE ***The mating gears can have different pitch diameters and number of teeth, but the number of teeth per inch, or diametral pitch (P), must be the same for the gears to mesh correctly.***

Rack-and-Pinion Gears A *pinion* is just another name for spur gear, and a *rack* is a linear gear. A rack is basically a spur gear unwrapped so that the teeth lay flat, as shown in Figure 1-20. The combination is used in many steering systems, and it is a great way to convert from rotary to linear motion. Movement is usually reciprocating, or back and forth, because the rack will end at some point, and the pinion can't push it in one direction forever.

FIGURE 1-20 Rack-and-pinion gears

Another common example of a rack-and-pinion gear is a wine bottle opener—the kind shown in Figure 1-21. The rack in this case is circular, wrapped around the shaft that holds the corkscrew. The handles are a pair of first class levers that end in pinion gears, and they go through a lot of travel when you push them down to give you the mechanical advantage needed to lift the cork out of the bottle easily.

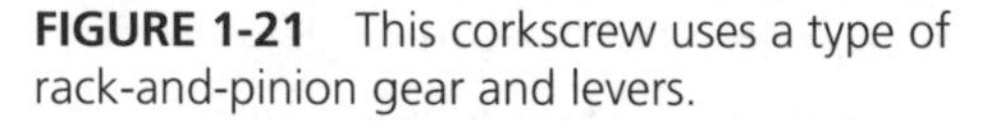

FIGURE 1-21 This corkscrew uses a type of rack-and-pinion gear and levers.

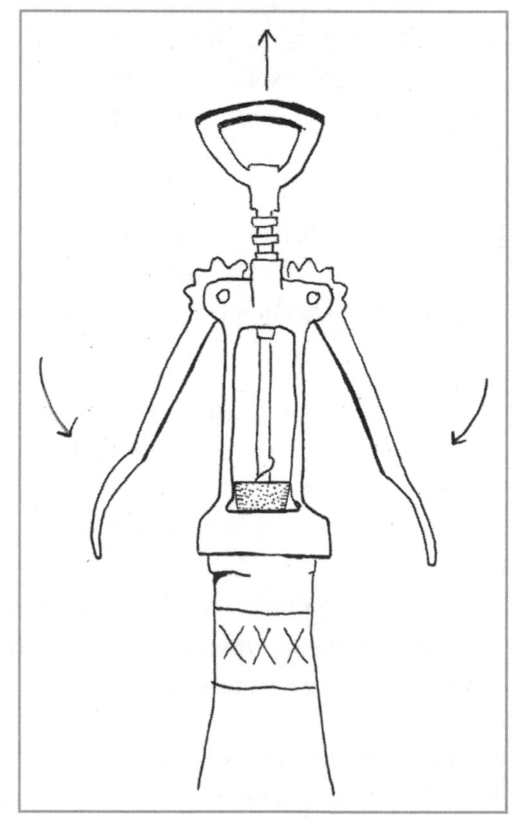

Bevel Gears *Bevel gears* mesh at an angle to change the direction of rotation. A *miter gear* is a specific kind of bevel gear that is cut at 45° so that the two shafts end up at a 90° angle, as shown in Figure 1-22.

Worm Gears *Worm gears* actually look more like a screw than a gear, as shown in Figure 1-23. They are designed to mesh with the teeth of a spur gear.

One important feature of the worm gear is the mechanical advantage it gives. When a worm gear (sometimes just called the *worm*) rotates one full revolution, the mating gear (sometimes called the *worm gear*) advances only one tooth. If the mating gear has 24 teeth, that gives the drive train a 24:1 mechanical advantage. (This is technically only true for single-lead worms; for a two-lead worm, two full revolutions are needed to turn a mating gear one tooth.) Of course, the mating gear will be moving very slowly, but a lot of times, the trade-off is worth it.

FIGURE 1-22 Bevel gears

Another great feature of worm gears is that the majority of the time, they don't *back drive*. This means that the worm can turn the worm gear, but it won't work the other way around. The geometry and the friction just don't allow it. So, a worm gear drive train is desirable in positioning and lifting mechanisms where you don't want to worry about the mechanism slipping once a certain position is reached.

Planetary Gears *Planetary*, or *epicyclic gears*, are a combination of spur gears with internal and external teeth. They are mostly used in places where a significant mechanical advantage is needed but there isn't much space, as in an electric screwdriver or a drill. You can even layer planetary gear sets to increase the

FIGURE 1-23 Worm gears

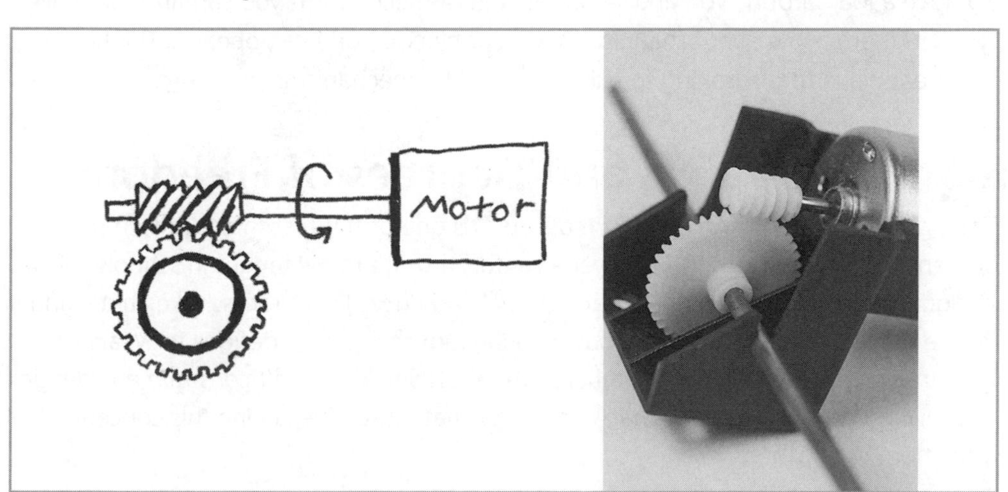

mechanical advantage. (howstuffworks.com has an excellent write-up on the topic, which you can find at www.science.howstuffworks.com/gear7.htm.)

Gear Ratios

Gears of different sizes transmit a mechanical advantage, similar to how pulleys work. As always, the mechanical advantage is the ratio of how much we put in to how much we get out.

The smaller of two gears in a set is usually called a *pinion*, and is the one being driven. Let's say we have a 20-tooth pinion attached to a motor shaft. Then a 100-tooth spur gear (of the same diametral pitch, of course) mates with the pinion to rotate an adjacent shaft. The pinion must rotate five times to turn the output gear once, so the mechanical advantage is 5:1.

When the gear train is being used to *magnify force*, the input gear will always be smaller than the output gear. This setup is great when you have a motor and need to multiply the work it can do by itself, or when you need to slow the motor's output to a speed that fits your application.

To use a gear train to *magnify speed*, reverse the gears so the big gear is the input gear. The gear train is at a mechanical disadvantage in this configuration, but since one turn of the input gear on the motor turns the mating gear five times, the speed of the output is magnified by five.

So, take a look around you and see what kind of mechanisms you can find that have gears in them. How about that old clock, your blender, or a can opener? The kitchen is a great place to go looking for all sorts of useful mechanisms.

Design Constraints and Degrees of Freedom

The principle of *minimum constraint design*[3,4] is one of the first things I teach my students. It has been around for over a century, but it's rarely taught in schools. Most designers and engineers learn it through trial and error. That process takes time, and if you're reading this book, you probably don't want to mess up designs for years to gain that experience firsthand. So here's the short version: *Don't constrain any design or moving part in more ways than necessary*. That's it. Let's examine this concept a little more in depth.

Degrees of Freedom

FIGURE 1-24 Coordinate system of axes and planes

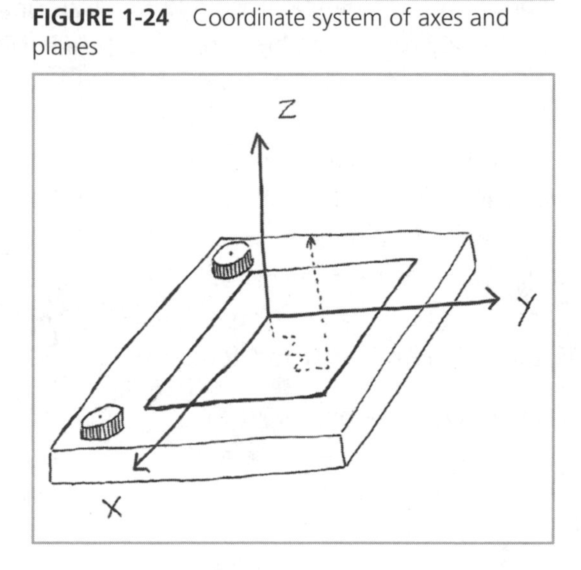

Every object has six different ways it can move: three straight line motions, called *translations*, and three *rotations*. This is usually shown on a coordinate system, as in Figure 1-24.

If you stand straight up and picture the origin (the middle where all the lines meet) of this coordinate system at your belly button, it will be easier to understand the movement. You can jump up and down (translation along the Z axis), shuffle side to side (translation along the Y axis), or walk forward and backward (translation along the X axis). Every linear movement is a combination of X, Y, and Z translations.

For example, if you walk forward diagonally, you are moving in X and Y. Remember the Etch A Sketch? It has two knobs: one that controls horizontal, or X motion, and one that controls vertical, or Y motion. To make a diagonal line, you spin both knobs at once. You could say that you are drawing in the XY plane, because your motion is part X and part Y movement. You can do this with your body if you walk forward and to the right diagonally. The axes in Figure 1-24 also define three planes: the XY plane, YZ plane, and XZ plane. Can you think of a way to move in the XZ plane?

In addition to these three translations, any object can spin around any of these three axes. If you spin around in place, you are rotating about the Z axis. If you bend forward and backward at your waist, you're rotating your body about the Y axis. And if you bend to the side, you are rotating about the X axis.

Rotations may be easier to picture on an airplane, where they have more specific names, as shown in Figure 1-25. When a plane tilts its wings with respect to the horizontal, it is rotating about the X axis, called *roll*. When then nose dips up or down, it's rotating about the Y axis, called *pitch*. And when it rotates around the Z axis to go into a turn, it's called *yaw*. Remembering the names of these three rotations isn't important. Just keep in mind that all movement is a combination of three translations and three rotations.

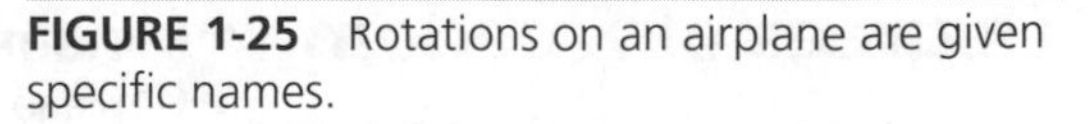

FIGURE 1-25 Rotations on an airplane are given specific names.

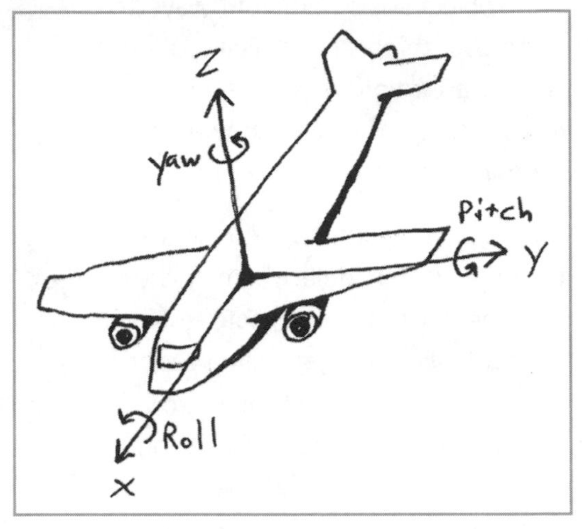

Before we move on, I want to introduce two more simple terms used to describe motion:

1. *Axial* refers to along an axis. Axial rotation is around an axis (either clockwise or counterclockwise). Axial load or force is applied parallel to an axis.

2. *Radial* refers to perpendicular to an axis. A radial load or force is applied perpendicular to an axis.

Figure 1-26 illustrates these concepts.

FIGURE 1-26 Axial and radial motion and forces

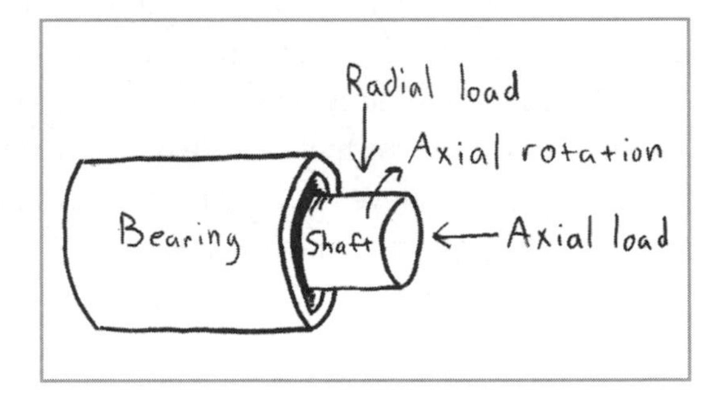

Minimum Constraint Design

Have you ever sat at a table that wobbled every time you rested your elbow on it? Most of us have, and we've also probably tried to stop the wobble by putting some napkins or coasters under the offending wobbly side. What you may not have realized at the time was that the table most likely had four legs. You see, a three-legged table can't wobble. Sure, it can fall over, but it can't wobble that annoying half an inch that the four-legged table can. That's because three points define a plane, and four points is one too many. Any time you have four points that are trying to coexist on a flat surface, you have issues. At least one of those legs will need to have some "give" in it. That's what you were adding when you squished the napkins under the wobbly table.

Using three points to define a plane is what is called good *minimum constraint design*. You can see this concept in tricycles made for young children who are not known for their superior balancing abilities, as well as in camera tripods.

Sometimes there are reasons to add more than the minimum number of constraints. For example, a car has four wheels. I just said that you only need three points to define a plane. What gives? All four wheels of the car do! Because the four points, or wheels in this case, are made of air-filled rubber, they can "give" and distribute the weight of the car evenly between them, without ever experiencing the four-legged table wobble. Also, a four-wheeled vehicle is less susceptible to tipping over than a three-wheeled one. This is what we call an example of acceptable *redundant constraint design*, because the fourth wheel is redundant. Keep this concept in mind when you are making anything. It is sure to save you loads of time reworking mechanisms, and I will point out good minimum constraint design in the projects throughout this book.

For good examples of minimum constraint design, check out any moving LEGO kit. These kits are designed well, with just enough parts to get the job done, and the parts themselves are made precisely to stick together or slide through each other with just the right amount of clearance. For bad examples, or overconstrained designs, try to assemble furniture from IKEA or any other budget retailer. Inevitably, the holes in the desk legs don't line up, the wooden pegs are too tight to go all the way in, or some other part is sized in a way that it creates an unacceptable redundant constraint and causes you a headache. Put the extra effort in on your own projects to avoid such scenarios.

Project 1-1: Rube Goldberg Breakfast Machine

Rube Goldberg was an engineer turned cartoonist who is best known for his cartoon series depicting complex contraptions that perform simple tasks in extraordinarily complex ways. In fact, the adjective *Rube Goldbergian* is defined as "accomplishing by complex means what seemingly could be done simply."[5]

Omega Engineering (www.omega.com), a company that specializes in automation equipment, uses Rube Goldberg cartoons in its ads as a comical way to show that its products are more efficient at automating tasks than the Rube Goldbergian way. These cartoons epitomize the idea that there is always more than one way to accomplish a task (see Figure 1-27). This is certainly true when it comes to making mechanisms, and it's important to realize that the first solution to a problem that comes to mind may not be the simplest or the best.

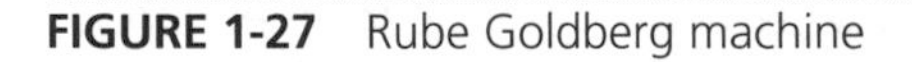

FIGURE 1-27 Rube Goldberg machine

The objective of this project is to build a Rube Goldberg machine that cracks an egg in no less than five steps. This can be done quickly and cheaply with material you find around the house, but the means to the end is limited only by your imagination and budget. I've included an example in case you're stuck, but I encourage you to ignore it and develop your own project. The idea is to get you working with your hands and making something to accomplish a specific task, without thinking too much about it.

The rules for this project are as follows:

- The majority of the egg and no more than half the shell should end up in the final receptacle.
- Limit yourself to a 3 × 3 ft area for the entire machine.
- Starting the machine is the only human interaction allowed. For example, this could be a button press, pushing a toy car over a ledge, or removing a stopper.
- From the time you initiate movement, your egg must be cracked in 5 minutes or less.
- Each step, or energy transfer, must be unique and contribute to the goal. For example, you can't have a golf ball roll down a ramp, spin five pinwheels, and then trigger a knife to cut the egg. That's boring. Also, the pinwheel spins don't contribute to the final goal of egg cracking.

I have assigned this project to my students at New York University (NYU) in the first class for the past few years, and it's always a hit. My favorite example of a successful Rube Goldberg machine to date is one that was designed to suck the egg out of the shell using a large syringe (check out the video at www.flickr.com/photos/fxy/3260972797/; credit Xiaoyang Feng, Mike Rosenthal, and Ithai Benjamin). You can browse the student pages from 2008 onward at http://itp.nyu.edu/mechanisms, as well as the rest of the Internet, to find other examples of Rube Goldberg machines. I've included a simple example here to get you started. So get to work!

Shopping List:

- 1 sheet 1/4-in thick clear acrylic, approximately 15 × 31 in (Ponoko.com stock)
- 1/4-in diameter, 36-in long wooden dowel
- Multitool with knife and file
- Mousetrap
- Paint-stirring stick
- Fishing line or other thin string
- Rubber bands
- Duct tape roll
- Small bowl
- Spoon or fork
- Egg(s)

Recipe:

1. See www.makingthingsmove.com for the link to download the templates from Thingiverse. This template was designed to be laser cut from the 1/4-in clear acrylic that Ponoko.com stocks and configured to fit on its P3 template. Register for Ponoko.com, and order the design cut out of your chosen material. You can also use the template to cut pieces out of cardboard or foam.

2. Use the knife on the multitool to score the wooden dowel into eleven 2 1/2-in long sections. Snap off the pieces at the score marks and file the edge. Set aside the remaining few inches.

3. Place all 2 1/2-in dowels in the back side template (refer ahead to Figure 1-29).

4. Loop rubber bands around the dowel just below the top-left dowel, as shown in Figure 1-28.

5. Tape the spoon or fork to the bottom-center dowel so that it can pivot when the duct tape roll falls and hits it.

FIGURE 1-28 Fixing the paint stick in place with rubber bands

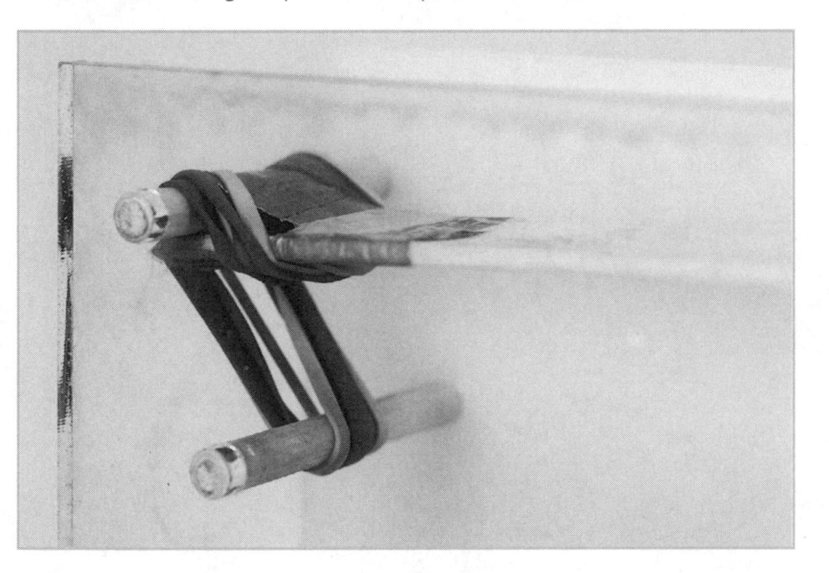

6. On the handle side of the fork or spoon, duct tape a ~2 ft length of fishing line.

7. Put the front side template onto the 11 dowels while sandwiching the string guide piece (see Figure 1-29).

8. Route the fishing line as shown by the arrows in Figure 1-29. Then tie it to the hole in the egg gate.

9. Slide in egg ramp pieces to slots in the front and back template. Secure with tape if necessary.

10. Fix the wooden paint stick by looping the rubber bands around the end and securing with duct tape (refer to Figure 1-28). Hold it in position by sliding the remaining 1/4-in wooden dowel length into the slot (see Figure 1-29).

11. Duct tape the mousetrap to the two far-right dowels in the orientation shown in Figure 1-29.

FIGURE 1-29 Final assembly of Rube Goldberg breakfast machine

12. Carefully set the mousetrap.

13. Place the egg gate in the egg ramp and position the egg behind it. The string should be tight enough that when the duct tape roll hits the spoon or fork, that small amount of movement in the string dislodges the egg gate.

14. Now it's showtime! Slide the wooden dowel out of the top slot. Watch the paint stick slap the duct tape roll, which lands on the spoon, which yanks the string, which pulls out the egg gate, and allows the egg to fall and trigger the mousetrap.

15. Cook the egg and enjoy your breakfast!

References

1. U.S. Bureau of Naval Personnel, *Basic Machines and How They Work* (New York: Dover Publications, 1971).

2. David Macaulay, *The Way Things Work* (Boston: Houghton Mifflin, 1988).

3. Lawrence J. Kamm, *Designing Cost-Efficient Mechanisms: Minimum Constraint Design, Designing with Commercial Components, and Topics in Design Engineering* (New York: McGraw-Hill, 1990). Also published in paperback by the Society of Automotive Engineers, Inc., 1993.

4. James G. Skakoon, "Exact Constraint," *Mechanical Engineering* magazine (http://memagazine.asme.org/Articles/2009/September/Exact_Construction.cfm).

5. Merriam-Webster Online Dictionary, "Rube Goldberg" (www.merriam-webster .com/dictionary/rube%20goldberg).

2 Materials: How to Choose and Where to Find Them

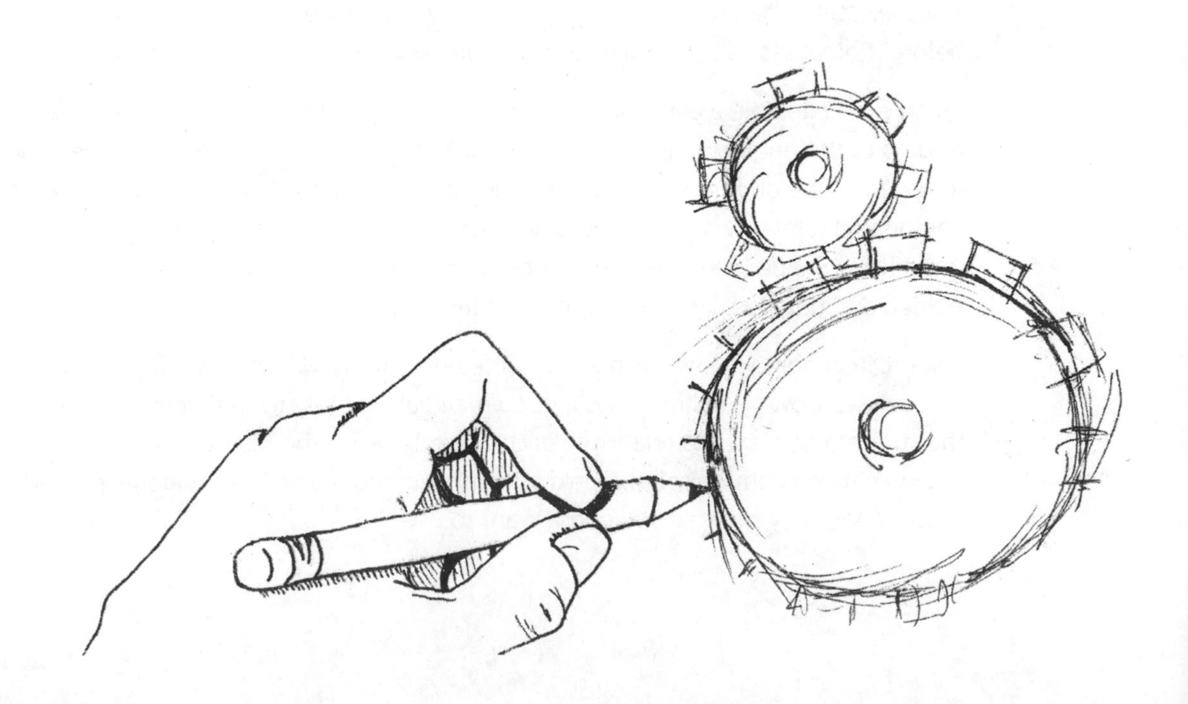

Once you've decided to make something that moves, you'll need to find parts and materials to build it. Aside from the usual constraints of availability and budget, how do you choose materials for your project? What's the difference between types of wood or kinds of aluminum, and are there any other options?

In this chapter, you'll learn about the various types of materials, how to use them, and where to find them. But first, we need to talk about how materials are described.

Describing Materials

In order to choose materials for projects, you need to learn how to describe materials and how strong they are. Each type of material is characterized by material properties.

Material Properties

A *material property* is just something about the material that's the same regardless of its size or shape. For example, density is a material property, but weight is not. Density is equal to mass divided by volume, so no matter how much stuff you have, that ratio stays the same. However, the more stuff you have, the more it will weigh, so weight is not a distinguishing material property.

Another useful material property is *yield strength*. A material that yields, or stretches, before it breaks is called *ductile*. One that breaks right away is called *brittle*.

As an example, take a paperclip and try to bend one of the legs just a little, so it returns to the original shape. This leg will have deflected, or deformed, but because it returned to the original shape, it hasn't actually yielded yet. Now take that same leg and bend it a lot—way out to the side. It stays in its new position, and returning it to the original shape would be hard (if not impossible). At this point, the paperclip has yielded and deformed in a way that is not temporary.

If you can imagine paperclips made of different materials yielding at different angles, you can see how yield strength is a good material property to use when comparing the strength of materials relative to each other. Look on the MatWeb site (www.matweb.com) to see the yield strength (among many other material properties) of just about any material you might want to use.

Material Failure: Stress, Buckling, and Fatigue

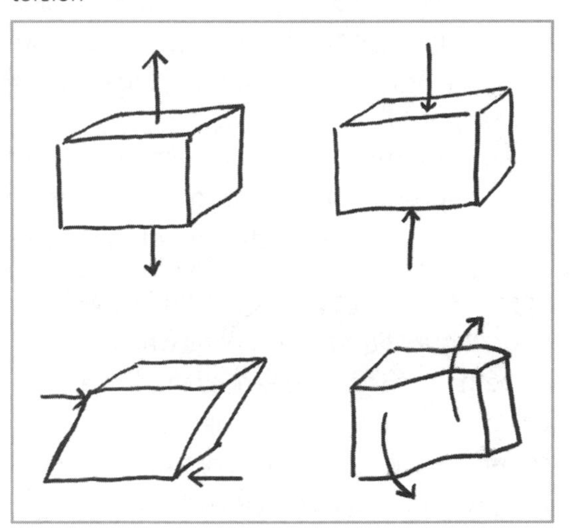

FIGURE 2-1 Tension, compression, shear, and torsion

The yield strength is the name given to the specific *stress* a material can experience before it gets, um, stressed out. Stress is just a force applied over a certain area, commonly expressed in pounds per square inch (psi). The stress at which a material actually breaks is called the *ultimate strength*.

There are actually four different kinds of stresses, and therefore four different ways a material can fail (see Figure 2-1):

1. **Tension** Tension is a fancy word for stretch. Think of the chains that hold up a child's swing on a playground. These chains are in tension when someone sits on the swing because they are being stretched. If someone very heavy sits on a swing designed for a two-year-old, the chain will break or fail in tension. The force or weight of the person divided by the cross-sectional area of the chain is the stress the chain feels in tension.

2. **Compression** Compression is a fancy word for squish. When you look at material properties on a website like matweb.com, you will sometimes see two different numbers for tensile strength and compressive strength. Consider this when you are designing support structures for your mechanisms—especially ones that could hurt someone if they break. There's a reason why buildings don't have foundations made of cheese: The compressive strength just isn't high enough. The force of the building divided by the area of the foundation is the total stress the cheese would feel.

3. **Shear** Shear stress is what's happening to the box in Figure 2-1, where the force is coming from the side of the box instead of in line with it. Try to avoid this situation in your designs, because the shear strength is only half the tensile yield strength in things like metal bolts.

4. **Torsion** Torsion is a fancy word for twist. The hex keys in Figure 2-2 failed in torsion when I tried to use them to unscrew a bolt that was glued too tight. The hex key twisted out of shape before I could get the screw unstuck.

Special cases of failure also include buckling and fatigue. *Buckling* happens when something is too long and skinny, like a column, and doesn't even get a chance to squish before it gives out. For example, you can probably balance your coffee mug on an empty toilet paper roll, but if you tried to balance it on one drinking straw, that wouldn't work out so well.

FIGURE 2-2 Allen keys that failed in torsion

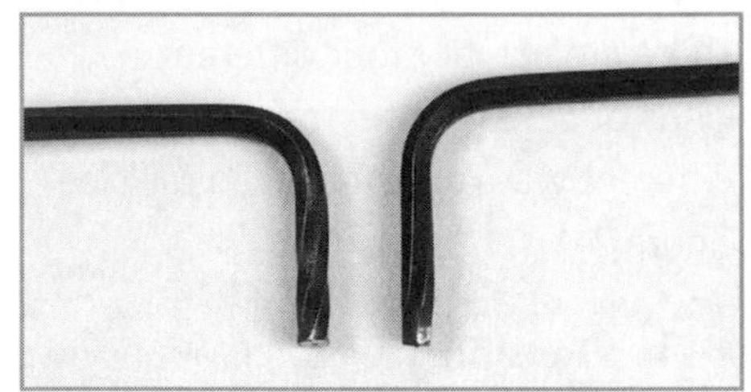

Fatigue failure is what happens when you bend a paperclip back and forth a bunch of times until it snaps. A single back and forth bend isn't enough, but after 20 or so, the paperclip gets stressed out and breaks from fatigue.

How to Tolerate Tolerances

The description of most raw materials will tell you what the tolerance is on the length, width, diameter, or some dimension of the part. So what is a tolerance?

The *tolerance* of a part dimension is the range of values a thing could actually have when you get it. For example, you may think you need a half-inch-diameter aluminum rod, but you don't really mean 0.5 in. That implies that you want a rod that is 0.500000 in, or perfectly 1/2 in. There are two problems with this:

1. You probably don't want a 1/2 in rod. You want one a little smaller or a little bigger.

2. No manufacturing technique is perfect, so there is no machine that exists that can make you a perfect 0.500000 in rod.

You need to figure out the range that's acceptable to you. Let's say you need this 1/2 in rod to fit in a hole in a wood block you measured with your calipers that has a diameter of 0.517 in. You need the rod to slide in and out and spin freely, so you want to leave some *clearance*, or space, between the rod diameter and the hole diameter (see Figure 2-3).

You look on the McMaster-Carr website (www.mcmaster.com) and find a 1/2 in rod that says the diameter tolerance is ±.005 in and the length tolerance is ±1/32 in. This means that your rod diameter could be as big as 0.505 in or as small as 0.495 in, and McMaster-Carr is not guaranteeing anything more specific than that. Luckily, either one of those will work for you in this case, because they are both smaller than your 0.517 in hole. The length tolerance is a wider range, but length matters less to you for your specific application than diameter, so you decide that anywhere in that ±1/32 in length range is just fine.

TIP ***When looking for just about anything—from raw materials to project supplies and components—McMaster-Carr (www.mcmaster.com) is a good place to start. It's like a giant online hardware store with an enormous selection of parts and materials paired with the best user interface for a website of its kind. It also has helpful descriptions of materials and uses, with detailed pictures of the majority of the products. I will mention it frequently in this book, and call it McMaster for short.***

FIGURE 2-3 Tolerances of a shaft and hole combination

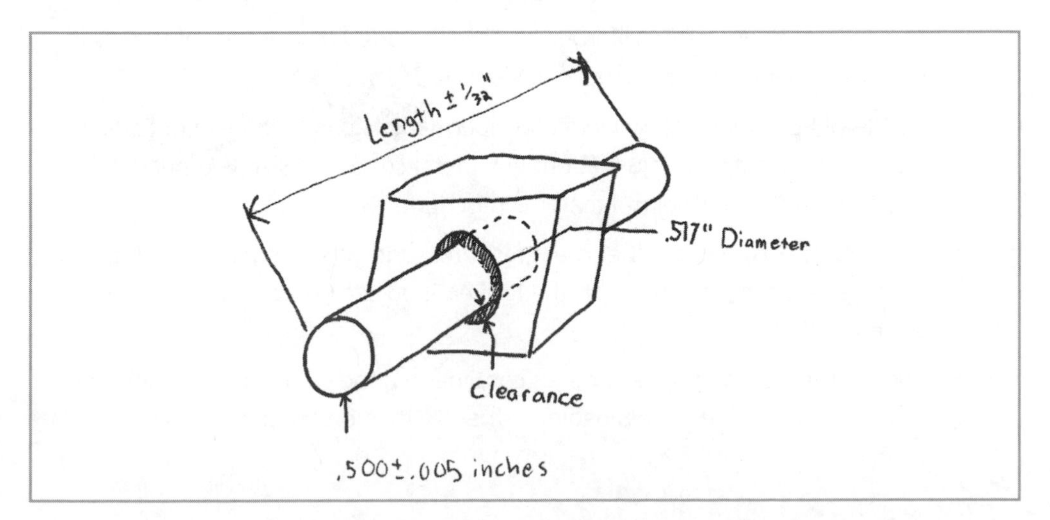

You will almost always find that more precise materials with tighter tolerances are more expensive. It takes more work to make a metal rod that is 0.5 ± 0.001 in than it does to make one that's 0.5 ± 0.010 in. If you don't need perfection, don't ask for it! You will pay the price if you do. Words like *precision ground* and *tight* will get you tighter tolerances and very straight, flat material, but only spring for them when you really need them. On the other hand, *oversize* parts guarantee they will always be bigger than the standard dimension you choose.

Material Types

There are three main types of solid materials: metals, ceramics, and polymers (plastics). Three other categories are useful to define: composites (including wood), semiconductors, and biomaterials.

Metals

Metals can be pure, as in the case of pure aluminum or iron, or *alloys*, like steel. An alloy is just a mix of two or more metals. Metals are generally good conductors of heat and electricity. They tend to be strong yet ductile, so you'll find them in a wide variety of shapes and sizes.

Your local home-improvement store should have a good selection, and you can look to more specialized metal distributors, depending on the sizes, shapes, and materials you need. In addition to our staple supplier McMaster, here are some other places to look:

- Metalliferous (www.metalliferous.com) sells a good assortment of metals in sizes and shapes common for hobbyists.
- At Metal Supermarkets (www.metalsupermarkets.com), you can get small pieces and quantities of just about any metal cut to the size you need, while you learn from their helpful material descriptions.
- OnlineMetals.com is another great resource, and you can get a 5% discount if you access the store through the Maker Shed site (www.makershed.com/v/metals/).
- Speedy Metals (www.speedymetals.com) has a good selection and good prices, but searching the site is less beginner-friendly than on the preceding sites.

Although we'll cover different types of metal here and talk about finding the raw materials, don't discount sets like Erector Set and VEX Robotics Design System for structural metal parts that are easy to configure. 80/20 Inc (www.8020.net/) and Item (www.itemamerica.com/) sell grown-up versions of these configurable structural systems, and MakerBeam (www.makerbeam.com/) is an emerging open hardware version of the same.

Steel

Steel has excellent strength, but this comes at the expense of high density. Out of the materials we'll cover, steel is one of the more difficult to work with in its raw form because drilling holes in it or shaping it takes patience. However, it's great for screws, shafts, and bearings, and as sheet metal.

Steel is a mix of mainly iron and carbon, with other elements like chromium, sulfur, and manganese thrown in to create different combinations of strength and other material properties. It conducts electricity and heat well.

Stainless steel resists rusting and is generally not magnetic, but it is difficult to sharpen. Plain or carbon steel will rust, but it is easier to sharpen, so it is often used to make cutting tools.

If you go to mcmaster.com and type "steel" in the Find box, you'll get a screen that shows a bunch of different shapes, dozens of alloys, over a dozen finishes, and something about specifications. So how are you supposed to make sense of all this?

First, realize that you don't need to find the perfect material—just something that satisfies your criteria. As a rule of thumb on McMaster, start with the characteristics that are most important to you, and then keep narrowing down the options until you reach a manageable number of options.

Let's assume you're looking for some thin, steel sheet metal you can bend and drill holes in for an automatic electro-mechanical toilet paper dispenser you have in mind. In this case, you start by choosing the Sheets, Bars, and Strips icon. You don't want anything fancy, so choose Plain on the next screen under Type. You also think you want something about 12 in long, so choose 12 in the Length section, but you're not picky about the other dimensions.

Now scroll down and read through some of the material descriptions. It looks like the first one, General-Purpose Low-Carbon Steel, is easy to bend, so you choose that one. You want your toilet paper dispenser to be about 2 in wide, so you choose that option. Uh-oh—unfortunately, the thinnest piece at that width is 1/8 in thick, which you won't be able to cut or bend by hand. So you back up and try choosing a thickness you know you can cut and bend first, and decide on 0.020 in. Phew! Now your job is easy, because there is only one option, a 12 × 8 in piece that comes in packs of five.

TIP ***Sometimes sheet metal is sold in a gage thickness, not a decimal thickness like 0.020 in. The higher the gage number, the thinner the material. For example, a 0.030 in thick steel is also called 22 gage, and a 0.060 in thick sheet is about 16 gage. To convert from one to the other, check eFunda's handy online calculator at www.efunda.com/designstandards/gages/sheet_forward.cfm.***

Local metal shops are another great resource, but they may not have convenient online stores. Don't be afraid to pick up the phone and ask questions about what size/thickness/gage/alloy of steel would be best for your project. Most suppliers will be happy to help you—after all, it is in their best interest to have an informed customer. Also check out plumbing pipes, fixtures, and electrical conduit as sources of steel tubing that are much less expensive than more precision extruded material.

Aluminum

Aluminum is less dense than steel and is known for having a great strength-to-weight ratio. It's much easier to bend and drill than steel, and won't rust like plain steel.

A good multipurpose aluminum alloy is 6061. If you want something strong but still lightweight, alloy 7075 is about twice as strong as 6061 and just as light. It's actually stronger than a few steel alloys, and is used in aircraft and aerospace mechanisms. This means, of course, that it's generally more expensive, so choose the alloy that is most appropriate for your application.

NOTE ***The naming schemes for alloys of most metals (aluminum 6061 and 7075, 303 stainless steel, and so on) are confusing and not necessary to understand or memorize. I'll mention the most common alloy names, but other than those, just read the descriptions of the material options you have once you have narrowed down your options using other variables that are important to you. McMaster often includes suggested uses in the descriptions, like "excellent for sheet metal work," "use for gears," or "light structural applications." Follow that advice, or give the store you are ordering from a call for help with selection.***

Aluminum extrusions are made by pushing hot metal through a shaped orifice, just as you did with the Play-Doh Fun Factory when you were little, only the metal is heated to a much higher temperature to be formable. Extrusions can be flat bars, L-shaped angle stock, or C-shaped channels.

Shelving standards (the rails you can nail into a wall, and then attach brackets and shelves to) are readily available at home-improvement stores and can be used as structural components. Aluminum angle and shelf brackets are good for mounting motors and joining pieces together.

Aluminum is a good conductor of heat, but it's a poor conductor of electricity relative to copper.

Copper

Copper is a good conductor of electricity and very cheap, so it's used a lot in wires, printed circuit boards, and other electronic equipment. It is also used extensively in the plumbing industry. Copper tubes can be assembled by brazing (see Chapter 3) to create small art pieces, sculptures, or stands. Copper is relatively soft, so you will rarely see it used for structural or high-strength parts, but you will see it a lot in sculpture and decorative work. When exposed to air for long periods of time, it will get a light-green coating, or patina, as on the Statue of Liberty.

These are two common copper alloys:

1. **Brass** Mix copper and zinc together, and you get brass. It's stronger and more durable than copper, and good at resisting corrosion from the atmosphere and water (including saltwater). It's not generally used for structural parts, but is common in furniture and architectural design.

2. **Bronze** Mix copper and tin together, and you get bronze. The alloy is harder and stronger than copper, and more corrosion-resistant but softer than brass. Bronze is also used in bushings because of its relatively low friction. (If you don't know what a bushing is, no worries—you'll learn in Chapter 7.)

Silver

Silver is actually a better conductor than copper, but it is much more expensive, so it is used only in very high-end electronics. It is also soft and easy to form, making it a favorite of jewelry makers.

Mixing silver with a bit of copper yields sterling silver, an alloy that is even easier to work with. Silver is not used structurally because it is too soft and too expensive.

Ceramics

Ceramic compounds fall between metallic and nonmetallic elements (oxides, carbides, and nitrides). Examples include clay, glass, diamonds, precious stones like amethyst, and your favorite coffee mug.

Ceramics are hard, but if you've ever dropped a coffee mug and seen it shatter, you know they are also brittle. Unlike metals that stretch and yield before they break, ceramics just break. They are good insulators of electricity and heat, and resist high temperatures and corrosion well. For our purposes, you might see them used only as insulating spacers (standoffs) that keep electrical components safe.

Polymers (Plastics)

Plastics and rubbers are types of *polymers*. Foams, like neoprene, also fall into this category. They have low densities, so they are relatively light.

Thermoplastic materials can be molded and remolded when heated, and will return to original form. This quality makes thermoplastics, like soda bottles, recyclable.

Thermoset plastics cannot be remolded and thus can't be recycled. Examples are bowling balls, football helmets, and epoxies.

As with all materials, plastics come in a variety of shapes and sizes. The following are some common types and uses (thanks to Peter Menderson and his materials class notes and resources at http://itp.nyu.edu/materials/):

- ABS is tough and impact-resistant. A lot of small appliances are made from this. It's softer than acrylic and easy to machine. It's what LEGOs and the original printing material used for Makerbot's CupCake CNC are made of.
- Acrylic (trade name Plexiglas) is a clear, hard plastic commonly used in laser cutters and model making. You can cut thin sheets just by scoring it with a hobby knife and then snapping it apart.
- Delrin is tough, easy to machine, and low friction (although not as low as Teflon). It is used commonly in gears and bearings.
- Nylon is similar to Delrin and good for general-purpose wear applications.
- PETG bends easily and is a cheaper alternative to polycarbonate.
- Polycarbonate (trade name Lexan) is shatter-resistant, has excellent clarity, and has high-impact strength. It is porous, so it will absorb moisture from the air.
- Polyethylene comes in a wide range of grades and properties. It also vacuforms well (see Chapter 9 for a description of vacuum forming).
- PVC plastic is the white material usually used for plumbing pipes and fittings. Although easy to saw, cut, and drill, it is particularly environmentally unfriendly because of toxins released in its manufacture and disposal. You cannot laser cut it, and you should use safety gear (mask, goggles) even when cutting or drilling it.
- Styrene is easy to machine, low cost, and flexible. It comes in thin sheets that can be cut with a sharp knife.
- Teflon is slippery, so sheets and tubes are used for bearings and sliding surfaces.
- Rubber parts are used as shock absorbers and stoppers, as well as in O-rings and gaskets to seal mechanisms from the elements.

The type of plastic you're looking for, and how much you need, will help determine where you can find it. As always, starting at the McMaster site is convenient. However, plastics are so popular that you can likely find what you need at a local arts-and-crafts store like Pearl (www.pearlpaint.com) if you're looking for common shapes and sizes like sheets or rods.

Foams, rubbers, and plastics for molds and casted parts fall in this section as well. Insulation foam (the harder, pink and blue kind) and RenShape foam are popular for prototyping. Smooth-On (www.smooth-on.com) makes dozens of materials for prototyping, sculpting, and model making. The Compleat Sculptor (www.sculpt.com), based in New York City, is one of Smooth-On's distributors, and a one-stop shop for everything you could ever need for molding and casting. We'll talk more about this process in Chapter 9.

For more do-it-yourself (DIY) versions of shaping your own plastic parts, check out ShapeLock Design Plastic and Sugru. Both of these products are plastics that you can mold and shape by hand, and then let harden at room temperature. Depending on your application, don't overlook Tupperware or toy sets like LEGO and K'NEX for structural parts and housings.

Composites

Composites are similar to alloys in that two or more materials are mixed together to combine the favorable characteristics of each. However, metal alloys mix only types of metal, while composites can mix materials across different material groups, such as fiberglass and carbon fiber. For example, composites are being used increasingly in the newest airplanes, including the Airbus A380 and Boeing 787. Composites are lighter weight but still strong, so the plane uses less fuel to cover the same distance. Rapid prototyping companies like Solid Concepts (www.solidconcepts.com), which we'll talk about more in Chapter 9, can use composites of aluminum, nylon, and glass to create small functional parts for prototype designs.

Wood

Wood is actually a natural composite, made of strong *cellulose fibers* (think of them as straws) held together by a stiff material called *lignin* (think of this as the glue around the straws).[1] It's relatively easy to work with and generally low cost.

In general, harder woods are stronger, but more difficult to work with. Soft woods like balsa are very light and easy to work with, but also very weak and split easily. Wood tends to split *along* the grain (that's why those disposable wooden chopsticks pull apart so easily).

Composite woods like plywood, medium-density fiberboard (MDF), chipboard, oriented strand board (OSB), particle board, and Masonite are popular due to their low cost and availability. All are combinations of wood chips or particles and binding agents.

- Plywood is made of alternating layers of large wood chips at right angles to each other that are held together with glue, so it is stronger than the sum of its parts and tends not to split like pure woods. Aircraft-quality plywood, available from hobby stores, is higher quality than home-construction plywood.
- MDF is made with fine sawdust combined with a wax and resin binder, and formed under pressure, which makes it stronger and denser than plywood, but also heavy.
- Thin Masonite is very popular to use with laser cutters. It is more environmentally friendly than some of the other composites because it is made from natural materials and doesn't used formaldehyde-based resins to bond the wood particles.
- Bamboo is even more environmentally friendly. Because it grows so fast, it's easily renewable. A few companies make bamboo plywood (try www.plyboo.com for a sample), but it may be a few years before prices come down and bamboo becomes more popular with hobbyists.

Natural woods have a wide range of properties as well. Maple, cherry, and oak are very hard. Spruce and balsa are very soft. They also come in a variety of shapes and sizes, from sheets to more common structural forms like 2 × 4 boards.

The lumber section of your local home-improvement store, arts-and-crafts stores, and hobby stores are good places to find wood materials. Look for wood with few or small knots, dings, or cracks. Look down the board length to check for warp, and avoid wet boards. You may also find scrap wood outside these stores, and some of that might be suitable for your projects.

To construct structures from wood, make through-holes (holes that go all the way through the piece) whenever possible, and use nuts and bolts instead of screwing or nailing directly into the wood and risking a split edge. The farther from an edge the hole is, the less likely the wood is to split.

Wood glue works very well when two pieces of wood are glued along the grain (*not* end to end!). When glued along the grain, the glue has a chance to seep in and grab onto the straw-like fibers, and will actually reinforce the wood. When glued end to end, the glue does not have a chance to seep in and grab anything; it just makes a very weak joint. Keep in mind that wood glues do not accept stains, so be careful not to leave any glue smudges on your work if you plan to stain the final piece. Natural woods accept stain much better than composite woods, so keep this in mind if aesthetics are important to you. See Chapter 3 for more on wood screws and glues.

Paper and Cardboard

Paper, cardboard, and foam-backed boards are also made of wood. Although most of you might not consider paper an engineering material, you can actually do a lot with some thick card stock. Just check out the paper animation kits at Flying Pig (www.flying-pig.co.uk).

Paper or cardboard is also great for quick prototyping to get your head around your ideas—kind of like LEGOs, but easier to cut. You can even cut out pieces of paper or cardboard first, and then glue them to your final work material to use as a cutting guide. Get a good X-Acto (utility) knife, a cutting mat (from an artists' supply store, not a kitchen store), and a metal ruler to use as a straight edge. You can substitute old magazines for a cutting mat, but the mat makes a great work surface in general, so it's a good investment.

To find paper and cardboard, your local arts-and-crafts store is a good place to start, followed by office-supply stores. You can find posterboard and index cards at most drugstores.

Foam Core

Foam core is made of stiff foam sandwiched between two stiff paper boards. It's lightweight, easy to cut, easy to glue, and easy to find. Your local arts-and-crafts store, as well as just about anywhere with a stationery or school supplies section, will carry it.

Semiconductors

Semiconductors have electrical properties halfway between electrical conductors (metals) and insulators (ceramics). These materials have made integrated circuit chips possible, and enabled the electronics and computers we take for granted every day.

You won't work directly with semiconductors in this book, but you will use components made from them (like transistors), so it helps to know what they are.

Biomaterials

A *biomaterial* is a nonliving material used in a component that interacts with the human body. *Biocompatibility* is the ability of a material to avoid rejection by the body.[2]

Some common biomaterials are titanium, stainless steel, PMMA (the pinkish translucent plastic used in dental devices like retainers), Teflon, and silicone. For our purposes, these will come into play only if you are designing wearable technology or something that people will interact with for long periods. In those cases, you need to pay attention to biocompatibility to avoid problems with skin allergies and to enhance comfort. For example, if you're designing a metal frame knee brace that generates energy when you walk, realize that up to one in four women have metal contact allergies (less than half that many men), and you should put a biocompatible pad of neoprene (or something similar) on any part of the frame that rests on skin.

Project 2-1: Different Diving Boards

In this project, we'll put a fixed weight on the end of each of four different materials and see how they behave based on their material makeup and geometry.

Shopping List:

- Four ~12 in long strips of different materials, such as the ones used for Figure 2-4:
    - Wooden paint stirring stick
    - Brass 1/16 × 1/4 in cross section
    - Brass 1/32 × 1/2 in cross section
    - Steel 0.020 × 1/2 in cross section
- Three C-clamps
- Scrap 1 × 3 wood board or hardcover book
- 4 oz clay blocks (or other small, fixed weights)
- Duct tape

FIGURE 2-4 Material testing

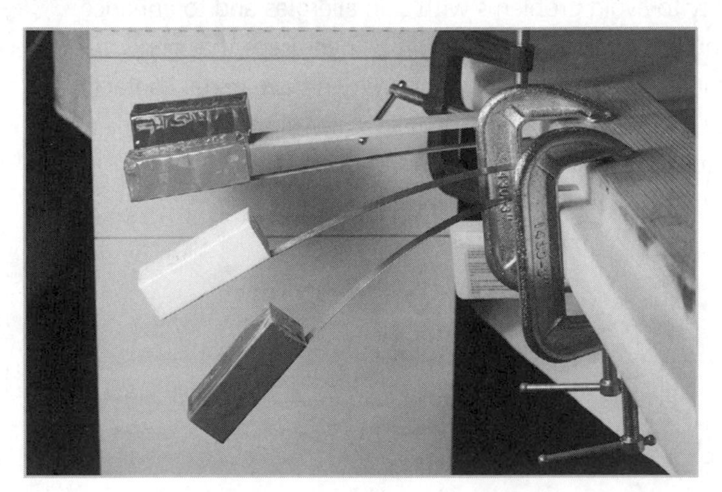

Recipe:

1. Allow about 10 in of each material to hang off the edge of a table. Clamp the wooden paint stick to the edge.

2. Clamp the remaining three materials underneath the book or scrap wood piece.

3. Duct tape the clay blocks to the ends of each material.

4. Notice how small changes in geometry as well as material affect how far the clay droops at the ends of the diving boards. Always keep material and orientation in mind when designing structures for your projects!

References

1. Michael R. Lindeburg, *Mechanical Engineering Reference Manual for the PE Exam, 12th Edition* (Belmont, CA: Professional Publications, 2006).
2. Buddy Ratner, Allan Hoffman, Frederick Schoen, and Jack Lemons, eds., *Biomaterials Science: An Introduction to Materials in Medicine* (San Diego, CA: Academic Press, 1996).

Screw It or Glue It: Fastening and Joining Parts

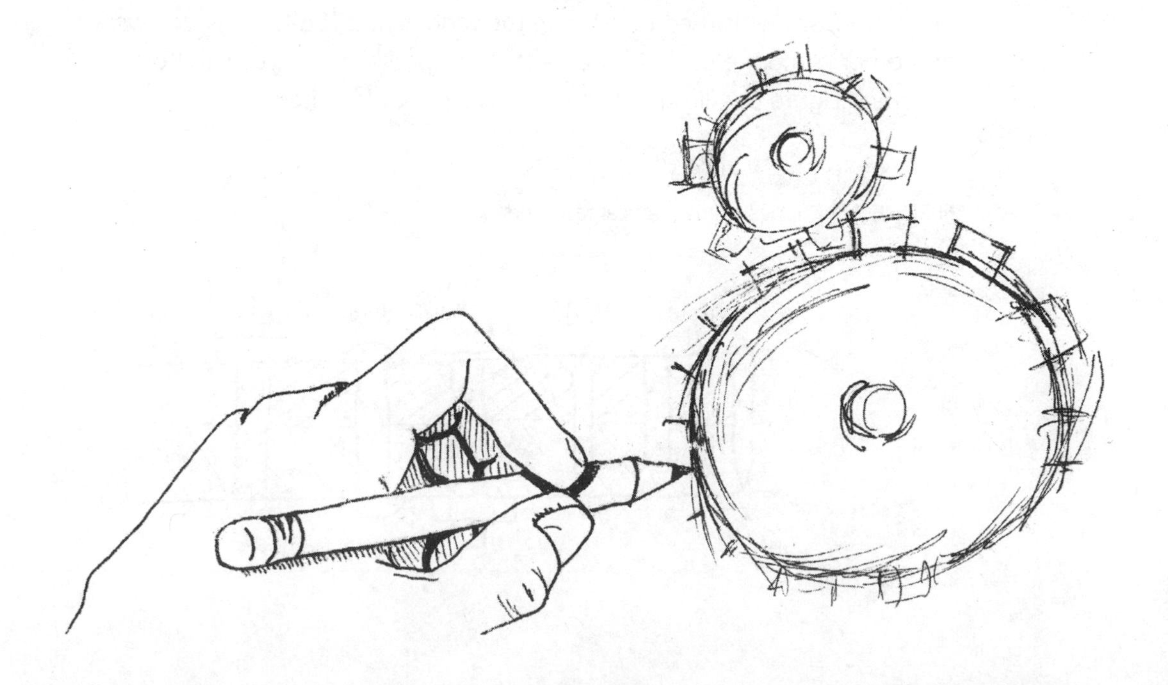

When you first build a project—whether it's for your home, an art installation, or a robot competition—something always goes wrong. Sometimes the cause is dead batteries, broken connections, or a dog using your project as a chew toy. But the problem I see most often is the use of inappropriate materials and fasteners that don't hold up over the course of your mechanism's intended lifetime.

Usually, your mechanical devices will have a base, a skeleton, or some other kind of central structure that will need to be put together. Knowing how to do this efficiently will save many headaches later on. There are two main ways to join components to each other: nonpermanent joints and permanent joints. I recommend using nonpermanent joints whenever possible, because they allow you to take things apart without damage. However, I will discuss permanent joints for use in situations where they are the only option.

This chapter will give you a general understanding of various ways to put things together. In later chapters, I'll talk more about specific situations, such as attaching things to motor shafts.

Nonpermanent Joints: Fasteners

Nonpermanent joints are practical and quick, and they come in a variety of designs. The term *fastener* is used to describe the various nuts, bolts, nails, screws, washers, and other components that have been developed over the years to hold things together. Figure 3-1 illustrates the various types of fasteners.

FIGURE 3-1 Types of mechanical fasteners

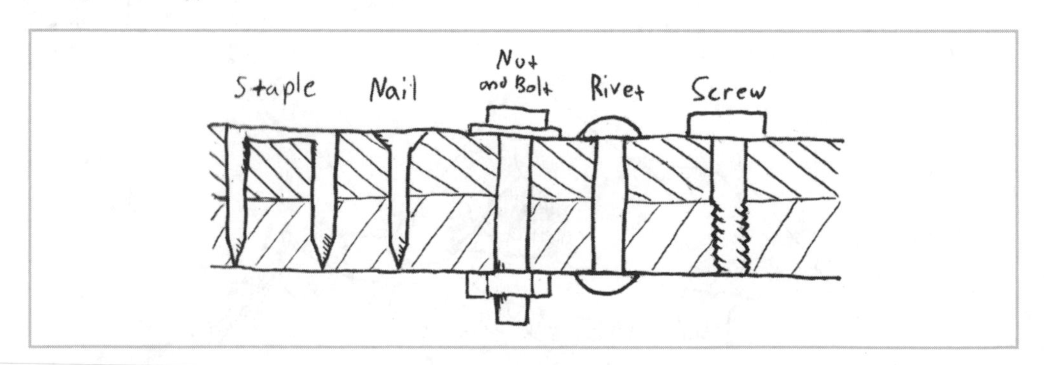

When first building a project, you can bet you'll need to take it apart at some point. Nonpermanent joints make this easy to do.

NOTE ***From Velcro to magnets and cable ties to hose clamps, there are dozens of ways to fasten parts together. Finding the perfect solution is less important than finding something that works for your particular project and the intended life of your mechanism. More fastening techniques and components are entering the market daily, so there's a very good chance you'll find something that works for you.***

Screws, Bolts, and Tapped Holes

For the purposes of this book, the words *screw* and *bolt* are interchangeable. Some people use *bolt* when the component is paired with a nut, and *screw* for a component that threads into a tapped hole in another piece instead of using a nut. However, the distinction is not generally agreed upon and is not important here.

Major Diameter and Threads per Inch

Figure 3-2 shows the characteristics of a screw. The most important ones are *major diameter* (outside diameter) and *threads per inch*. The diameter can be measured in inches or millimeters. In the inch system, screws with a diameter of 1/4 in or more are labeled with the diameter first, then the threads per inch. For example, a 1/4-20 screw is one with 1/4 in major diameter and 20 threads per inch of the screw shaft. For some reason, screws with a diameter less than 1/4 in are given a number. For example, a 4-40 screw has a diameter of 0.112 in. To convert from screw number to decimal diameter, use this formula, or refer ahead to Table 3-1.

$$\textit{diameter (inch)} = (\textit{screw number} \times 0.013) + 0.073$$

Most screws come in a standard (coarse) pitch as well as a fine pitch (more threads per inch). The *pitch* is the distance between threads. In the metric system, screws are labeled with the diameter first, then the pitch. So a 3mm diameter metric screw with a standard pitch of 0.05mm is called an M3×0.05. We'll focus on the imperial system in this book, but the metric system is good to know when you come across components (like some motors) that have holes in metric sizes.

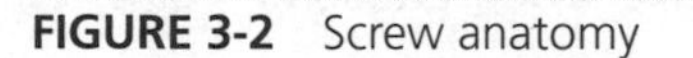

FIGURE 3-2 Screw anatomy

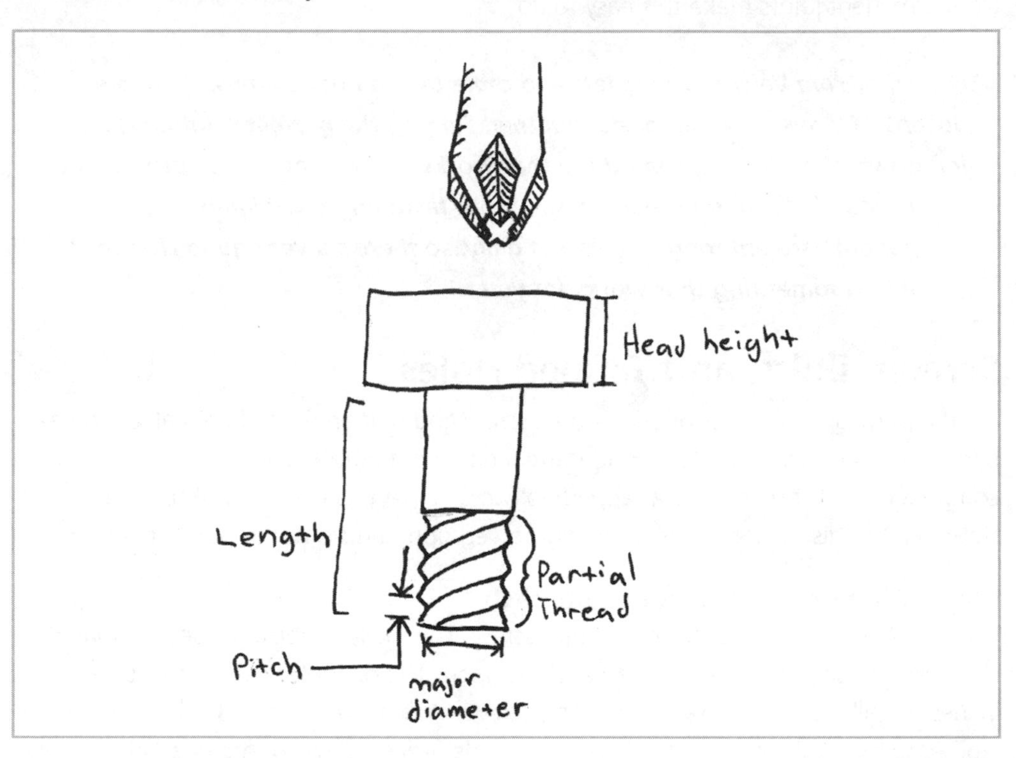

The easiest way to join two materials with a screw is to drill a *clearance hole* through both of them, insert a screw that is longer than their combined thickness, and use a nut on the opposite side to sandwich the pieces together, as illustrated in Figure 3-3. As a rule of thumb, make sure the screw extends all the way through the nut and sticks out at least two or three threads past it. The right clearance hole gives you just enough room to put the screw through, but not so much that it's sloppy. A *close fit* is standard, but a *free fit* is a little larger and will give you more wiggle room for poorly aligned parts.

Sometimes it won't be possible to use a screw and nut due to space or other constraints. In this case, you join the two parts by screwing directly into one of them.

FIGURE 3-3 Anatomy of bolted joints: using a clearance hole with a nut on the end (left) and screwing into one piece that is tapped (right)

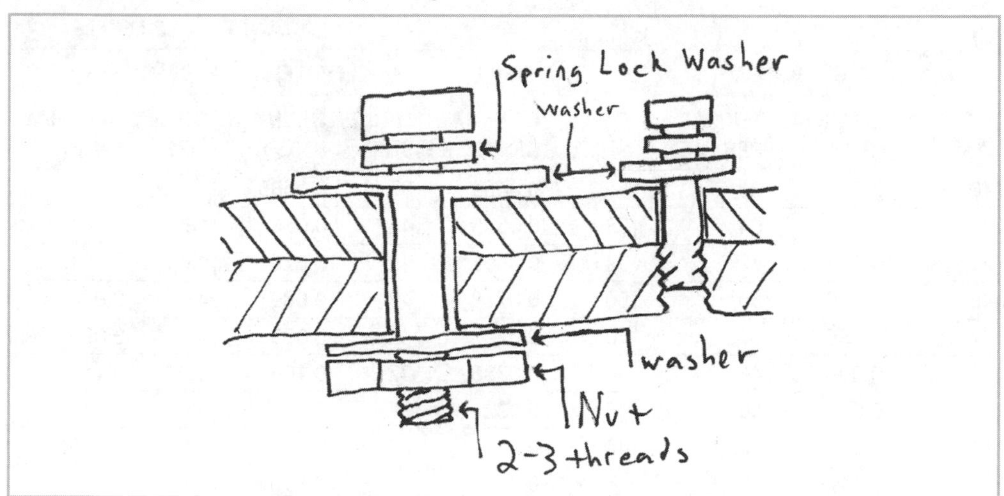

With wood and most plastics, it is best to drill a pilot hole in the piece that you'll drive the screw into to avoid splitting or cracking and make it easier to install the screw. A *pilot hole* is just a hole that's slightly smaller than the screw's major diameter and makes it easier to install.

You can find the right size pilot hole for a corresponding screw by looking it up on McMaster. For example, for part 90031A153, a wood screw, the pilot hole size in soft wood is 1/16 in. In metal, you *must* drill a pilot hole equal to the tap drill size listed for your screw. Then use a tap that matches your screw size to create the threads in the piece you are screwing into. This is not as complicated as it may sound.

For a well-designed joint, use Table 3-1 to determine the clearance hole size that matches your screw. The table also lists the tap drill sizes. (The table information is from www.stanford.edu/~jwodin/holes.html and www.efunda.com/DesignStandards/screws/tapdrill.cfm.)

TABLE 3-1 Screw sizes and tap drill table

SIZE OF SCREW			TAP DRILL		CLEARANCE HOLE DRILLS: CLOSE FIT		CLEARANCE HOLE DRILLS: FREE FIT	
NO. OR DIA.	DECIMAL (INCH)	THREADS PER INCH	DRILL SIZE	DECIMAL (INCH)	DRILL SIZE	DECIMAL (INCH)	DRILL SIZE	DECIMAL (INCH)
#0	0.06	80	3/64	0.0469	52	0.0635	50	0.07
#1	0.073	64	53	0.0595	48	0.076	46	0.081
#1	0.073	72	53	0.0595	48	0.076	46	0.081
#2	0.086	56	50	0.07	43	0.089	41	0.096
#2	0.086	64	50	0.07	43	0.089	41	0.096
#3	0.099	48	47	0.0785	37	0.104	35	0.11
#3	0.099	56	45	0.082	37	0.104	35	0.11
#4	0.112	36	44	0.086	32	0.116	30	0.1285
#4	0.112	40	43	0.089	32	0.116	30	0.1285
#4	0.112	48	42	0.0935	32	0.116	30	0.1285
#5	0.125	40	38	0.1015	30	0.1285	29	0.136
#5	0.125	44	37	0.104	30	0.1285	29	0.136
#6	0.138	32	36	0.1065	27	0.144	25	0.1495
#6	0.138	40	33	0.113	27	0.144	25	0.1495
#8	0.164	32	29	0.136	18	0.1695	16	0.177
#8	0.164	36	29	0.136	18	0.1695	16	0.177
#10	0.19	24	25	0.1495	9	0.196	7	0.201
#10	0.19	32	21	0.159	9	0.196	7	0.201
#12	0.216	24	16	0.177	2	0.221	I	0.228
#12	0.216	28	14	0.182	2	0.221	I	0.228
#14	0.242	20	10	0.1935	D	0.246	F	0.257
#14	0.242	24	7	0.201	D	0.246	F	0.257
1/4	0.25	20	7	0.201	F	0.257	H	0.266
1/4	0.25	28	3	0.213	F	0.257	H	0.266
5/16	0.3125	18	F	0.257	P	0.323	Q	0.332
5/16	0.3125	24	I	0.272	P	0.323	Q	0.332
3/8	0.375	16	5/16	0.3125	W	0.386	X	0.397
3/8	0.375	24	Q	0.332	W	0.386	X	0.397
7/16	0.4375	14	U	0.368	29/64	0.4531	15/32	0.4687
7/16	0.4375	20	25/64	0.3906	29/64	0.4531	15/32	0.4687
1/2	0.5	13	27/64	0.4219	33/64	0.5156	17/32	0.5312
1/2	0.5	20	29/64	0.4531	33/64	0.5156	17/32	0.5312

Project 3-1: Drill and Tap a Hole

As an example, let's drill and tap a piece of aluminum for a 4-40 screw.

Shopping List:

- Scrap piece of aluminum
- Clamp (like McMaster 5031A6) that is wide enough to fit your scrap aluminum piece and worktable
- Center punch or other hard, sharp, pointed object
- 4-40 screw (any length and head style will do)
- Small tap handle (McMaster 25605A63)
- 4-40 tap (one that is designed to fit into the tap handle)

TIP ***The chamfer** **is the tapered part at the front of the tap that helps guide the tap into the hole. Choose a taper chamfer (McMaster 25995A125) versus a plug or bottoming chamfer if you have the choice, because they are easier to start.***

- Tapping fluid or WD-40
- 0.089 in diameter drill bit, also known as wire gauge size 43 according to Table 3-1 (any good drill bit set should have this size, or you can buy it individually, and choose the standard jobber's length if you have a choice; McMaster 30585A57).
- Drill (any hand drill will do, cordless or otherwise)
- Small rounded file or countersink tool (like McMaster 2742A26)
- Safety glasses

Recipe:

1. Put on your safety glasses and clear your workspace.

2. Use the center punch to make a mark where you want to drill. This is not strictly necessary, but will prevent the drill bit from wandering when you start drilling the hole.

3. Clamp the aluminum down to your worktable. Install the 0.089 in drill bit in your drill, and drill a hole either partway through the material (blind hole) or all the way through (a through hole is much easier to tap). You will need to clamp or hold down your scrap aluminum while doing this, depending on its size.

4. Clear off any metal chips or burrs from the hole with your file or countersink tool.

5. Spray or squirt some tapping fluid or WD-40 on the hole. Although not strictly necessary, this will make your job easier and decrease your risk of breaking the tap. It's much harder to tap a dry hole.

6. Install the tap in the tap handle, just as you would insert a drill bit into a drill.

7. Place the end of the tap in the hole you just drilled, and position the tap perpendicular to the material (see Figure 3-4). *The tap must stay perpendicular to the material surface the whole time or it will break*. Carefully turn the tap handle clockwise one or two turns until you feel the little teeth on the tap start to bite into the aluminum. From this point on, turn the handle one-half turn clockwise, then one-quarter turn counterclockwise. This backing up is necessary to cut the aluminum in small pieces so the chips don't build up, plug up the tap, and cause it to break.

CAUTION ***Taps are made of material that is very strong and sharp, but very brittle. They will break surprisingly easily if you twist too hard. Don't try to correct for misalignment once you've started to tap the hole.***

FIGURE 3-4 Ten-step tapping procedure

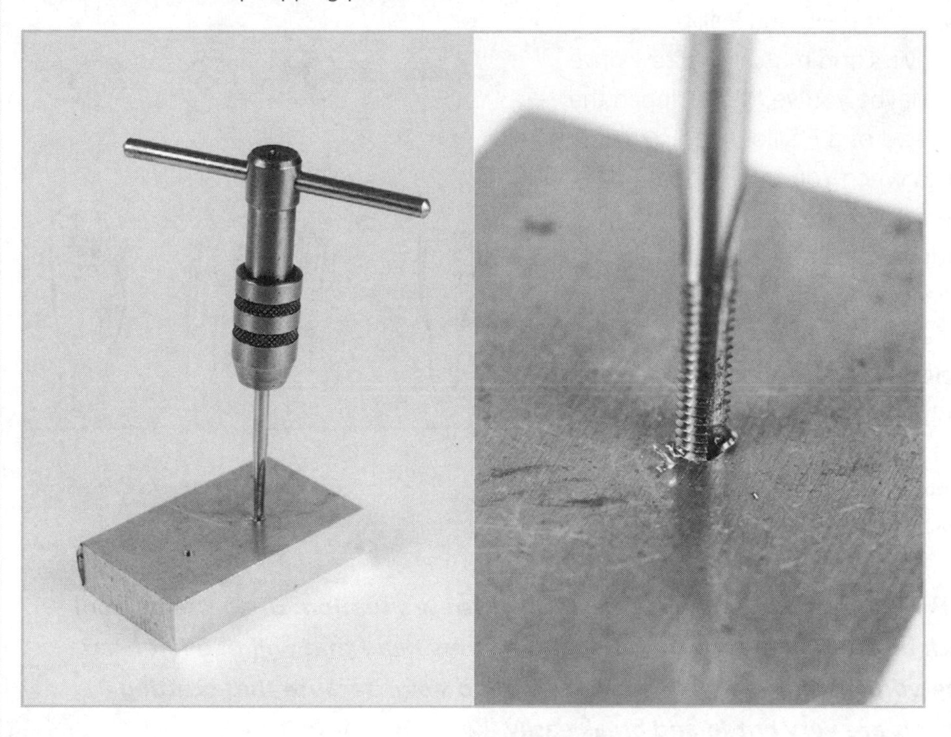

8. Once you've gotten to the bottom of the hole, or through the material, unscrew the tap all the way back to remove it.

9. Wipe, rinse, or blow away any metal chips and excess tapping fluid.

10. Try screwing the 4-40 screw into the hole. Ta da! It should twist in easily.

Head and Drive Styles

The next important considerations with screws and bolts are the *head style* and *drive style*. Figure 3-5 illustrates some common drive styles.

You are probably familiar with the standard flat head and Phillips head screwdrivers and matching screw drive styles. Maybe you've even stripped the screw head of a Phillips head screw. This happens when you try to tighten or loosen a screw that is stuck, and the screwdriver slips out, squishing the material on the head so much that it becomes impossible to tighten or loosen at all. If this happens, you're screwed. Avoid this problem by being particularly careful with Phillips head screws, or avoiding them altogether by using socket cap or hex head screw drive styles.

FIGURE 3-5 Drive styles of common screws

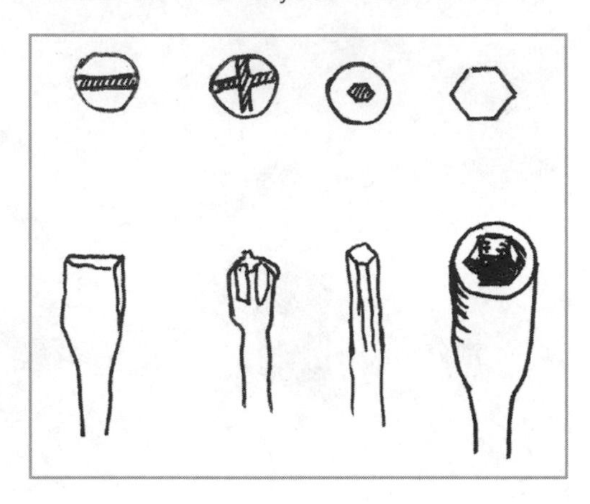

TIP ***There is one way out of a stripped screw situation. Use a Dremel tool with a cutting wheel to cut a slot in the screw head that will fit a flat head screwdriver. Use your safety glasses and go slow, because these cutting wheels are very brittle and break easily.***

Socket head screws are slightly less convenient because you need to have a different Allen wrench (also called an Allen key or a hex key) for each screw size, as shown in Figure 3-6. However, these types of screws are much less likely to strip. This is especially true when you're trying to undo a screw you accidentally glued in place because you *thought* you were done. Socket head cap screws are designed to resist tension in the joint; button heads are not—they are designed to look nice. Hex head screws are more popular for heavy-duty applications, or for when you can't get at the screw head with a conventional screwdriver but can get to it from the side with a wrench.

The head style you use will vary depending on your application and convenience. Wood screws commonly come in flat head styles that you can drive right into the wood. Socket cap screws and machine screws come in a variety of styles. Some of the most common head styles are shown in Figure 3-7.

FIGURE 3-6 Allen (hex) key sets

NOTE ***The length for flat head screws is measured from the flat top. The length for all other styles is measured from under the head.***

If you want your screw to sit flush with the material, instead of on top of it, you will need to use a *countersink* or *counterbore* (see Figure 3-8). McMaster (among others) sells special countersink and counterbore drill bits for this purpose.

FIGURE 3-7 Head styles of common screws

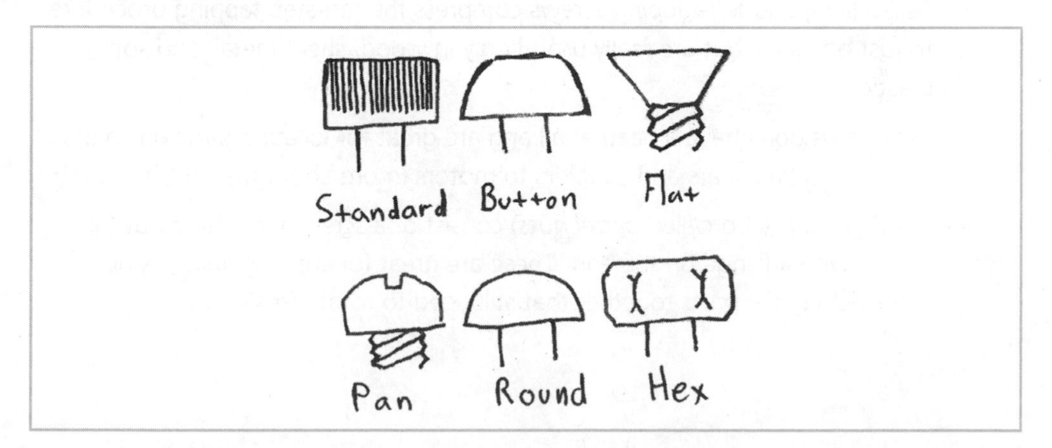

Material

The last thing to worry about when choosing screws is material. The vast majority of the ones you use will be steel. Choose stainless steel if you don't want the screws to rust. A plain steel screw with a zinc-plated or black-oxide finish will also protect from rust and might be cheaper than the stainless steel option.

FIGURE 3-8 Counterbores (left) and countersinks (right) allow screws to sit flush with the surface.

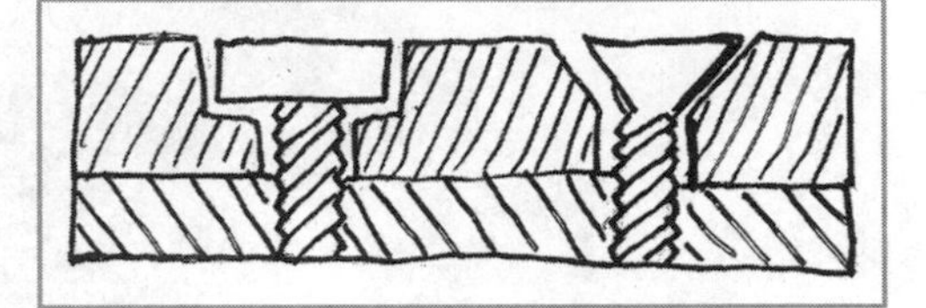

Threaded Rods and Speciality Screws

Threaded rods, or all-thread, are like long screws with no head. You can get them full or partially threaded and in many different lengths and sizes. One can act as a shaft to align multiple parts, then sandwich them together with nuts on each side. They are also used as push-pull rods to create small motions, like steering a rudder on a model airplane.

There are more types of screws than I have pages to write about them, but a few deserve a quick mention (see Figure 3-9):

- *Shoulder* screws have a smooth cylindrical shoulder under the head before the threads start that is great to use as a spacer or shaft.
- You can use *U-bolts* to create a loop on an otherwise flat surface.
- *Eyebolts* serve a purpose similar to U-bolts, but need only one mounting hole (versus the two you need for U-bolts).
- *Self-drilling* and *self-tapping* screws compress the ten-step tapping procedure to just one step, but are really useful only in wood, sheet metal, and soft plastics.
- *Set* screws don't have a head at all and are great for locating parts on shafts and connecting gears and couplers to motors (more about that in Chapter 7).
- *Binding posts* (also called *barrel nuts*) consist of a screw and a long nut (or barrel) with a flange on the end. These are great for creating linkages or sandwiching flat parts together that still need to rotate freely.

FIGURE 3-9 Specialty screws

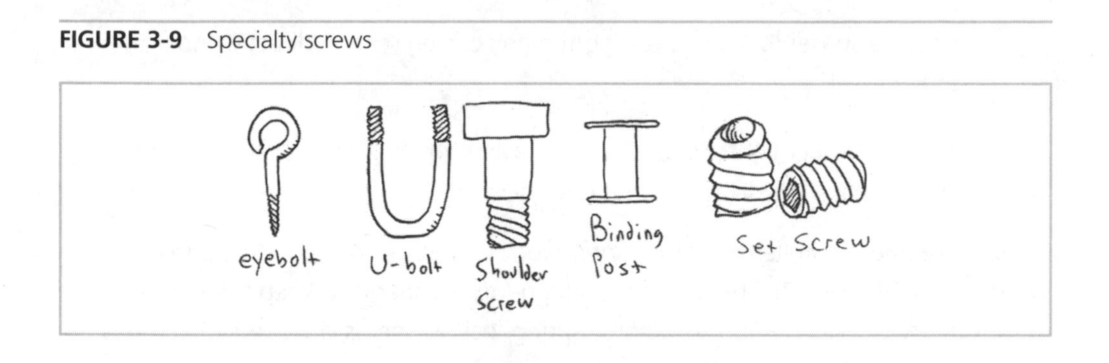

Nuts

Nuts thread onto the end of screws and are the easiest way to form a bolted joint. Choose a nut with the same thread size and pitch as your screw or it will not fit. The following are some types you might use in your projects:

- The most common nut is called a *hex nut*.
- Next most common in usefulness is a *locknut*. Locknuts have a nylon insert or deformed thread that makes them vibration-resistant and very secure. However, they are generally not reusable.
- *Wing nuts* are great if you need to assemble and disassemble connections without tools.
- *Tee nuts* are ingenious little components that you can hammer into a hole in wood and soft plastics to create metal internal threads with little effort.
- *Rivet nuts* do the same job as tee nuts in sheet metal assemblies, but require a special installation tool and are not easily removable once installed.

Washers

Washers serve at least four purposes:

1. Allow you to avoid marring the base material during installation
2. Spread the screw fastening force over a larger area

3. Act as a spacer to avoid stressing the tiny curved section directly under the screw head

4. Indicate proper tightness of the bolt—when the washer is snug and stops spinning, it's time to stop turning the screwdriver

For a well-designed bolted joint, use *spring-lock washers* in addition to standard washers (see Figure 3-3). The job of the spring-lock washer is to keep the joint tight even if the screw vibrates loose. When a spring-lock washer is compressed, it looks just like a fat washer with a gap in it, or a C shape. However, if the screw begins to loosen—either through wear and tear or vibration—the spring-lock washer springs up as the C shape untwists to fill the tiny gap created.

Nails and Staples

Nails are more permanent than screws but can still be removed relatively easily. The following are types commonly used in projects:

- *Double-headed nails* are good to hold something together temporarily and much easier to remove than common nails.
- *Finishing nails* have small heads designed to sink into the material, so never use these if there's a chance you'll need to take one out.
- Small nails, called *upholstery tacks*, can secure fabric stretched over a frame, as used in furniture designs.

Nail guns make short work of this job. Similarly, a staple gun, or just a normal stapler opened up, makes short work of stapling material together to temporarily secure it.

Nails and staples are often the lazy way out of designing good, nonpermanent joints. I recommend using them only for temporary holding or early prototyping stages, not for general use. Removing wrongly installed nails and screws usually renders them unusable, damages your base material, forces you to line up your pieces again, and wastes more nails or screws. This cycle can get repetitive and tedious. Try designing your projects to use nuts and bolts from the early stages, and resort to nails and staples only if you have no other choice.

Pins

If you're ever put together furniture from IKEA, or used an exercise machine at a gym with a weight stack, you've dealt with pins used for fastening and alignment. You can find wooden dowel pins at any arts-and-crafts store, but metal dowel pins and spring pins are more common to use for aligning parts.

If you make a hole in a part just slightly smaller than the pin diameter, you can hammer or press in the pin, and friction will keep it in place. If you make a hole just slightly bigger, your pin will slide in easily and be removable.

Retaining Rings

Retaining rings can be used with pins and shafts to stop them from sliding all the way through holes or to create pivoting joints. Figure 3-10 shows an example of a retaining ring used in a piece of gym equipment.

FIGURE 3-10 Retaining rings on gym equipment keep the shaft of a pivoting joint in place.

You need to carve a slot in a shaft to hold retaining rings in place, but this is easy to do if you have access to a lathe. See Chapter 9 if you want to know more about lathes.

Permanent Joints: Glues, Rivets, and Welds

Permanent joints are used when two pieces are designed to go together and never need to come apart. This includes welds, rivets, glues, and epoxies. Although a glued joint is quick and easy, permanent joints can cause headaches if something goes wrong and you need to take things apart. Permanent joints are a last resort. Consider using nonpermanent joints first.

Adhesives

Adhesives come in many different forms, from the common white Elmer's glue we all used in elementary school to two-part epoxy. They can take anywhere from a few seconds to a week to dry and reach full strength. Some are designed to join similar materials, while others can be used on dissimilar materials as well. (Go to www.thistothat.com for some good advice on which adhesives to use based on what you are gluing together.)

As a rule of thumb, make sure both surfaces are clean, and give them texture with some sandpaper or a file to give the glue more surface area to grab.

Wood Glues

Wood glues should be used to join wooden pieces along the grain. When used this way, the bond is very strong—sometimes stronger than the wood itself.

Common white multipurpose Elmer's glue will work for wood, but yellow wood glue is better suited for most applications. With yellow glue, you have about 15 minutes from when the glue leaves the bottle to when it starts to dry.

The next step up is Titebond II, which you should use if you need water resistance, a tackier working material, and a faster setting time.

Finally, there are polyurethane foam glues (like Gorilla Glue) that react with moisture to expand and fill gaps and crevices and dry to form extremely strong joints.

Epoxies

The classic 5-minute epoxy is a favorite of hobbyists. Epoxy is a glue that comes in two parts and is activated only when these two parts are combined. This mixing can be done by hand with a popsicle stick, or more conveniently through an applicator gun and mixer nozzle.

Epoxy can be used to bond many different types of plastic, metal, and composites. It dries hard and can be sanded or painted.

You can find epoxy at just about any hardware or home-improvement store (and, of course, online at McMaster). Look for epoxy putty in the plumbing section of hardware stores. Propoxy, FastSteel, and QuikSteel are common brands (see Figure 3-11). It's a putty-based two-part epoxy that hardens like steel after being mushed together and exposed to air for about 20 minutes.

Plastic Glues and Solvents

Glues made specifically for plastics can be particularly effective because they can react chemically with the plastic to melt the two pieces into a strong joint. Weld-On is a well-known brand for bonding acrylic, and is used extensively in the architecture and

FIGURE 3-11 Epoxy putty—the great project saver

modeling industries. Other glues, epoxies, and solvents that cite specific plastics as their target materials work great as well.

CAUTION ***Weld-On is pretty nasty stuff. Open the windows, use a fan, and wear a respirator mask if there is not good airflow where you work. Avoid direct inhalation. Use gloves to avoid skin irritation. Wear splash-proof safety goggles. The fumes can irritate your eyes even if you don't spill.***

Threadlockers

When screws are used in metal and are exposed to vibrations, shock, or varying loads, it is common to use a *threadlocker*, which is a type of glue used to fix the screw into the tapped hole or nut. A threadlocker also seals the bolted joint from fluid leakage and helps prevent corrosion.

The Loctite brand name has become synonymous with this application. It comes in dozens of varieties, so read the description to make sure you choose the best one for your application.

Super Glue

Most kinds of super glue come in small or one-use tubes and dry within seconds. The gel-based ones are the easiest to use because they are thicker and don't drip or run easily.

In general, these glues are weaker than epoxies and should be used only as a last resort or for a temporary fix. Avoid using them in applications in which they are under force or pressure. Super glues are better suited to fixing your sunglasses than for holding moving mechanisms together.

You can find super glue just about anywhere, from your local drugstore to McMaster.

Hot Glue

A hot glue gun and some glue sticks definitely deserve a place in your toolbox. Hot glue can be used for just about anything, from gluing cardboard to insulating exposed metal on electrical components. It's also inexpensive and easy to find at any arts-and-crafts store.

CAUTION ***Be careful when the glue gun is plugged in so you don't burn yourself or the table on which you're working.***

One disadvantage is poor vibration tolerance. Hot glue tends to separate from the base material if much vibration happens, or if anything attempts to pull the two things apart with much force. Also, try not to glue something that melts when it gets hot, or you'll end up with hot goo.

Tape

Common translucent office tape or double-sided tape is not that useful for our purposes, but the double-sided foam tape you can get at hardware stores can be handy. You can use this tape to mount components that don't have mounting holes, like small motors, or to hold something in place while you drill holes to fasten a component more securely.

> **CAUTION** ***Double-sided foam tape is very sticky and hard to remove from surfaces once attached. Make sure you've decided where the components you are using should be before using it. If you do need to peel off a connection and redo it, scrape off as much of the tape as you can with a hobby knife, and use Goo Gone to remove the rest.***

Duct tape is easy to tear and relatively strong. Its fancier cousin gaffer tape is similar, but leaves no sticky residue when removed.

Rivets

A blind or pop rivet consists of a small, flanged metal tube with a metal rod running through it and a ball at the end (see Figure 3-1). A rivet installation tool pushes the rivet into a predrilled hole and pulls the rod back into the tube, and the resulting distortion of the ball flares outward. The rod breaks off in order to form another flange on the back side of the material. These flanges sandwich two or more pieces of sheet metal or other thin materials together. You can also find a rivet tool attachment for a hand drill that works in a similar way. Once installed, a rivet is not removable, except by drilling it out to create a hole bigger than the initial hole used before riveting.

Rivets are used primarily when there is access to only one side of the joint, or when there is no room for a screw-and-nut combination. They are also used for aesthetic reasons to keep a surface relatively flat looking, and are used in mass production of commercial parts after a design has been finalized. You can find all kinds of rivets and rivet tools online at McMaster.

As with other permanent fastening methods, I recommend avoiding rivets unless they are a last resort, because you will not be able to easily redo and adjust your designs.

Welding, Brazing, and Soldering

Welding, brazing, and soldering are ways of joining metal with metal by using heat. They are all permanent joints and need to be reheated, cut, or both in order to reverse them. Welding melts two similar metals together, sometimes using filler rod of a similar metal. Brazing and soldering both use heat to melt *dissimilar* metals, as a kind of glue between two pieces.

Welding

In welding, two metals are joined together by melting them along a seam or at a spot, sometimes by using a similar metal to fill the voids. The two main types of welding are gas welding and arc welding. Gas welding uses a blow torch that combines a fuel (commonly acetylene gas) with oxygen to produce a flame that melts the two pieces of metal you are welding together. Arc welding uses a DC or AC electrical current that is converted to heat to melt the materials. There are two common types of arc welding:

1. Tungsten inert gas (TIG) welding uses a pointy metal electrode to initiate the arc between it and the materials to be welded. You can also supply a filler material in this arc as you go along a weld seam, but this method takes considerable coordination.

2. With metal inert gas (MIG) welding, a wire is fed at a constant rate through the welding tip while the materials you are welding heat up. This method still takes practice to master, but is slightly easier to learn than TIG welding.

NOTE ***If you want to learn more about welding, look for classes at your local art or community center. For example, in the New York City area, check out the Educational Alliance (www.edalliance.org) and 3rd Ward (www.3rdward.com) sites.***

Welding is useful for heavy mechanism work, but isn't as useful at small scales. It also has a lot of overhead (equipment, safety supplies, and so on) compared to other fastening techniques. That said, you will definitely feel more hardcore if you learn to weld, and it always helps to have another fastening technique at your disposal when the other options just won't work. For example, one of my former students made Skybike, a bike you ride upside down, by welding together parts of old bicycle frames (see http://itp.nyu.edu/~md1660/skybike.html). This project would have been much more difficult with any other fastening technique.

Brazing

Brazing uses a copper-zinc or silver-based alloy filler with a melting point above 800°F to glue two pieces of metal together. Although the melting point of the filler is lower than that of the metals being joined, the metal parts both melt a bit, and this fusion helps the joint strength.

Copper brazing is a popular way to join tubes for home plumbing systems. Silver brazing (sometimes inaccurately called silver soldering) is used extensively when making silver jewelry.

Braze welding is the term used for joining metals with a dissimilar filler rod. Although these joints are weaker than in traditional welding, the advantages are that you can join dissimilar metals and minimize any distortion from heat.

Soldering

Soldering traditionally uses a lead-tin filler with a melting point below 800°F, although lead-free solder is becoming more popular to reduce e-waste, in compliance with the Restriction of Hazardous Substances directive (RoHS).

Soldering is not a fusion process because the base materials don't melt, so these joints are the weakest we've talked about so far and are generally only for connections in electrical components and wires. Even with these components, you should solder only when there is no other way to make the connection. Solderless breadboards were created for prototyping purposes, and we'll talk more about using them in Chapter 6.

Forces, Friction, and Torque (Oh My!)

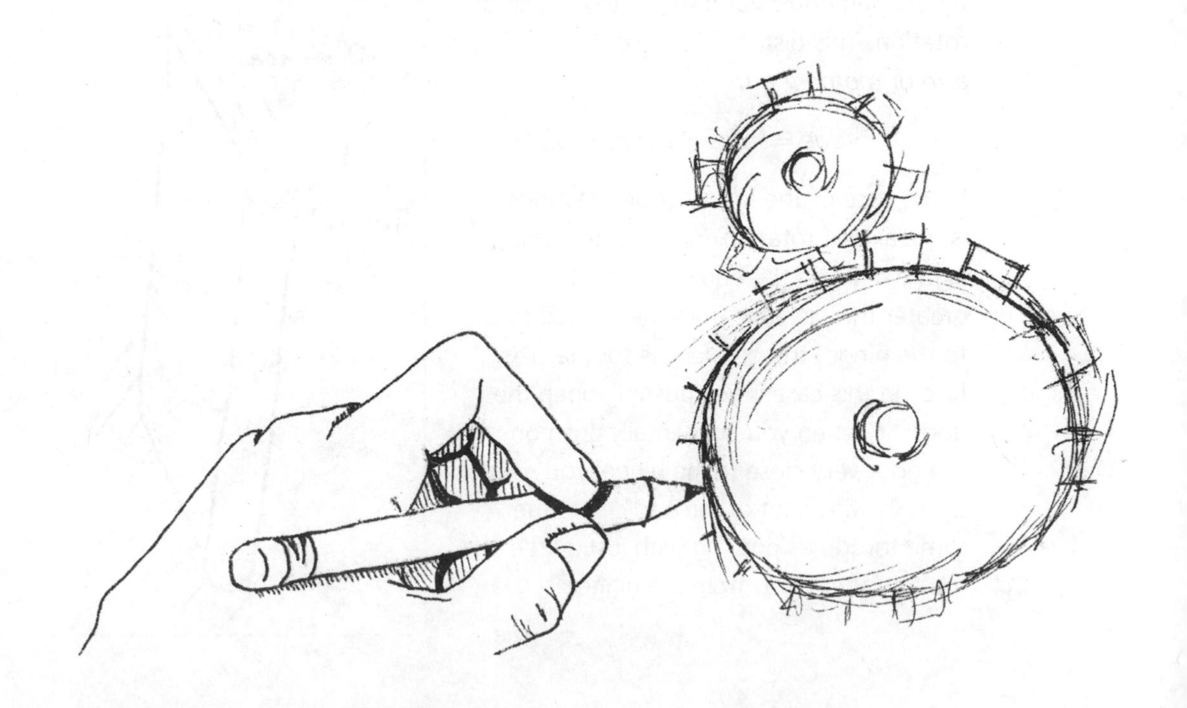

Have you ever tried to push open a door, only to realize you're pushing on the wrong side—the one closer to the hinge—and it's really hard to open? This happens because a door needs a certain amount of torque to open, and if you push too close to the hinge, you have to use a lot more force than if you push at the handle to create the same amount of torque. I suggest pushing on doors in the middle to avoid embarrassment.

In order to estimate torque, know where on the door to push, figure out if something will break, choose a motor, or pick a material for a project, it helps to think about the world around us in terms of numbers. So, before we get to examples of forces, friction, and torque, we need to review some math.

Torque Calculations

First, you need to understand the relationship between force and *torque* (also called *moment*). We talked about force in Chapter 1. Just as a force can be thought of as a push or pull, torque can be thought of as turning strength.

Torque is how hard something is rotated. More specifically, torque is force multiplied by the perpendicular distance to the axis of rotation. This distance is also called the *lever arm* or *moment arm*:

$$\text{Torque} = \text{Force} \times \text{Distance}\ (\perp)$$

In the case of the unruly door, the hinge is the axis of rotation. You can see from Figure 4-1 and the equation that the greater the distance from the applied force to the hinge, the greater the torque. The force in this case is you pushing open the door. So when you accidentally push on the door very close to the hinge, you need to push with a lot of force to create the same torque as pushing with just a little force farther away from the hinge.

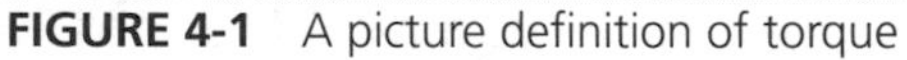

FIGURE 4-1 A picture definition of torque

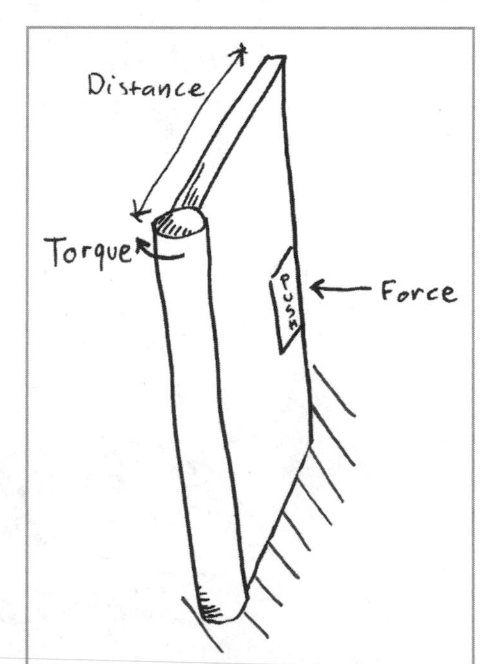

You can feel torque in action with a simple exercise. Grab a can of soup from your pantry and hold it in your right hand, all the way out to the side so your arm is parallel to the floor. The strain you feel in your shoulder is your muscles creating the necessary torque to support the soup can. Your shoulder is acting as an axis of rotation, and the torque is the force of the can (its weight) multiplied by the distance the can is from your hand to your shoulder (see Figure 4-2). If the can weighs 1 lb, and your arm is 2 ft long (d_1), the torque at your shoulder is 1 lb × 2 ft = 2 ft-lbs.

FIGURE 4-2 Shoulder torque when holding a can with your arm parallel to floor

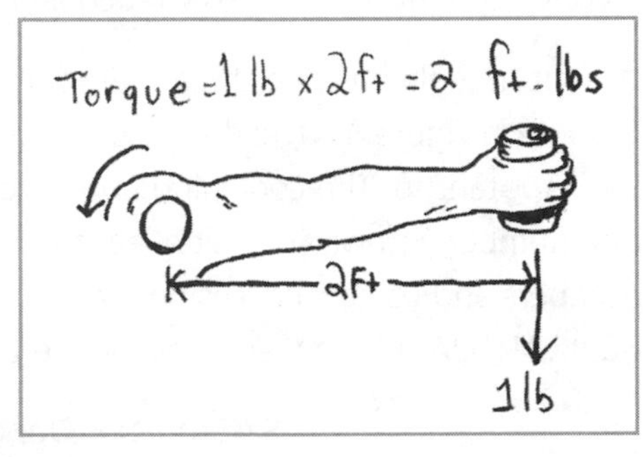

Your shoulder will get tired after a while in this position, so lower the can about halfway (see Figure 4-3). Now the torque on your shoulder is less, even though your arm is still the same length and the can weighs the same. Why? Because the force of the can is still pointed down (gravity always is!), but the perpendicular distance to the axis—your shoulder—is smaller (d_2).

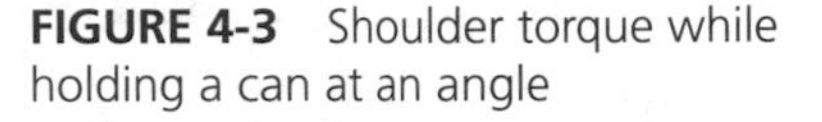

FIGURE 4-3 Shoulder torque while holding a can at an angle

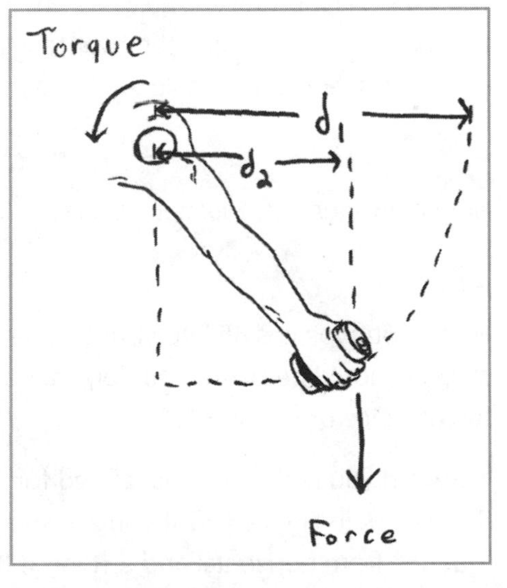

Torque always has units of *distance* × *force* (sometimes written as *force* × *distance*). Unfortunately, there are many ways of measuring distance and force, so torque can be in foot-pounds, ounce-inches, millinewton-meters, and so on. You can go to www.onlineconversion.com/torque and convert from any measurement you come across to one you prefer. There are even smart phone apps that do this for you. For example, if you find a motor that lists its torque in millinewton-meters, and foot-pounds make more sense to you, convert the motor torque to foot-pounds.

The last bit of math we need to review before going through some examples is the basic geometry of triangles you probably learned in high school. Remember sine, cosine, and tangent (abbreviated sin, cos, and tan)? Those three trigonometric properties help you to estimate torque when, as in Figure 4-3, the line of force isn't at a convenient right angle to what connects it to the axis of rotation.

A right triangle has one angle that's 90° (indicated by the box in the corner of the triangle in Figure 4-4), and the side opposite the 90° angle, the longest side, is called the hypotenuse. The cool thing about right triangles is that you can figure out any one number you want—a side length or angle—just by knowing any two other numbers and using sine, cosine, or tangent. To remember the relationships, think SOHCAHTOA. It's a mneumonic device to remember these formulas:

Sin (angle) = **O**pposite Side / **H**ypotenuse

Cos (angle) = **A**djacent Side / **H**ypotenuse

Tan (angle) = **O**pposite Side / **A**djacent Side

Use these relationships to solve for the distance of side X in Figure 4-4. All we know is that one angle is 45° and the hypotenuse is 2 ft. Since we want to solve for the side that is adjacent to the angle we know, we can use the cosine, like this:

$$\cos(45) = X / 2$$

Now rearrange the equation to solve for X:

$$X = \cos(45) \times 2$$

When you type cos 45 into your calculator, the answer should be 0.707. Multiply that by 2 to get the distance of side X = 1.4 ft.

FIGURE 4-4 Right triangle

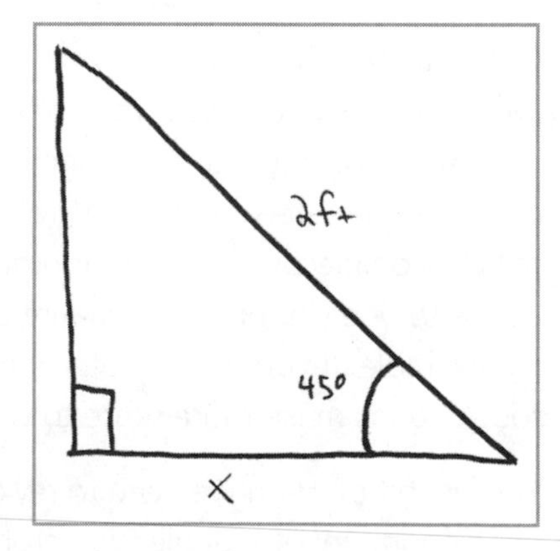

Now, did you realize you just solved for unknown distance d_2 in Figure 4-3? If you assume the arm is at 45° from horizontal and 2 ft long, that's exactly what you've done. In Figure 4-2, we already figured out that the shoulder torque when holding the can of soup straight out was 2 ft-lbs. Now figure out the torque when the can is held down at this 45° angle, as in Figure 4-3. The perpendicular distance we just solved for is 1.4 ft, multiplied by the weight of the can (1 lb), so the torque is 1.4 ft × 1 lb = 1.4 ft-lbs. That's why it's easier to hold the can lower than it is to hold it straight out.

Let's take this example one step further and assume that you want to make a human-sized puppet that can raise and lower a can of soup. If you put a motor at its shoulder, how strong would the motor need to be?

In order to design mechanisms that move, you first need to understand how to estimate if something will break when it's not moving, or *static*. Most of the time, you can isolate a static problem that represents the worst case of a *dynamic* (moving) one. In this case, it's when the weight of the soup can is farthest from the motor shaft, thus requiring the greatest shoulder torque. Since the weight of the soup can doesn't change, the highest torque will be needed when the distance from the can to the puppet shoulder is the highest. This happens when the puppet arm is parallel to the floor, as in Figure 4-2. Since we already solved for this maximum torque of 2 ft-lbs, you know to look for a motor that is at least that strong, and it will be able to handle all the other angles just fine.

Friction

Friction occurs everywhere two surfaces are in contact with each other. It's what makes door hinges squeaky and mechanisms noisy. High friction is sometimes a good thing when your mechanism interacts with the environment, such as the way friction allows your car tires to grip the road. However, friction is usually your enemy when it comes to making things move. It can rob your mechanism of power and decrease efficiency. Low friction is what we strive for inside mechanisms to make things run smoothly. Low friction is what makes nonstick cookware slippery and causes you to slide on ice.

So what is friction, and how can you design projects and choose materials to minimize it?

FIGURE 4-5 Forces on a box at rest

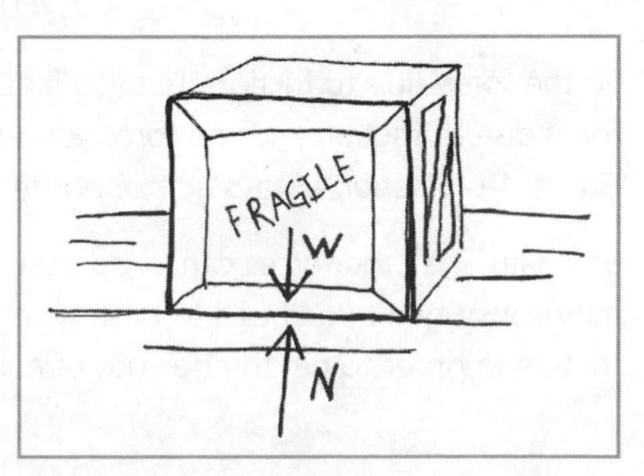

Friction is actually a force, just like your weight is a force. In fact, the force of friction is a percentage of any object's weight. Suppose you are trying to move a 50 lb box across a hardwood floor. At first, when the box is at rest, there are two forces acting on it, as shown in Figure 4-5:

1. **Weight (*W*)** The weight of the box is focused at the box's center of gravity and points down toward the floor, in the direction of gravity.

2. **Normal force (*N*)** This means the floor is actually pushing up with the same amount of force as the box weighs in a direction normal (or perpendicular) to the floor. This might seem like a made-up force, but think of what would happen if you set the box down on a bed of marshmallows. Those marshmallows would squish until they were compressed enough to support the weight of the box. The floor doesn't need to squish, because it's already strong enough. (If your floor needs to squish to support a 50 lb box, you should get yourself a new floor!)

The normal force (N) equals the weight (W) of the box when it's standing still. The fancy term for this is *static equilibrium*, which is when all the forces acting on an object cancel out so the object doesn't move.

Since the box is heavy, you decide to sit on the floor and push it with your feet instead of trying to lift it. How hard do you need to push? You must push just hard enough to overcome the friction between the 50 lb box and the hardwood floor. This force is a fraction of the normal force. In equation form, the last sentence looks like this:

$$\text{Force Due to Friction } (F_f) = \text{Fraction } (\mu) \times \text{Normal Force } (N)$$

The fraction, or *coefficient of friction*, commonly represented by the funny-looking Greek letter μ (pronounced *miu*), is less than 1. This means that you'll need to push sideways with less than 50 lbs of force to move the box.

For example, let's assume the coefficient of friction for our cardboard box on a hardwood floor is 0.4. Our equation looks like this:

$$F_f = 0.4 \times 50 \text{ lbs}$$

So the force due to friction that you'll need to overcome is just 20 lbs. If you put the force due to friction and the force you're pushing with into the diagram, it looks like Figure 4-6. *Friction always acts opposite the direction of movement*.

So again, if all the forces cancel out, the box will be in static equilibrium, and it won't move anywhere. You need to push with just slightly more force than the force due to friction in order to get the box out of equilibrium and moving.

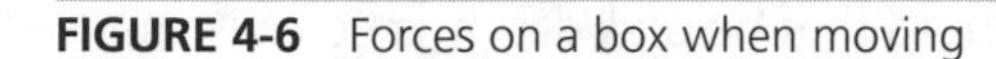
FIGURE 4-6 Forces on a box when moving

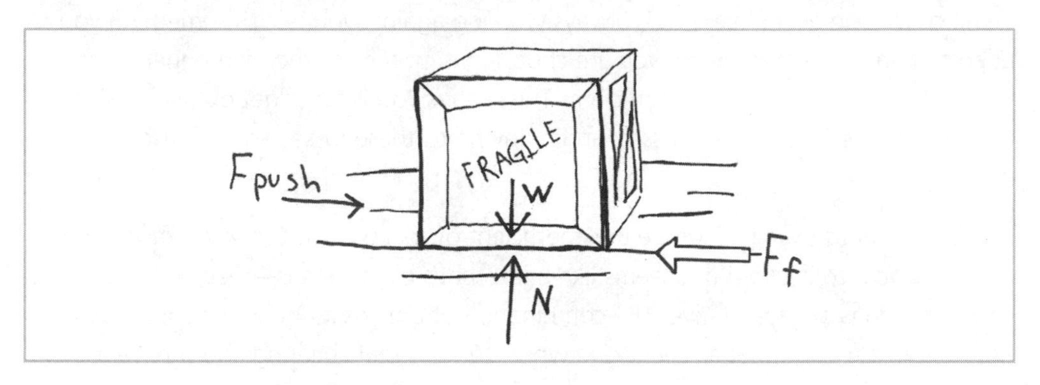

In the previous example, we just assumed a coefficient of friction. But what if you don't know the coefficient of friction and need to measure it? Let's use the same 50 lb cardboard box, but this time on a piece of plywood. Put the box in the middle of the plywood, and then start lifting up one edge of the plywood until the box just starts to slip. Use a protractor or other angle-measurement tool. The coefficient of friction is just the tangent of the angle where the box starts to slip. In equation form, it looks like this:

$$\mu = \tan \theta$$

The location of angle θ is shown in Figure 4-7. If the box starts to slide down the plywood at an angle of about 22°, that matches the coefficient of friction of 0.4 we assumed earlier.

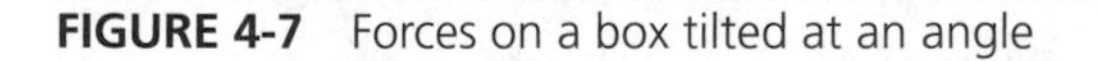
FIGURE 4-7 Forces on a box tilted at an angle

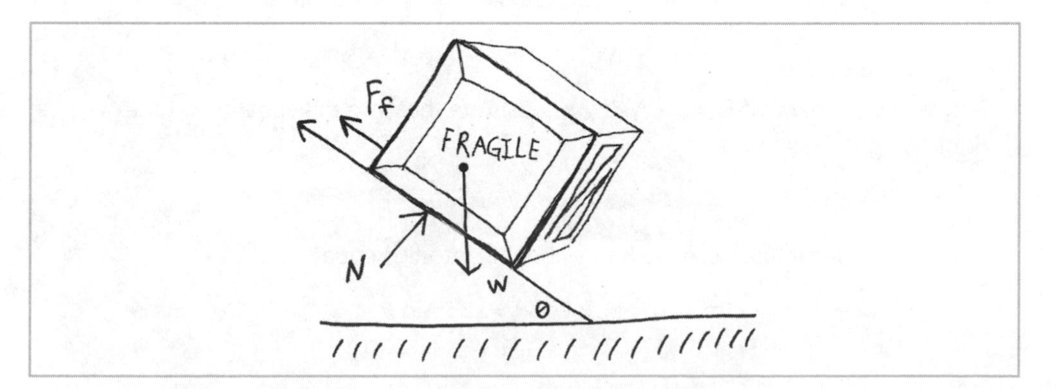

All this talk of sliding boxes is great, but how does that help us design mechanisms? Think of friction as a force that is always working against you, so it's something you need to compensate for when you are choosing a motor or other components for your project. Nothing is 100% efficient. This means you'll never get out everything you put in. There are always losses, and many times these losses are because of friction.

In the preceding example, where the coefficient of friction was 0.4, *40% of the input force was lost to friction*! If it seems like a lot, that's because it is—well, at least in this case. Friction is always relative. The combination of a box sliding on plywood might have a lot of friction, but a roller-skate wheel turning in its bearing might have a coefficient of friction of 0.05, for only a 5% loss.

The Coefficient of Friction

Some fancy math goes into making this simple equation to estimate the coefficient of friction. The force arrows in Figure 4-7 are not all at perfect right angles, like those in Figure 4-6. In order to cancel out the forces and calculate them, they need to be pointing in the same direction. To do this, we split up the weight force (W) into two components: one that is parallel to the plywood (opposite the force of friction) and one that is parallel to the normal force (N).

You can see that the angle at the top of this force triangle is the same as the angle the plywood is lifted off the floor (angle 1 on the left side of Figure 4-8). So we use some trigonometry and figure out that the component of the weight parallel to the plywood is $= W \times \sin\theta$, and the component parallel to the normal force (N) is $= W \times \cos\theta$. We can put these forces back in our diagram, as on the right side of Figure 4-8. Now we can start canceling out forces. We know these values:

$$N = W \times \cos\theta \qquad \text{and} \qquad F_f = W \times \sin\theta$$

But we also know that $F_f = \mu \times N$. If we substitute that into the equation on the right, we get this:

$$\mu \times N = W \times \sin\theta$$

We can also substitute what we know about N into this equation:

$$\mu \times W \times \cos\theta = W \times \sin\theta$$

You can see here that the W values cancel out, so cross those out. To simplify the equation, realize that $\sin\theta / \cos\theta = \tan\theta$. So, the equation boils down to a simple $\mu = \tan\theta$. Too easy!

FIGURE 4-8 Breaking the weight into components to solve for friction (left) and revised arrangement (right)

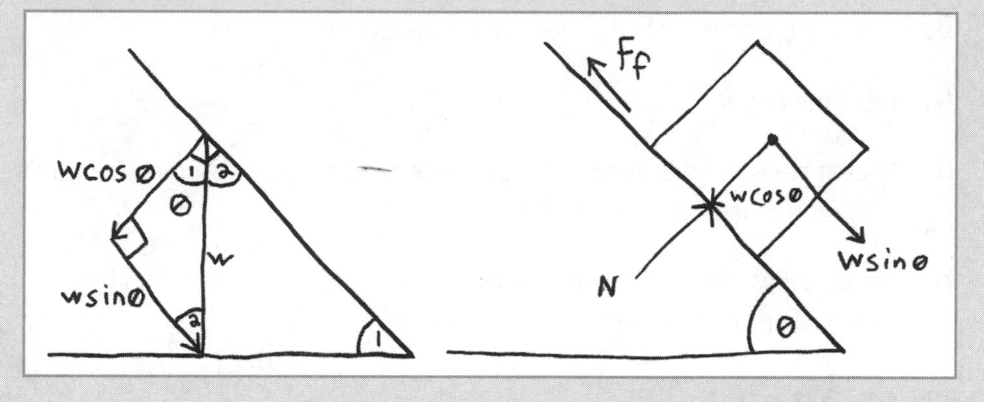

Project 4-1: Estimate the Coefficient of Friction

Let's do a quick test to get a feel for the coefficient of friction between different materials, so you know what 5% and 40% losses look like.

Shopping List:

- Two small objects made of different materials (a 4 oz clay block and iPhone 2G with aluminum back are used here)
- Scrap wooden board
- Protractor
- Calculator

Recipe:

1. Position your protractor at the pivot point of the scrap wood board, as shown in Figure 4-9.
2. Place the first object (the clay block) on the board.
3. Increase the angle until the block just starts to slide.
4. Read the angle.
5. Repeat steps 2 and 3 three times, and take the average measurement. In this test, I found the clay slipping at 45°.
6. Repeat steps 2 though 5 with the second object (the iPhone). I found the iPhone slipping at just 12°.

FIGURE 4-9 Friction testing

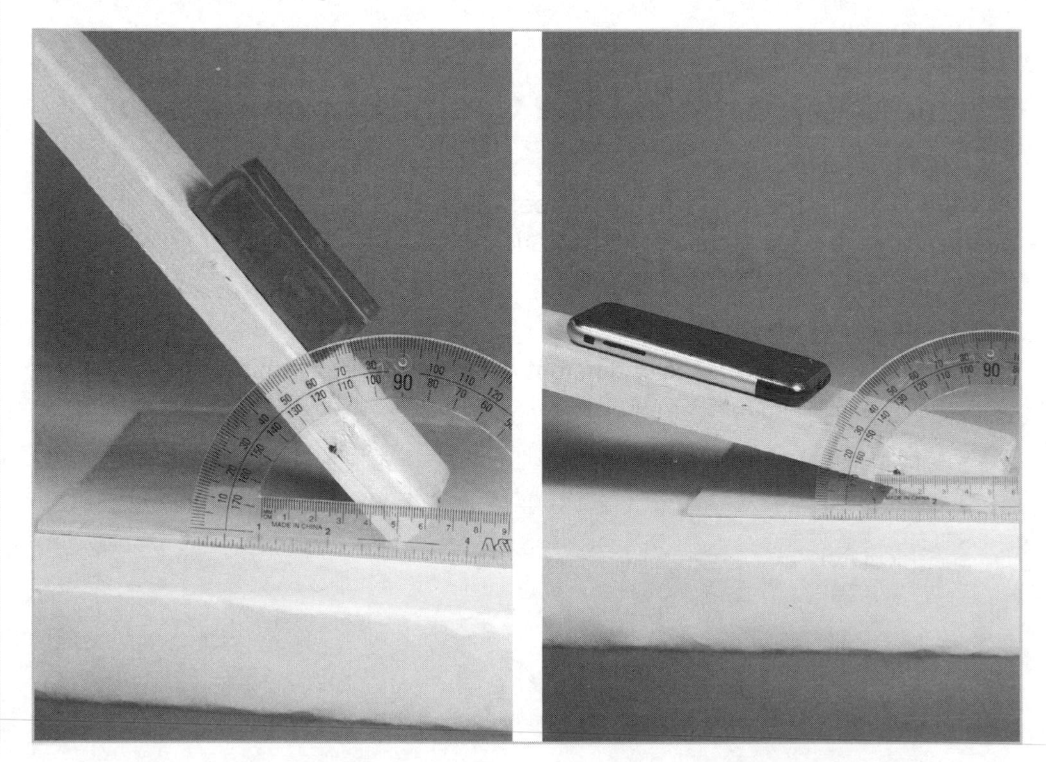

7. Break out your calculator. To find the coefficient of friction between the wood and clay, take the tangent of 45° (remember $\mu = \tan \theta$). You should get 1. This means that if you tried to push a clay block across the floor (as we did earlier with the box example), you would need to push with a force equal to its weight!

8. Do the same calculation with 12°. You should get 0.21. Since the iPhone is slipperier than the clay, it slides more easily. If you push an iPhone across a wooden floor, you need to push with a force equal only to 21% of its weight to get it to move.

Reducing Friction

Now that you know that friction is the enemy, let's look at a couple of ways to decrease friction: clearance between parts and lubrication.

Clearance

In Chapter 2, we talked about tolerances of materials and parts. So now you know that a 1/2 in shaft won't fit in a 1/2 in hole very well if they are both 0.50000 in. You need to leave a little clearance between parts that move relative to each other.

Clearance is just a fancy word for space. You need to leave space around your 1/2 in shaft for it to move, so you may want to drill out a hole that's around 0.515 in to give it some room to spin. There is no magic correct amount of clearance—it will depend on the size of your parts, their surface finish, and whether you want them to spin or stay put. For example, think of LEGOs. Some parts, like the axles, slide right through the holes in other parts. But the little gray stoppers, gears, and wheels you put on the axles are harder to slide on. Also, once you slide them to the right spot, they generally stay there. This is because there is less clearance between the axle and the gear than there is between the axle and the hole through the LEGO piece.

These differences in clearances between the parts allow you to constrain the motion of your LEGO parts just enough, but not too much. This follows the principle of minimum constraint design we talked about in Chapter 1. Using clearance appropriately is an excellent way to practice this principle.

Lubricants and Grease

It's usually a good idea to lubricate things that move. It keeps friction lower, which increases efficiency by allowing more input power to transfer to the output. It also helps keep mechanisms quiet.

A lot of bearings, motor gearboxes, and other components come with grease already in them. For quick, multipurpose fixes, WD-40 is a good light lubricant; the company claims it has over 2,000 uses. Another well-known brand that's a better lubricant is 3-IN-ONE. From bike chains to squeaky scissor hinges, a drop of this stuff will do the trick.

There are dozens of types of oils and greases available. Grease is thicker than oil and tends to stay where it's put. Oil can be runny. If your chosen multipurpose lubricant doesn't do the trick, try looking on McMaster for your application—for example, search for "gear grease"—to find a specific recommendation. Beeswax reportedly works well with moving wooden parts.

Free Body Diagrams and Graffiti Robots

One of the biggest problems my students have is figuring out how much motor torque they need for a certain project. It's impossible to size a motor accurately without doing at least a little analysis of what you want the motor to do. Do you want your motor to drive a small mobile robot quickly? Or maybe you want to use your motor to lift a heavy weight or pull on a rubber band?

Being able to sketch free body diagrams is a handy skill when you're trying to determine how to mount something or what kind of motor torque you need, and any other time you need to choose a component based on strength or torque. A *free body diagram* is like a simplified snapshot of all the forces and moments acting on a component. You can also use these diagrams to figure out the forces and moments you don't know. The *body* referred to is just one object or component of a system.

Here is what is included in a free body diagram:

- A sketch of the body, free from any other objects, with only as much detail as necessary (most of the time, a dot is all that's needed)

- All of the contact forces on the object:
  - *Friction* always acts in the opposite direction of motion.
  - The *applied force* is just what it sounds like—the force you apply to something. This could be a push or a pull (or a kick or a yank or a …). This can also be a force that something else applies to your object, such as the upward force that chains apply to the seat on a playground swing.
  - The *normal force* acts perpendicular to the surface of contact. This is what stops you, or your chair, from falling through the floor and travelling through the center of the earth.
  - *Drag* is the force that impedes an object when moving through air, water, or other fluids. Drag increases as an object moves faster. We generally ignore drag for things moving slowly in air.
- All of the noncontact forces on the object:
  - *Gravity* acts on all objects on the earth, and pulls them toward the center of the earth. This shows up in free body diagrams as the weight of the object.
- All of the moments (torques) on the object

For the object to be in static equilibrium, three conditions must be met:

1. All sideways (horizontal) forces must cancel out.
2. All up and down (vertical) forces must cancel out.
3. All moments (torques) must cancel out.

If these three conditions are not true, you have a moving object.

Let's walk through a real example to put this all in context. One of my former students built a graffiti-drawing robot that suspended a spray paint can from two motorized spools, as shown in Figure 4-10.

FIGURE 4-10 Graffbot original concept (credit: Mike Kelberman)

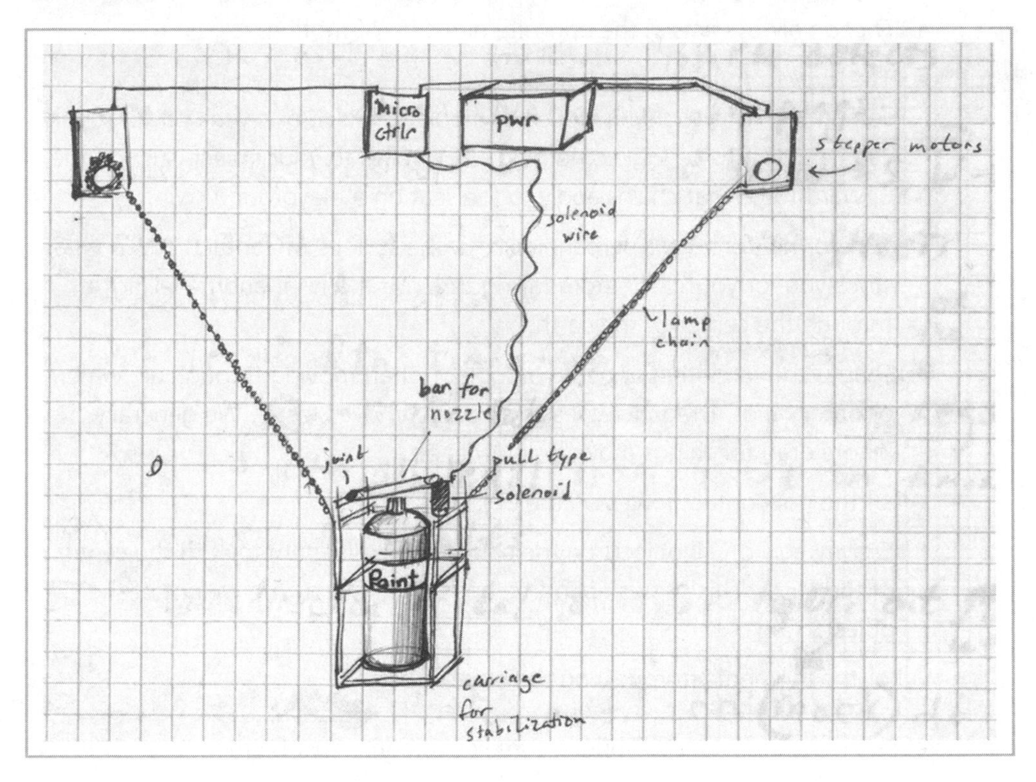

A motor and controller drive each spool, so working together, the spools can move the spray paint can in any shape possible on the wall below them. In order to figure out how hard the motors must work, let's draw a free body diagram of the paint can platform, as on the left side of Figure 4-11.

The only forces in this free body diagram are the weight of the paint can platform (W) and a force pulling up from each rope (F_1, F_2) to the motor spools. We can use what we know to find these forces so we can size the motor properly. So we need to use SOHCAHTOA to get all the forces pointing in straight lines.

On the right side of Figure 4-11, each rope forms a triangle with an imaginary horizontal line. If the can is hanging at 45°, part of the rope force is pulling sideways, and part is pulling up. By breaking the rope force into horizontal and vertical components, we can see that the two sideways forces ($F_{1\text{-OUT}}$ and $F_{2\text{-OUT}}$) are equal and

FIGURE 4-11 Free body diagram of paint can platform in Graffbot (left) and rope forces broken up into components (right)

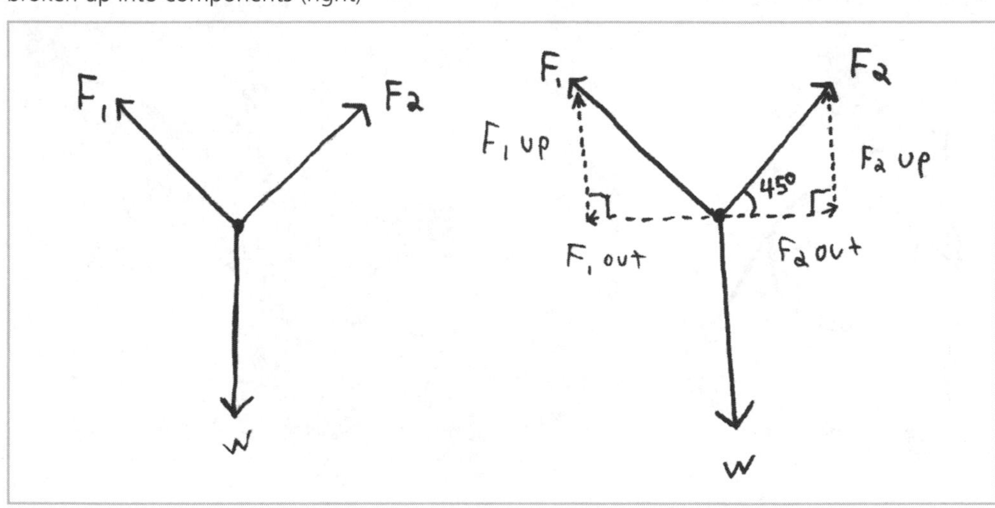

opposite, and therefore cancel each other out, satisfying our first condition for static equilibrium. That leaves us with just two up forces ($F_{1\text{-UP}}$ and $F_{2\text{-UP}}$) and one down force (W). In order for the system to balance, the up forces need to equal the down forces to satisfy the second condition for static equilibrium:

$$F_{1\text{-UP}} + F_{2\text{-UP}} = Weight$$

So if the paint can platform weighs 10 lbs, each F_{UP} force must be 5 lbs. There are no torques on this paint platform, so we can ignore the third condition for static equilibrium.

Now we're getting closer. We know the F_{OUT} forces cancel, and each F_{UP} force is 5 lbs. But we don't know the actual forces, F_1 and F_2, on the ropes that go to the spools. For the triangle on the right in Figure 4-11, recall from our earlier conversation that the sine of an angle equals the opposite side over the hypotenuse (the SOH part of SOHCAHTOA):

$$\sin 45° = F_{1\text{-UP}} / F_1$$

This gives us F_1 = 7 lbs. So we learned that it takes more force to pull something up at an angle like this than it takes to pull something up straight. If the angle gets smaller, so that the paint can platform starts closer to the motor spools, the force on the ropes will increase.

FIGURE 4-12 Free body diagram of Graffbot spool (left) and original Graffbot spool drawing (right) (credit: Mike Kelberman)

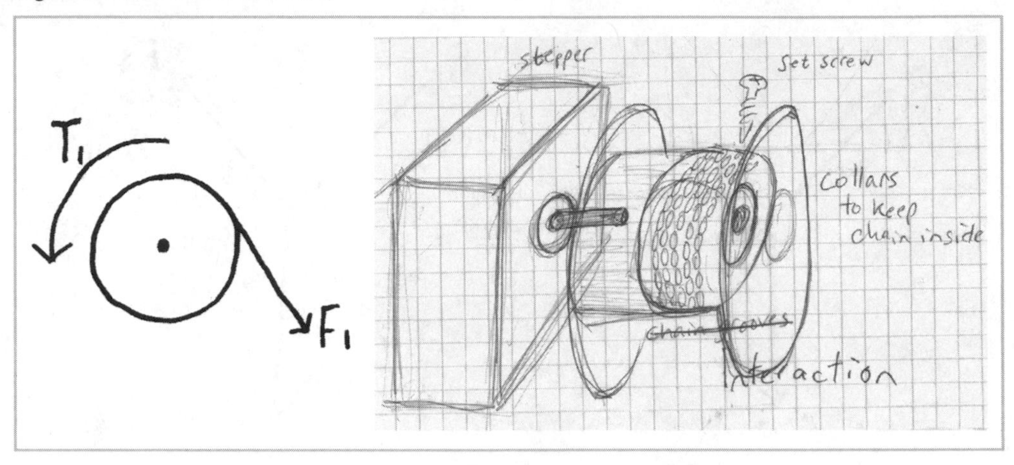

In order to choose the correct motor, we need to take this one step farther and draw a free body diagram of the spool, as shown in Figure 4-12.

We can neglect the weight of the spool and motor, because the holding force of the screws that mount them to the wall cancel it out. That leaves just the force from the rope and the torque from the motor that is in line with the spool shaft to resist the spool from being unwound. Remember that torque is just force times distance? The distance here is from the edge of the spool to the center of the spool. If the diameter of the spool is 4 in, then the radius is 2 in, and we can solve for the unknown motor torque:

$$\textit{Torque } (T) = 7 \text{ lbs} \times 2 \text{ in} = 14 \text{ in-lb}$$

We now know that we need a motor that can turn with *at least* 14 in-lbs of torque to get this Graffbot moving. Each motor needs to be this strong to control the spray paint can platform.

How to Measure Force and Torque

You can measure your weight (the force you exert on the ground) by standing on a common bathroom floor scale. But what if the object you need to weigh doesn't fit on a scale, or you need to measure the pulling force from something like a rope? And how do you measure torque?

Measuring Force

The simplest way to measure force, if you're trying to weigh something, is to use a scale. Some scales are mechanical, using weights and springs to turn or balance a dial; some are electrical. Tools for measuring force come in all different shapes, sizes, and price ranges. Throughout this book, we'll use force-measurement tools that are readily available and affordable.

Mechanical Options

The most affordable option is the standard bathroom scale. This is a smaller version of the scale you stand on at the doctor's office. The kitchen scale, its smaller cousin, is used to measure lighter objects like ingredients for recipes and is more accurate. These are mechanically based scales, which are easy to use. They typically have a needle that comes to rest on a dial to indicate the weight. With these scales, the object pushes on a base to measure a force.

To measure pulling force, you can use a luggage scale or spring scale. You can also find these at sporting good stores sold as fish scales. Mechanical luggage scales go for under $10 and look kind of like the scales at grocery stores to weigh produce. You can purchase spring scales, which are literally just a spring attached to a hook, for even less. Most spring scales have a housing that indicates the pulling force based on how much the spring stretches. These generally work only for a small range of forces, like 5 to 20 lbs. So you need to have a good idea of what you're measuring before you choose how to measure it.

Electrical Options

Bathroom and kitchen scales also come in electrical versions. Instead of a system of springs and levers underneath the platform, these use sensors to detect weight and display it digitally on a screen. You can use these kinds of sensors directly if you need to integrate them into a project, but they are not as plug-and-play as the mechanical options.

- Force-sensitive resistors (FSRs) are used to measure low forces. Their accuracy is not great (±5%–25%, depending on the application), so they are more useful for measuring relative weights or as a sensor to indicate whether something is being squeezed or sat on. An example is the SparkFun (www.sparkfun.com) sensor SEN-09375, which goes up to 22 lbs.

- Flexiforce pressure sensors are more accurate —about ±2.5%—but are about twice as expensive as FSRs. However, at around $20, they're still on the low end of the price scale for force measurement options. An example is the SparkFun sensor SEN-08685.
- Luggage/fish scales also come in digital versions. A company called Balanzza makes popular versions that start at around $15. MakerBot Industries used a digital fish scale to measure the pull force of its plastruder motor for the CupCake CNC machine, as shown in Figure 4-13. See Project 4-2 for how it works and to learn how to make your own version.

The next step up on the price scale is a big one. Higher-end force measurement tools use fancy electronics for precise measurements. They generally come in two varieties: handheld and button types that can be integrated into projects.

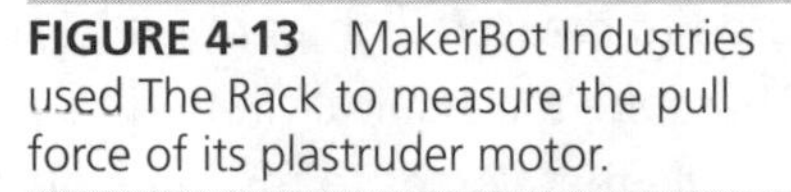

FIGURE 4-13 MakerBot Industries used The Rack to measure the pull force of its plastruder motor.

- Digital force gauges, like McMaster's 1903T51, start at around $370. You can attach a hook or a plate to the measurement end to measure force when pulling or pushing.
- Load cells, like the MLP-100 from Transducer Techniques (www .transducertechniques.com/), start at around $300. Unless you're a whiz at electrical engineering, you'll also need the $400 display to read the force. However, if you need accuracy and a sensor you can integrate into a project, these are ideal.

Measuring Torque

Measuring an unknown torque directly can get expensive. You can set a torque wrench to a certain number and tighten a bolt just the right amount, but torque wrenches are not really made for measuring an unknown torque. Torque wrenches start at around $100 and climb up steeply from there.

You can clamp a torque gauge onto a motor shaft or a screw head and read the resulting torque in real time. However, at $580 for a gauge like McMaster's 83395A29, this is not a very accessible option either.

Luckily, we can use the fact that *torque = force × distance* to our advantage here. All we need to do is measure the force and the distance from the axis to the point where it is being applied, and we get torque!

Project 4-2: Measure Motor Torque

When you buy a motor, it will usually come with a list of specifications to tell you everything you want to know about it. However, sometimes you're stuck with a motor that doesn't have a data sheet. Here, we will use an adaptation of MakerBot's Rack (Figure 4-13) to measure motor torque indirectly by measuring motor force.

Shopping List:

- DC motor (Solarbotics GM9 shown)
- Shaft collar, gear, or other component that fits the shaft and has a set screw hub; replace the set screw with a regular long screw
- Luggage scale, spring scale, or fish scale
- Two C-clamps to hold the motor and scale
- Ruler
- Optionally, epoxy putty and small hook (like one for hanging pictures)

Recipe:

1. Fix the screw and shaft collar or gear to your motor shaft.
2. Fix the motor to the edge of your work table with a clamp (see Figure 4-14).
3. If necessary, use the epoxy putty and small hook to create an attachment on the scale that can hook around the screw.
4. Fix the luggage scale to your work table with a clamp.

FIGURE 4-14 Measuring motor torque

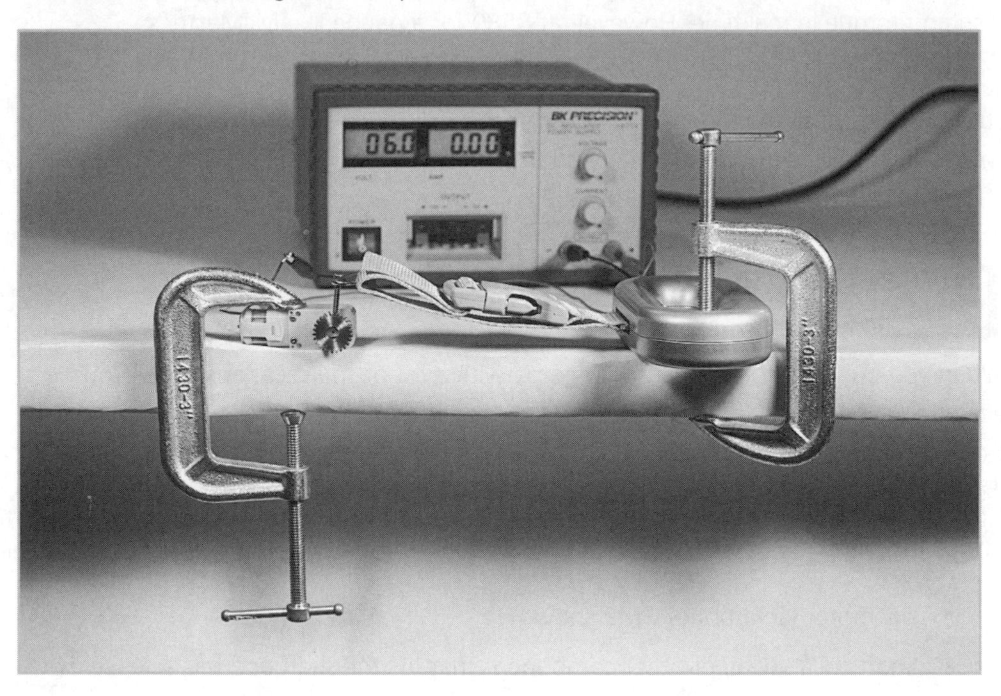

5. Turn the luggage scale on if it's digital.
6. Connect the end of a strap or hook to the screw as close to the screw head as possible.
7. Power on your motor.

NOTE ***If you're testing a motor for which you don't have a data sheet, it's best to use a benchtop power supply so you can find out the acceptable voltage range for your motor (see Chapter 5). Alternatively, you can try batteries that add up to the voltage your motor needs. Six volts would be a good guess for most small DC motors.***

8. Your motor will try to turn, but the connection to the luggage scale will stop it. Read the luggage scale as the motor is stalled like this, and record the number. Do this quickly, before your motor heats up from working too hard!

9. Turn off the power supply.

10. Measure the distance from the center of the motor shaft to the location on the screw that your luggage scale was attached.

11. Multiply the distance you found in step 10 by the force reading you got in step 8, and voila, you have the torque of the motor for a given voltage.

Mechanical and Electrical Power, Work, and Energy

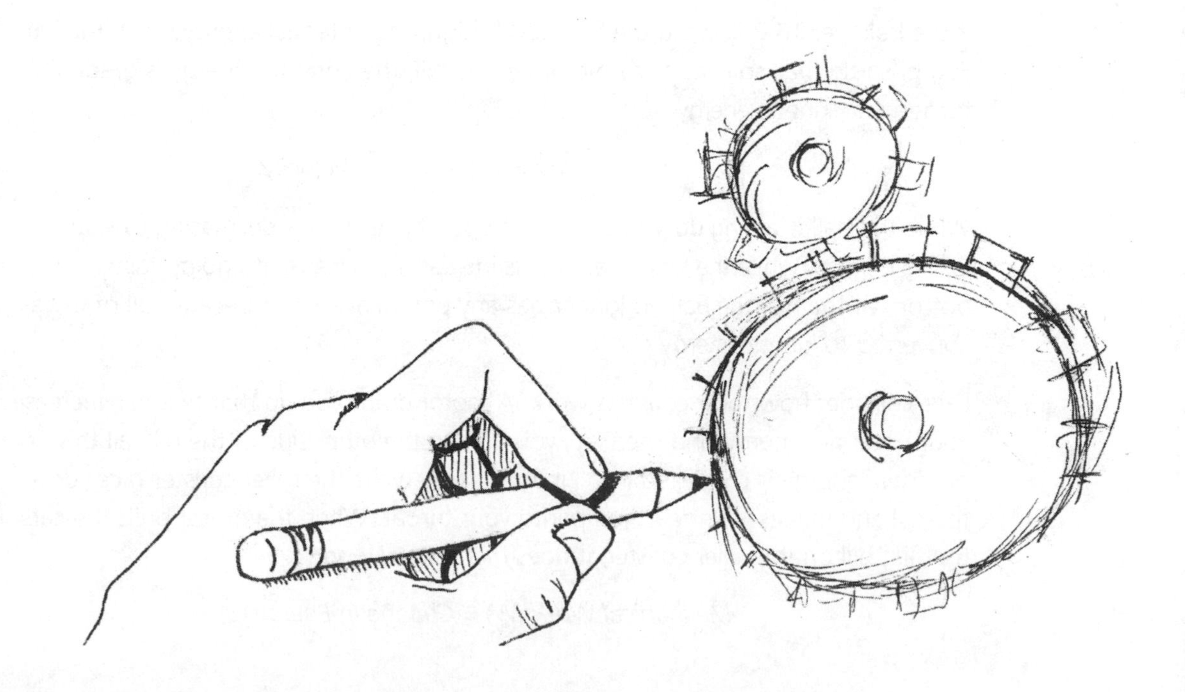

All things that move need some source of energy. This energy may be as simple as using the force of gravity to create movement (how apples fall from trees), or as complex as the internal combustion engine in a gas-powered car. A person can also supply power by cranking a handle or pedaling a bike. Our bodies turn the chemical energy from the food we eat into mechanical energy so we can walk, run, and jump. Motors turn electrical energy into mechanical energy so we can make things move and spin.

In this chapter, we'll discuss how power, work, and energy are related, identify sources of power, and highlight some practical examples of putting these sources to work.

Mechanical Power

Mechanical energy is the sum of an object's potential and kinetic energy. *Potential energy* is how much energy is stored in an object at rest. *Kinetic energy* is the energy an object has because of its motion.

For example, a ball stopped on top of a hill has potential energy equal to its weight multiplied by the height of the hill.

$$\text{Potential Energy} = \text{Weight} \times \text{Height}$$

If the ball weighs 2 lbs, and the hill is 20 ft high, the potential energy is 40 ft-lbs. If you push the ball so it starts rolling down the hill, the potential energy is gradually turned into kinetic energy.

$$\text{Kinetic Energy} = 1/2 \times \text{Mass} \times \text{Velocity}^2$$

While the ball is rolling down the hill, the potential energy is decreasing (because it's losing height), while the kinetic energy is increasing (because it's going faster). At the bottom of the hill, the ball no longer has any potential energy, because all of it was converted to kinetic energy.

Let's consider how roller coasters work. A motor drags you up that first hill, increasing your potential energy, and then lets you go. On the other side of the hill, all this potential energy is converted into kinetic energy while the roller coaster races down the hill and makes your heart jump into your throat. When the motor pulls the cars up the first hill on the roller coaster, it does *mechanical work*.

$$\text{Mechanical Work } (W) = \text{Change in Energy } (E)$$

Think of *energy* as the capacity for doing work. In this case, the mechanism dragging the roller-coaster cars up the hill took the cars from zero potential energy to a lot of potential energy.

Now suppose the roller-coaster cars weigh 1,000 lbs all together, and they're dragged to a height of 200 ft. By changing their elevation, the dragging mechanism did 1,000 lbs × 200 ft = 200,000 lbs-ft of work!

We can also define work as force multiplied by distance:

$$\text{Work } (W) = \text{Force } (F) \times \text{Distance } (d) = \text{Energy } (E)$$

This should look familiar from Chapter 1, where we talked about simple machines. *Work* is just the amount of energy transferred by a force through a distance. For our roller coaster, the dragging mechanism carried the 1,000 lbs of coaster cars up 200 ft, so the work equals 1,000 lbs × 200 ft = 200,000 lbs-ft, which is the same answer as from the potential energy method. The potential energy method and the mechanical work method are two ways of thinking about the same situation.

Mechanical power is the rate that work is performed (or that energy is used):

$$\text{Power } (P) = \text{Work } (W) \,/\, \text{Time } (t) = \text{Energy } (E) \,/\, \text{Time } (t)$$

In the United States, mechanical power is usually measured in horsepower (hp). This curious unit is left over from the days when steam engines replaced horses, and equals the power required to lift 550 lbs by 1 ft in 1 second—the estimated work capacity of a horse. One horsepower also equals approximately 746 watts, or 33,000 ft-lb per minute. You will often see motors and engines rated in horsepower.

Up until now, we've talked about work and power only in straight lines, but what about power for a rotating motor? You might remember from Chapter 1, when we talked about bicycles, that the speed of something spinning is called *rotational velocity*. You just saw that work has units of *force* × *distance*, and luckily, as you learned in the previous chapter, so does torque! So in this case, you can think of torque as work being performed in a circle. Here's the equation form:

$$\text{Power } (P) = \text{Torque } (T) \times \text{Rotational Velocity } (\omega)$$

Electrical Power

You will likely need to use electricity at some point in your project, unless your creation is powered directly by wind or a human (or a hamster). Just as a ball on top of a hill will roll from a higher potential energy position to a lower one, electricity flows from a high potential source to a lower one. The high potential is called the *power source* (or just *power*). The low potential is called the *ground*, which doesn't necessarily mean the ground you're standing on, but it can—that's where the term comes from.

Lightning starts out as a powerful charge just looking for a place of lower energy to discharge, so it finds the fastest path to the ground. When you walk across a carpet in your socks and pick up positive charge, you get a shock when you touch a doorknob for the same reason—your high charge is looking for a lower energy place to go. Your charge jumps to the doorknob and creates a spark, just like a tiny lightning bolt. Metals are good conductors of electricity, so electric charge flows through them easily. Since the doorknob is metal, possibly attached to a metal door that is hung on a metal frame, it provides a great place for your charged-up energy to travel. Your sister is also a great place to discharge built-up static electricity. Since humans are about 60% water, and water is very conductive, your high charge will escape using her body as the ground. (Note that sisters do not always appreciate the shocking result of this grounding experiment.)

For the projects in this book, the ground will not usually be the earth (or your sister), but a strip of metal that is isolated from the power source so it stays at a lower potential energy level. Sometimes the ground is just the negative terminal of a battery.

The relative difference between any high-energy and low-energy point is called the *voltage* and is measured in volts (V). If we compare electricity to the flow of water, voltage is like water pressure that comes from a pump, and wires are like pipes. The amount of voltage, or water pressure, gives an idea of how much work a power source can do. For batteries, you can read this voltage "pressure" on its label. Find a standard AA battery and you'll see 1.5V marked somewhere on it.

As an experiment, take an AA battery and your multimeter (like the SparkFun item TOL-08657). Make sure the black probe is in the hole marked COM and the red probe in the hole marked HzVΩ. Turn the dial to V and hit the yellow button to turn it on.

FIGURE 5-1 Using a multimeter

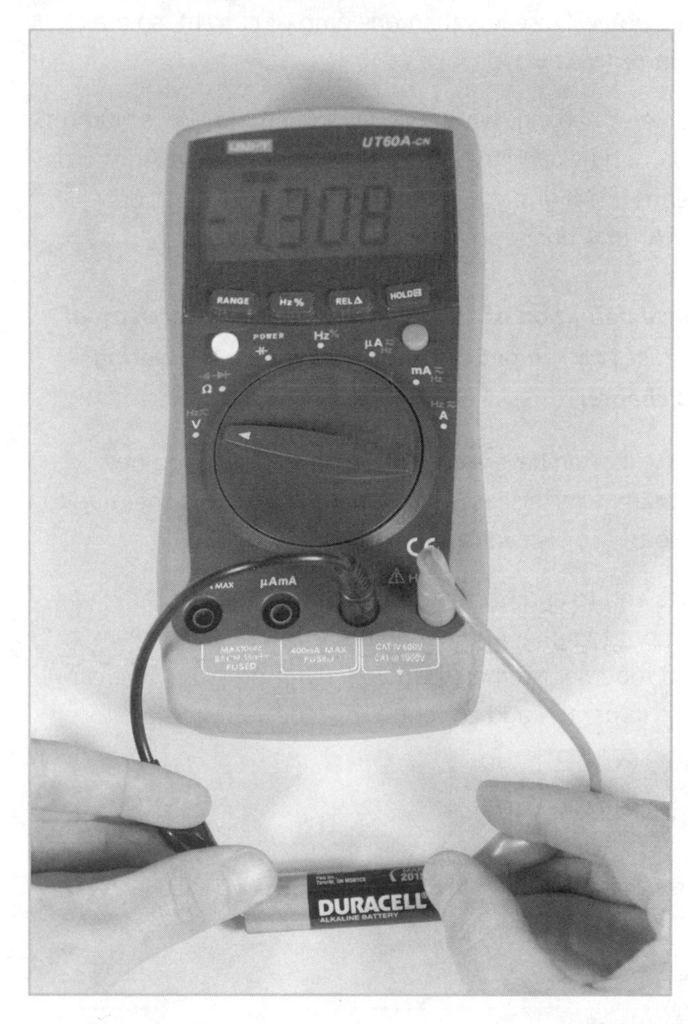

Touch one probe to the (+) side of the battery, and one to the (–) side, as shown in Figure 5-1. New batteries may read up to 1.6V. If it's an alkaline battery, it will read closer to 1.3V when drained. If you put the red lead on the (–) side of the battery, you'll get a negative number (as shown in Figure 5-1). The reverse is true: if you put the red lead on the (+) side, you'll get a positive reading. If you never use your multimeter for anything else, at least you know how to test for dead batteries now.

Current (I) is the amount of electrical energy passing through a point in the circuit. It is measured in amperes (A), or amps for short. Using the water analogy, the current at one point in a wire is like the amount of water flowing past one point on a water pipe. Most components in this book will use less than one amp of current, so are rated in milliamps: 1,000 milliamps (mA) = 1A.

Sometimes batteries have a current marking, which will say something like 3000mAh for a size AA (see www.ladyada.net/make/mintyboost/process.html). The mAh stands for milliamps × hours. This means it will give you up to 3,000mA for 1 hour, or 1,500mA for 2 hours, or 750mA for 4 hours—get it?

NOTE ***This is the technical definition of mAh, but the internal chemistry of the battery will limit how fast you can get current out of it (see "Powering Your Projects" later in this chapter).***

Motors and other components will often have a current rating that tells you how much current they need to operate, which is just as important as hooking them up to a battery or power supply set to the correct voltage.

If you connect two AA batteries end to end, so the (+) sides are facing the same way, you get 3V. When batteries are put together end to end like this, we say they are in *series*. Their voltages are added together, so there is more "water pressure" we can put to work. However, batteries in *parallel* add currents while the voltage stays the same. Figure 5-2 illustrates these concepts.

There are two flavors of electrical current:

1. *Direct current* (DC) is a constant flow of electricity from high energy to low energy. A battery supplies DC.

FIGURE 5-2 Batteries in series and parallel

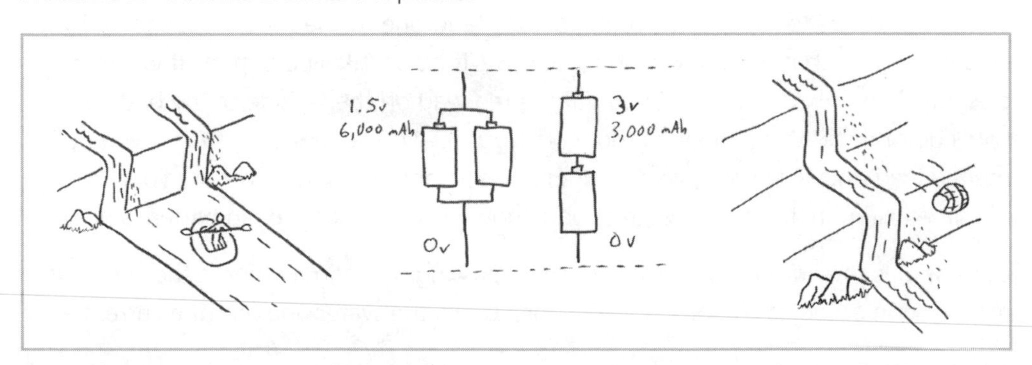

2. *Alternating current* (AC) comes out of our wall sockets in the United States. It is like a wave of power that fluctuates between 0V and 120V 60 times per second.

AC power is useful when electricity must travel long distances, like from the power plant to our homes. Some appliances, like fans and blenders, use AC power directly to run AC motors. Otherwise, for our purposes, DC power is most useful. Most modern electronics use AC-to-DC converters (also called AC adaptors or DC power supplies) that convert the 120V AC from American wall sockets into around 5V to 12V DC that our electronic components can use.[1] The converter is in the bulky black box on the charging cords for your laptop and cell phone.

FIGURE 5-3 A simple circuit

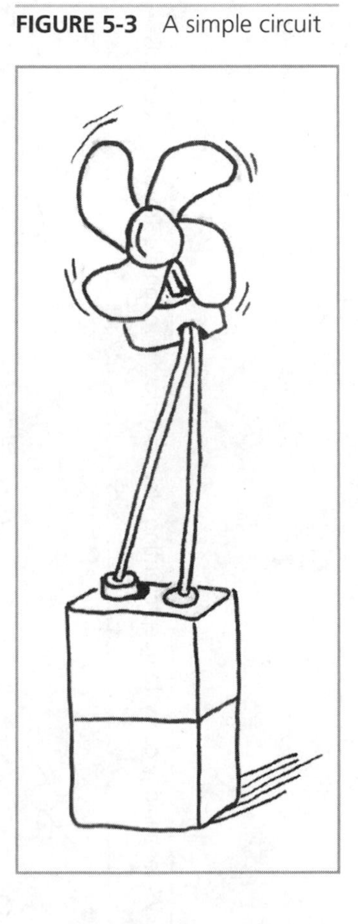

A *circuit* is a closed loop containing a source of power (like a battery) and a *load* (like a light bulb or motor), as shown in Figure 5-3. Current flows from the positive (high energy) terminal of the battery, through the light bulb or motor, back to the negative (low energy, or ground) terminal of the battery. If you put stuff, like a light bulb or a motor, in the current's way in a circuit, it has no choice but to travel through the light bulb or motor until it reaches a ground.

The *resistance* (*R*) of the load is measured in ohms (Ω) and represented by a squiggly line in circuit diagrams, as shown in Figure 5-4. You can think of resistance as a transition to a skinny pipe put in line after a fat pipe. All the current, or water, still must go through it, but it resists the flow, so there is higher voltage, or water pressure, before the transition than after it (see Figure 5-4). *Electricity always follows the path of least resistance to ground.*

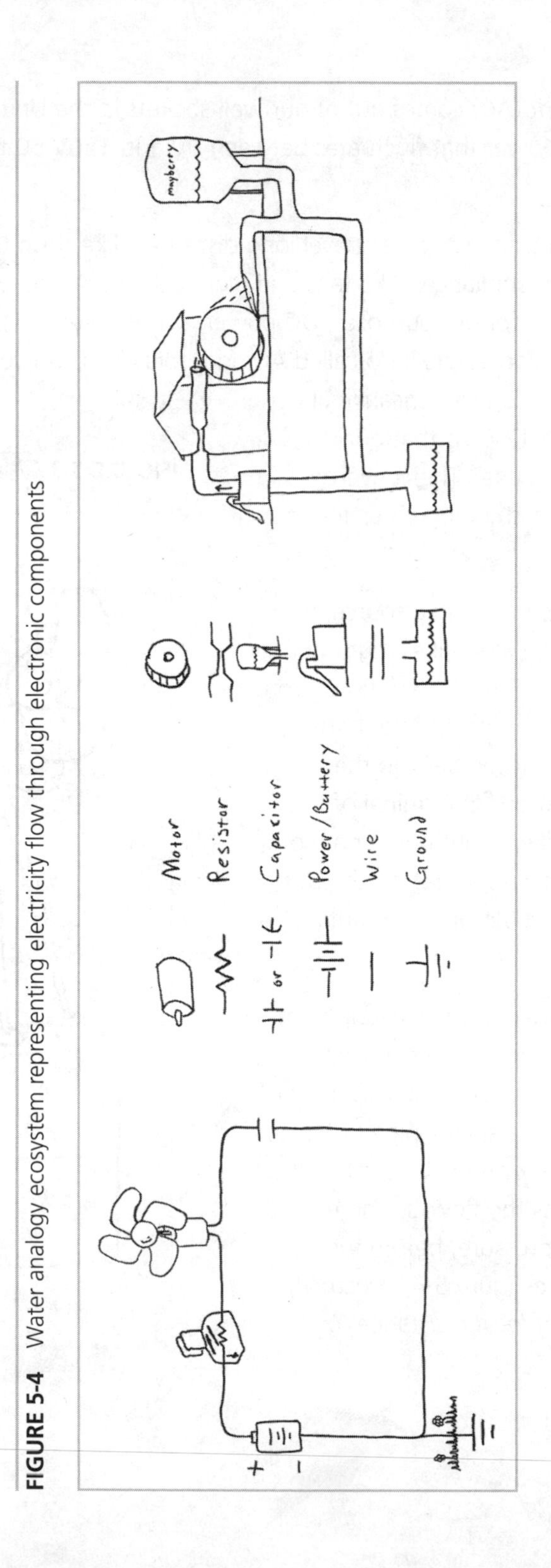

FIGURE 5-4 Water analogy ecosystem representing electricity flow through electronic components

A load (light bulb, motor, and so on) resists the flow of electricity by turning the electrical energy into some other form—light in the case of the light bulb and motion in the case of the motor. In an ideal circuit, all the electrical energy is converted into other forms of energy like this. In reality, some energy is always lost, most of the time as heat.

Technically, you can't lose energy. It just gets turned into forms that are not useful. For example, heat is a form of energy, but when our circuit has extra power that causes our motor to heat up, that heat is not useful to us, so we say that the energy is lost. Some appliances, like a toaster, take advantage of this. A toaster is like a big resistor that just takes the current from the wall and turns it into heat to toast our bread.

Voltage, current, and resistance are related by Ohm's law:

$$\text{Voltage } (V) = \text{Current } (I) \times \text{Resistance } (R)$$

Batteries actually have internal resistance, which just means that they're not 100% efficient, and this internal resistance makes the batteries heat up. Cheaper batteries tend to have higher internal resistance, which is a direct power loss.[2]

Electrical power (P), measured in watts (W), is the combination of current and voltage:

$$\text{Power } (P) = \text{Current } (I) \times \text{Voltage } (V)$$

For example, a 60-watt light bulb needs 0.5A at 120V. The more work you need to do, the more power you need. Compact fluorescent light (CFL) bulbs are environmentally friendly because they use much less power than standard incandescent light bulbs to produce the same amount of light. A 13-watt CFL can produce as much light as a standard 60-watt bulb.[3] The CFL bulbs are more efficient, which allows them to turn more of the input energy into light energy.

Now we can bring everything together. In the first section of this chapter, you learned that mechanical power is measured in horsepower. So, 746 watts equals 1 horsepower, and power equals torque times rotational velocity, and power also equals current times voltage—what does all this mean? It means you can calculate some important values. For example, if you have an electric motor rated in watts, you can figure out the torque for a given velocity (or figure out the velocity at a given torque).

If your eyes are starting to cross and you're thinking of throwing this book into the closet, don't worry. *It's not important to remember all these equations.* However, it is important to know that work, energy, mechanical power, torque, rotational velocity, and electrical power are all related to each other with simple mathematical relationships. And you can use them to figure out how to make things move!

Powering Your Projects

Remember that energy can't be created or destroyed; it just changes form. *Transduction* is the conversion of one form of energy to another. It follows that anything that converts energy from one form to another is called a *transducer*. For example, a motor is a transducer that changes electrical energy into kinetic energy, or motion. Light bulbs and light-emitting diodes (LEDs) are transducers that change electric energy into light and heat. Our bodies are transducers that change chemical energy into mechanical energy. Some of my students have created stationary bikes that power televisions and rocking sculptures that generate electricity by turning a motor. The number of ways you can convert one kind of energy to another is as endless as your creativity.

Back in Chapter 1, I defined a mechanism as an assembly of moving parts. Now you know that moving parts have kinetic energy. That energy needs to come from somewhere, right? Luckily, there are many energy sources we can use to make things move. Not all of them are practical for small-scale work, so we'll focus on the ones that are. The electricity we get from the wall socket comes from other sources like burning coal (and possibly wind or hydropower) that are not directly useful to us. But all we need to know at this point is that the wall socket provides a source of AC power.

To determine your preferred power source for a given project, consider these questions:

- Is your project actually mobile, like a robotic car? If it needs to be truly mobile, you may choose batteries or another power supply small enough to lug around.
- Will your project move but stay in one place, like a painting rotating on the wall? If so, you can use a wall outlet, but you probably need to convert the AC power to DC power. You can do this with AC adaptors like the ones found on your cell phone or laptop computer chargers.

You can also generate your own power with a wind turbine or by pedaling a bike. We'll talk about how much power to expect from these alternative energy sources later in the chapter.

Prototyping Power: The Variable Benchtop Supply

A variable benchtop power supply (see Figure 5-5) is ideal for *prototyping*—the initial phase of working out an idea. SparkFun's (www.sparkfun.com/) TOL-09292 is one example, and Marlin P. Jones & Associates (www.mpja.com) also has a good selection.

FIGURE 5-5 A variable benchtop power supply (image used with permission from SparkFun Electronics)

These power supplies are plugged into a wall outlet and act as power converters from AC to DC. They are expensive and large—hence the name benchtop supply. However, they are ideal because you will undoubtedly work with components that need different amounts of power: some motors want 3V, others want 24V, some want low or high current, and so on.

CAUTION ***Be careful with these things. They can source a lot of power.***

These types of power supplies allow you to vary voltage and supply the required current while you are testing your project. A typical supply ranges from 0V to 30V and 0A to 3A. This flexibility can be a huge time-saver, so you may want to start here, even if you plan to go mobile for the final product. Once you get all the power questions ironed out in the testing and prototyping phase, you can choose batteries or a fixed power supply and be on your way.

Look for benchtop power supplies that are *regulated*. These ensure that the voltage won't drop as current increases.[1] This is important and will save you a lot of frustration. You'll also see supplies listed as *switching* or *linear*. The switching type supply is more efficient, so usually more expensive, but worth it if you have the option.

NOTE ***Voltage is something that's set, but current varies with varying load. For example, a 3V DC motor might come with a data sheet that says something like "no load current: 40mA, stall current: 450mA." We'll cover details on motor data like this in the next chapter. For now, know that a motor with nothing attached to it (no load) doesn't have to do a lot of work so it isn't thirsty for current. As soon as you attach something to the shaft, it is loaded and will need more current to overcome the additional strain. If you load the motor to the point where it stops spinning (with your fingers or a pair of pliers), it will be really thirsty and draw the most current. This maximum current is called the*** **stall current.** ***When you choose your batteries or other power supply for this motor, you should make sure the current rating is high enough to supply at least this stall current. Actually, the current rating of the power supply can be as high as you want; the motor will take only what it's thirsty for. Using a benchtop supply during prototyping is a great way to find out how thirsty your mechanisms are and avoid draining batteries.***

Mobile Options: Batteries

Batteries are great when you need your project to be mobile, but not so great when you are prototyping and testing. There's nothing more frustrating than troubleshooting a mechanism that's not working right, or not working at all, and finding out many hair-pulling hours later that all you needed were fresh batteries! I recommend prototyping with a variable benchtop supply, but when it comes time to go mobile, batteries are where it's at.

Unfortunately, battery technology isn't advancing as fast as we might hope. Batteries are relatively heavy, costly, and large compared to some of the other components we'll talk about. If you plan to go mobile with your project, make sure you have accounted for the weight and size of the batteries within your mechanism.

CAUTION ***Avoid shorting batteries. This is when the positive (+) and negative (–) ends are accidentally connected, causing the power that the battery generates to flow back through itself! This will kill your battery and potentially cause a spark or even a fire. Keep your work space clear of wires and metal objects—like wrenches or screwdrivers—that could act as metal bridges and short your batteries. Also avoid using or storing batteries above 80°F. They won't work as well, and higher temperatures will shorten their life.***

The easiest way to incorporate batteries into your project is to use an off-the-shelf battery holder or snap, as shown in Figure 5-6. These holders can accommodate coin cells (like those in watches and calculators), 9V batteries, and up to eight AA, C, or D size batteries. All Electronics (www.allelectronics.com) usually stocks a good assortment in different sizes and configurations for less than $1 each. Many come with holes predrilled for easy mounting. All common cylindrical batteries are 1.5V (more like 1.2V for rechargeable batteries), but the larger and more expensive they are, the more amp-hours of current they provide.

FIGURE 5-6 Assorted battery holders

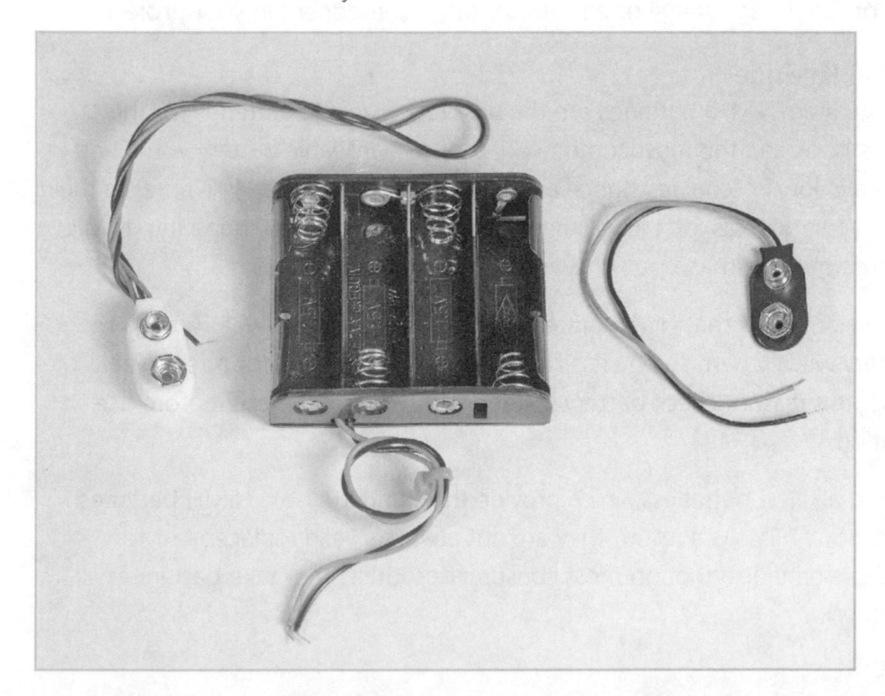

All batteries are not created equal. To help you find the best battery for your application, read on. We'll look at different kinds, including rechargeable batteries, and highlight the pros and cons of each.[2]

Zinc

Zinc batteries are super cheap, and come in all the common sizes (D, C, A, AA, and AAA). They can be rejuvenated a few times before their life is over with a special charger. However, their internal resistance is high, and they don't last very long.

Alkaline

Standard alkaline batteries (think Duracell CopperTop) last three to eight times longer than zinc batteries. They also cost twice as much, and standard ones can't be recharged. There are rechargeable alkalines, but you need a special low-current recharger that can't be used on other rechargeable batteries. They also make high-tech alkaline batteries that last two or three times longer that standard alkalines, but they're significantly more expensive.

Alkaline technology was not designed to handle the high levels of current that some motors require, so these batteries may not provide power fast enough. They also provide a decreasing voltage as they drain, which can cause problems if you're already on the edge of the voltage range of a motor or other component in your project.

Nickel-Metal Hydride

Nickel-metal hydride (NiMH) batteries are the best rechargeable batteries for the projects in this book and the most common kind of rechargeable battery you'll find in stores. They have low internal resistance and can be recharged over 400 times in their life. They don't last quite as long as alkaline batteries, but they are improving. NiMH batteries charge quickly (in an hour or two).

The main disadvantage is that NiMH batteries don't hold a charge well. A fully charged battery will discharge all by itself just by sitting on a shelf for a few weeks or even days. For this reason, check battery voltage with your multimeter before use to avoid frustration.

Unlike zinc and alkaline batteries, which provide the standard 1.5V, NiMH batteries provide only about 1.2V. As a result, they are not always a valid replacement for nonrechargeable batteries (though most consumer products that take batteries will work with either).

Nickel-Cadmium

Nickel-cadmium (NiCad) batteries are an older rechargeable battery technology. Some sources say they show a memory effect, which leads to diminished capacity if you fail to drain a battery completely before charging it to full. While the memory effect is debatable, these batteries more likely suffer from voltage drop. This means that if a battery is repeatedly used only partway before recharging, it will start delivering lower and lower voltages. This is true, but it's also true of all rechargeable batteries.

Once charged, NiCads maintain their voltages reliably until they are almost completely empty (unlike alkalines). They won't self-discharge as quickly as NiMH batteries do during periods of disuse. Like NiMH batteries, NiCad batteries are only 1.2V.

A downside is that cadmium is highly toxic and does not belong in a landfill. Disposing of NiCad batteries in the trash is illegal in many countries and states, so special recycling is necessary. For these reasons, properly charged NiMHs are usually a better choice.

Lead-Acid

Lead-acid batteries are found in your car and also come in smaller sizes like 6V, 12V, or 24V versions that power motorcycles, computers, and boats. They are useful to us because of their size. However, they are notoriously heavy. The trade-off for their weight is that these batteries are cheap and last a long time. Most lead-acid batteries can be recharged with a special high-current battery charger. Look for one that is sealed (denoted SLA for sealed lead acid or VRLA for valve-regulated lead acid) to avoid accidents with leaking battery acid.

Gel cells are just lead-acid batteries with a jello-like filling instead of a liquid acid. These are generally safer and cleaner. Look for the SLA batteries used in SADbot in Project 10-3 for reference.

Lithium, Lithium-Ion, and Polymer Lithium-Ion

Lithium-type batteries are the rechargeable batteries most commonly used in laptop computers and portable electronics. They're relatively expensive, but pack a lot of power for their size, and they will retain a charge for many months. The little coin cell batteries in watches and calculators are also sometimes lithium cells, but they aren't rechargeable.

SparkFun's PRT-00339 is a polymer lithium-ion battery (LiPo for short). This is currently the most advanced battery technology with the highest energy density (*energy density = energy / volume*).

These batteries also need a special charger, of course, and you can get a simple one at SparkFun (PRT-08293).

Plug-In Options

The advantage of having a project that doesn't need to move around is that you can have a big, heavy power supply, or you can plug right into the wall. The following are a few ways to turn AC power from the wall into DC power we can use.

Computer Power Supplies

If you've ever looked inside a desktop computer, you probably noticed a big, boxy looking thing, like the one shown in Figure 5-7, that was making some noise. This box is actually the power supply for the computer, and the noisy thing was probably the fan inside the power supply used to keep it cool. If you think your computer's power comes from the wall socket where you plug it in, you are only partly right. Your computer wants DC power, not the AC power from the wall, so this power supply does the conversion and has some smarts built in that regulate the flow of power and avoid overloads.

FIGURE 5-7 A computer ATX power supply (image used with permission from SparkFun Electronics)

SparkFun sells a computer power supply (TOL-09539), and you can get the accompanying ATX connector breakout board (BOB-09558) to take the mess of wires coming out of the box and give you a useful 3.3V, 5V, or 12V DC supply to power your mechanisms. Both of these can be had for under $35, so if you have a power-hungry immobile project, the combination is a very practical option.

Although you need to plug these power supplies into the wall, they are small enough to be pseudo-mobile. The CupCake CNC from MakerBot (www.makerbot.com) has one of these in it, and you can still carry the whole thing to a party if you want to.

Power Converters

AC adaptors, AC-to-DC converters, and DC power supplies are names for the power cables that usually have large black boxes (wall warts) at the point they connect to the wall. You probably have one to charge your cell phone (see Figure 5-8). They take the AC power from the wall and step it down to a nonlethal DC power in a range we can use. The scary high voltage and current from the wall stays inside the plastic lump and gets dissipated by the heat, and we get nice, smooth, usable DC power out of the other end.[4]

FIGURE 5-8 AC adaptor (image used with permission from SparkFun Electronics)

All Electronics usually has a large and affordable selection of these kinds of power supplies. They typically range from 5V to 12V and come in a variety of current options, from about 300mA to 3A. You can also find supplies with selectable voltage settings, like TM03ADR4718 from Herbach and Rademan (www.herbach.com). You can plug these into your solderless breadboard to power your circuits. If you don't know how to do this or what a solderless breadboard is yet, that's perfectly okay. You'll learn about solderless breadboards in Chapter 6.

Alternative Energy Sources

One problem with generating energy is that you must either use it or store it immediately. You don't need to worry about that with batteries or power supplies you plug into the wall. The power companies that feed our homes will just charge more money if we use more electricity, and batteries supply power through a chemical reaction until they are drained.

Alternative energy sources provide ways to generate electric power, but you need to use that power right away or convert it to a form that you can use later. Even if you use it right away, the unsteady flow of power from something like a solar panel on a cloudy day or wind turbine might need to be smoothed out before it is useful to power mechanisms.

Capacitors for Energy Storage

There are many ways to store energy, but only a few that are practical for our purposes. Gasoline and food store energy in chemical bonds, dams store potential energy in the elevated water, and flywheels store energy in the movement of a heavy spinning wheel. However, we're mostly interested in storing electric potential energy so we can directly use the electricity in our projects.

We've already talked about one way to store electric potential energy: batteries. You can use an alternative energy source, such as solar power, to charge your rechargeable batteries instead of using the energy directly. Another electrical component, a *capacitor*, can be useful for storing energy as well as smoothing out unsteady flows of energy.

A capacitor is like the water tower in our water analogy (see Figure 5-4). When there's plenty of water around, it gets pumped up to the water tower and stored for later use. When there's a shortage and the pump stops bringing in water, the water tower can drain immediately and supply the water it was storing. Capacitors store electrical energy like water towers store water. Similarly, when electric current is flowing into one side of a capacitor, it takes in all the energy and stores it. As soon as there is no current flowing in, the capacitor discharges immediately, until there is no stored up energy left. (See http://electronics.howstuffworks.com/capacitor1.htm for details of how capacitors work.)

A capacitor is a little like a battery because it stores electrical energy and has two connections. But unlike a battery, it doesn't create energy (it only stores energy), so it's much simpler. A capacitor is made of just two conductive plates close to each other but separated by something nonconductive. There are two kinds of capacitors: ceramic and electrolytic (see Figure 5-9). Ceramic ones don't care which way you put them in a circuit, but electrolytic ones definitely do! Make sure the gray stripe or minus sign is on the ground side of the current flow.

FIGURE 5-9 Ceramic capacitors (left) and electrolytic capacitors (right)

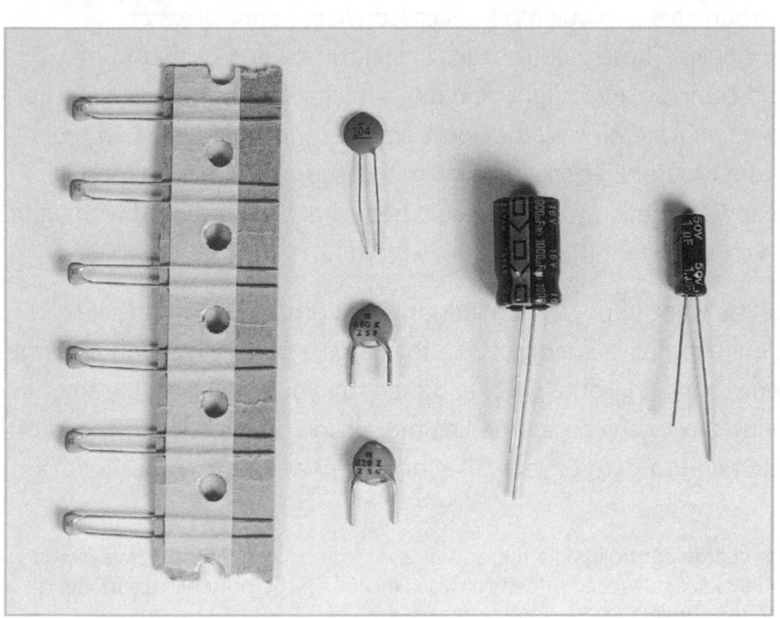

The amount of energy a capacitor can store is called its *capacitance* (*C*) and is measured with a unit called a farad. We normally deal with tiny capacitors that are measured in millionths of a farad, or microfarads (µF). Compared to other means of energy storage such as batteries, the energy density of capacitors is low.

The energy a capacitor can store is calculated as follows:

$$Energy = 1/2 \times Capacitance\ (C) \times Voltage\ (V)^2$$

This gives you energy in a unit we haven't talked about yet, called a joule. A joule is a standard unit equivalent to the ft-lbs we measured mechanical power in earlier.

The advantages of using capacitors over batteries are that they charge much faster, are very efficient, and give you high power over a short amount of time when they discharge. The disadvantages are that they can get very big and expensive for a capacitor equivalent to just an AA battery because of their poor energy density. For example, you would need a soda-can-sized capacitor just to hold enough charge to light a standard flashlight for a minute or so.

So, how do we use capacitors? The easiest way to integrate these into your alternative energy projects is through ready-made modules called *charge controllers* or *energy-harvesting* modules. These modules take unsteady power, like that from a solar panel or a hand crank on a flashlight, and use it to charge a battery or capacitor that releases the power in a steady way that looks smooth and consistent to motors. Part 585-EH300A from Mouser Electronics (www.mouser.com) is one such module that can filter unsteady input energy and release it between 1.8V and 3.6V with up to 1A of current for a very short time.[5]

If you have a motor you want to power continuously, you probably want to set up a circuit that allows your alternative energy source to charge a battery through a charge controller like the one shown in Figure 5-10, and then run your motor off the smooth battery power through the charge controller. See the section on decoupling capacitors in Chapter 6 and the Wind Lantern Project 10-2 for more ways to use capacitors.

FIGURE 5-10 A solar charge controller like the one shown (from Silicon Solar) allows you to charge a battery through solar panels, and then run your motor off the battery. This model is used in the SADbot project (Project 10-3).

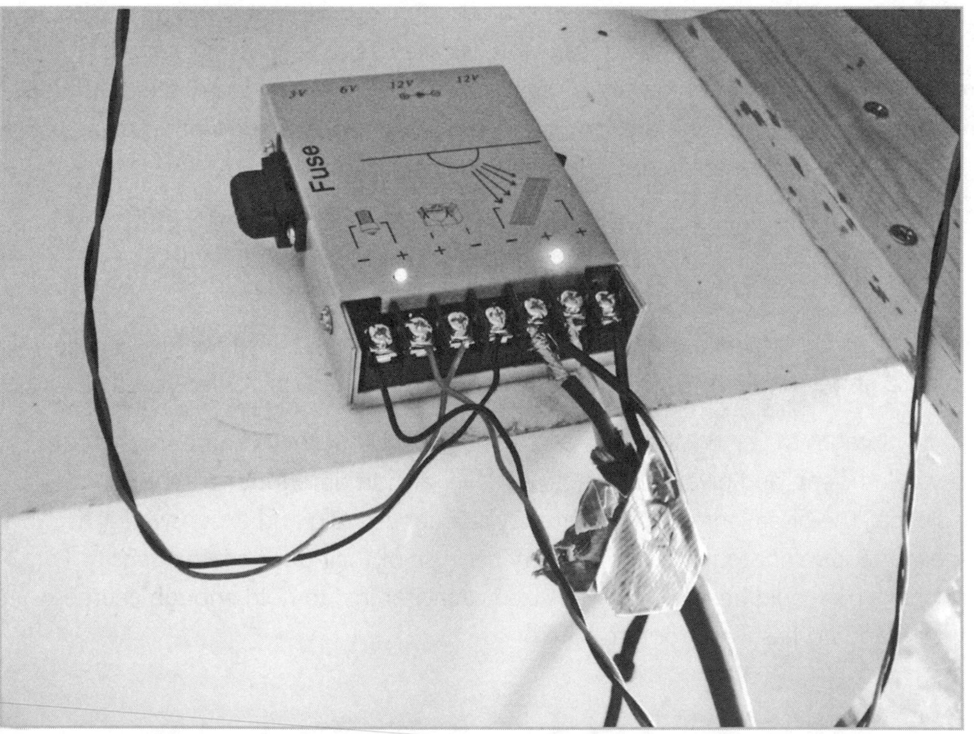

Solar Energy

Solar cells convert light energy to electric energy. The amount of energy you get depends on the area of the cell, so you need a pretty big cell to power the motors we'll talk about in this book. SparkFun's huge solar cell (TOL-09241) can provide 5.2 watts of power (8V at 650mA current) in direct sunlight, and that's enough to drive most of the smaller motors we'll talk about in the next chapter. They call it huge, but at about 7 × 8.5 in, it's smaller than a sheet of notebook paper. The solar cells used in SADbot in Project 10-3 are 13.5 × 18 in and they are from Silicon Solar (www.siliconsolar.com/).

Wind, Water, and Other Fluids

A *fluid* is anything that flows. It could be air, water, or maple syrup. Historically, wind and water power were used directly by using the circular motion from a water wheel or a windmill to grind flour or cut wood. Nowadays, fluid power is mostly used to generate electricity as a cleaner alternative to fossil fuels.

You can try harnessing the power of the wind to make your own wind turbine to create small amounts of electricity. For example, try connecting a small wind turbine to a small electric motor like in the Wind Lantern project in Chapter 10 (Project 10-2). When we give electricity to electric motors, they give us motion (as discussed in Chapter 6). However, if we give an electric motor motion by spinning the shaft, we get electricity. One company, Duggal Energy Solutions, created a streetlight powered by a small vertical-axis wind turbine and solar cell called the LUMI-SOLAIR (www.lumisolair.com). This is a commercial product, but there are also quite a few hobbyist sites devoted to making your own power using wind (for example, see www.gotwind.org and www.windstuffnow.com).

Compressed gases are another way you can use fluids like air and carbon dioxide to do work for you. Potato guns and bottle rockets are good examples of movement created by harnessing compressed air. *Pneumatics* is the name of the field concerned with using pressurized gas to create mechanical motion, but these devices are usually driven by an air compressor that works on electricity to begin with (more on this in Chapter 6).

Unless you have a large dam nearby or a river running through your backyard, hydropower will not be very useful for you in terms of powering projects in this book. I have seen a concept for a showerhead that uses a mini-hydroelectric generator to power an electronic display that shows users how much water they are consuming, but it has not been tested (see www.epmid.com).

Hydraulics is the technical name of the field concerned with creating movement from compressing fluids, but is usually reserved for heavy machinery like bulldozers and excavators. We'll cover this a bit more in Chapter 6.

Bio-batteries: Creating Power from Food

You can make a battery out of any fruit or vegetable that's acidic: potatoes (see Figure 5-11), tomatoes, onions, lemons, oranges, and so on. A bio-battery functions on the same basic principles as a traditional battery. When two strips of different metals (typically copper and zinc) are inserted into an acid solution (in this case, the acidic moisture inside the food), an electrochemical reaction takes place, which generates a potential difference (voltage) between the metal pieces. In a bio-battery, you can use a galvanized nail or other metal (zinc is the coating in anything galvanized) along with

FIGURE 5-11 Potato power (credit: Kaho Abe)

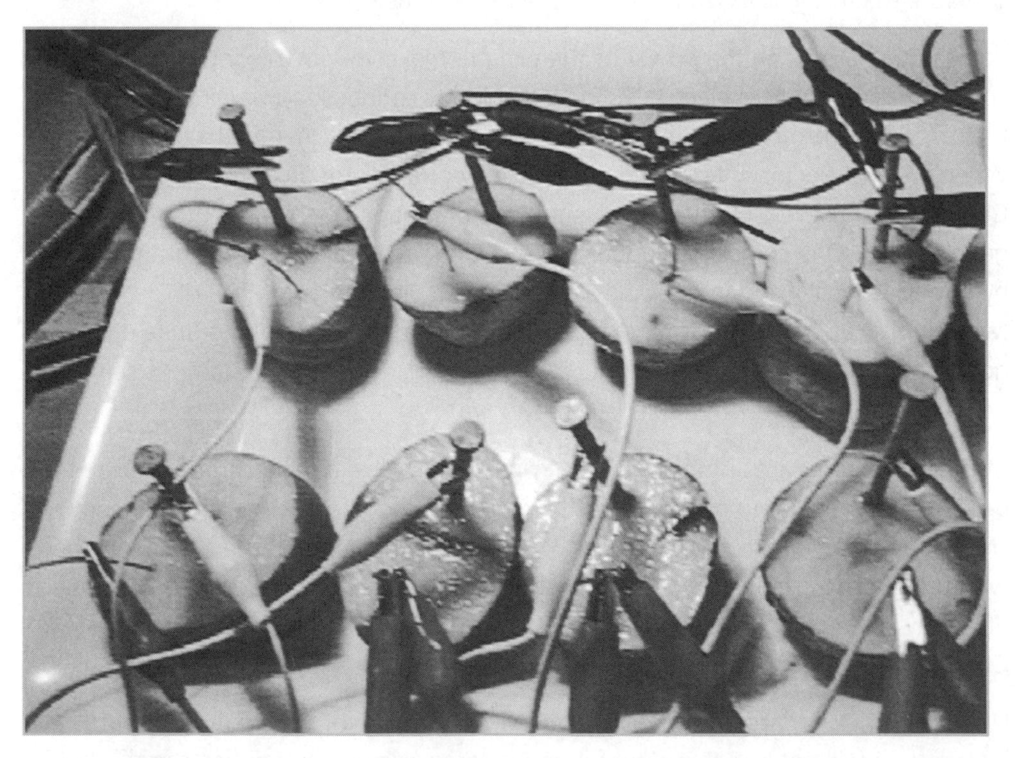

a penny for the copper. A pair of electrodes like these inserted into a potato will generate around 1V at a very small current (just a few milliamps). Individual bio-cells can be added in series to generate higher voltage, and in parallel to generate more current.[4]

Your basic LED needs about 10mA at 2V just to light up, so you'll need a whole plate full of potatoes to get enough power. As far as making things move, even tiny pager motors won't start spinning until you feed them about 20mA, and they want more current as soon as you do anything with them (see www.solarbotics.com/products/rpm2/ for an example). So to get enough power for a motor, you'll probably need a whole sack of potatoes, or better yet, a whole garden.

Humans

The simplest way we can create energy is to use our mechanical energy of movement to create different kinds of movement. For example, hand-cranked mechanical toys like those from the Cabaret Mechanical Theatre (www.cabaret.co.uk) in the United Kingdom range from dancing goats to flapping owls. These are powered by only a hand crank interacting with all kinds of gears, springs, and cams inside the wooden mechanisms. We'll discuss mechanical toys and kinetic sculpture more in Chapter 8.

Wind-up toys have been around for hundreds of years. They have springs or elastic bands to store the wind-up energy we create and use it to make a bug crawl or a toy car move. Modern ones from Kikkerland (www.kikkerland.com) and Z Wind Ups (www.zwindups.com) are popular with kids as well as adults. The following section talks about how you can use springs to store energy like in these mechanisms.

We can also use our mechanical power to create electrical power to use or store for later in rechargeable batteries. Hand-cranked flashlights and radios have been around for years to eliminate the problem of dead batteries, especially in emergency or power-outage situations. Because of the low power requirements of LED technology, about 1 minute of cranking can give about an hour of light on one of these flashlights.

The problem with hand-powered mechanisms is that it's hard to create enough electrical power to do much more than light a few LEDs. Luckily, our legs are built to convert more energy than our arms or hands—after all, we do walk around on them all day. For example, you can find bicycle-powered blenders at Fender Blender (www.bikeblender.com) that will whip up a smoothie while you work out. Other companies are exploring this space, and a couple innovative products are currently being tested. Bionic Power (www.bionic-power.com) created a knee brace with a generator in it that the company claim generates 7 watts of electricity per leg when walking. It's generated as your leg swings in the air, so you don't need to expend extra energy. They aim to use this power to charge cell phones and radios for hikers and soldiers, but that kind of power is enough to run the small motors we'll talk about in the next chapter. Another small company, Lightning Packs (www.lightningpacks.com), uses the weight of a heavy backpack along with the up and down motion of walking to generate up to 7.4 watts.

Springs and Elastic Energy Storage

A spring is an energy storage device, since a spring has the ability to do work.[6] Springs have many different shapes and sizes. *Compression springs* are the most common ones you can find inside mechanical pencils and pens. These will squish a certain amount when you put a certain force on one end. This force and squish distance tell us the spring's *stiffness*:

$$\text{Stiffness } (k) = \text{Force } (F) \,/\, \text{Squish Distance } (x)$$

You can sort springs by stiffness on McMaster. We need this stiffness to tell us how much energy the spring can store. The energy storage depends just on the stiffness and the amount of distance the spring deforms:

$$\text{Energy } (E) = 1/2 \times \text{Stiffness } (k) \times \text{Distance } (x)^2$$

Although we normally think of springs as coils of wire, the same spring equations apply to things like diving boards. They are really just long, flat springs, similar to leaf springs in the suspensions of trucks.

Torsion springs are the kind you find in mousetraps and hair clips that keep them shut.

No matter what kind of spring we're dealing with, it stores *elastic energy*. *Elastic* just means that as soon as we stop pushing or jumping on it, the spring will return to its original state. We'll talk more about different kinds of springs and how to use them in mechanisms in Chapter 7.

Project 5-1: Mousetrap-Powered Car

In this project, we'll use the energy that a torsion spring can store to power a small car. Refer to Figure 5-12 as you step through the recipe.

Shopping List:

- Mousetrap
- 1/4 in diameter wooden dowel
- Multitool with knife and file
- Two eye screws that the dowel fits into (McMaster 9496T27)
- Monofilament fishing line
- Two old CDs
- Laser-cut hub (see www.makingthingsmove.com for links to the digital files for download on Thingiverse.com, and Ponoko.com if you want to buy them) or epoxy putty
- Wooden paint stirring stick

FIGURE 5-12 Mousetrap-powered car ready to roll

- Ping-pong ball
- Duct tape
- Hot glue gun (with glue)

Recipe:

1. Line the edges of the CDs with duct tape to give them some traction.
2. Cut a 4 in length of the wooden dowel with your knife and file any splintered ends.
3. Attach the laser-cut hubs to the CDs with hot glue. Alternatively, use epoxy putty to bridge the gap between the wooden axle and the hole in the CD.
4. Gently insert the wooden dowel into the laser-cut hub on one wheel. It should be snug. If it's too loose, wrap the dowel in a little duct tape and try inserting it again. If it's too big, use the file to sand down the end until it fits (see Figure 5-12).
5. Twist the eye screws into the side of the mousetrap opposite the "bait" hook. They should be as close to the edges as you can get them without splitting the wood.
6. Duct tape the mousetrap down to one end of the paint stirring stick.
7. Insert the wooden dowel with one wheel attached through the eye screws, and attach the wheel on the other side.
8. Duct tape the ping-pong ball under the other end of the paint stick, and your car should balance.
9. In order to stop the wooden dowel from sliding back and forth, wrap strips of duct tape just inside the eye screws around the dowel. The dowel should still spin freely.
10. Cut at 2 ft length of fishing line. Tie one end around the center of the wooden dowel. Tie the other end to the center of the mousetrap arm. Secure with duct tape if necessary.

11. Spin the wheels backwards (clockwise in Figure 5-12) while guiding the fishing line so it wraps around the dowel. When you get almost to the end, keep winding the axle as you lift the mousetrap arm and flip it over.

12. To set the mousetrap, bring the long hook over the arm and catch it on the "bait" hook. This takes a delicate touch sometimes. Watch your fingers!

13. Once you've set your mousetrap, you're ready to race! Set it down on the floor and use a pencil or other long object to trip the mousetrap. The fishing line attached to the arm will pull on the line wrapped around the axle and it will start to unravel. Your car should be able to go about 10 ft with this design. Now try some variations and see if you can get the car to go faster or farther!

References

1. Dan O'Sullivan and Tom Igoe, *Physical Computing: Sensing and Controlling the Physical World with Computers* (Boston: Thomson, 2004).

2. Gordon McComb, *The Robot Builder's Bonanza*, ed. Michael Predko (New York: McGraw-Hill, 2006).

3. US Environmental Protection Agency and US Department of Energy, ENERGY STAR site, "How Much Light?" (www.energystar.gov/index.cfm?c=cfls .pr_cfls_lumens).

4. Nicolas Collins, *Handmade Electronic Music: The Art of Hardware Hacking, Second Edition* (Routledge, New York: 2009).

5. Jeff LeBlanc, "ALDEH 300 Energy Harvesting Modules" (http://itp.nyu.edu/physcomp/Notes/ALDEH300EnergyHarvestingModules).

6. Michael Lindeburg, *Mechanical Engineering Reference Manual for the PE Exam, Twelfth Edition* (Professional Publications, Belmont, CA: 2006).

Eeny, Meeny, Miny, Motor: Options for Creating and Controlling Motion

The most important component of any moving system is an *actuator*, which is the thing that causes a mechanical system to move. Motors are the most common actuators, and as you'll learn in this chapter, there are many different kinds to choose from for your projects. We'll also cover a few other ways to create motion.

In previous chapters, you learned about force, torque, and power, so by now you have the tools to determine how strong your actuator must be for a specific task. We'll use that information, along with other project-specific requirements, to help us narrow down the available options. This chapter covers a lot of information, so take it slow and don't expect to understand everything on the first pass. Now that you've been warned, let's talk a bit about how motors work.

How Motors Work

Motors turn electrical energy into mechanical energy using coiled-up wires and magnets. When electricity flows through a wire, it creates a magnetic field around it. When you bring a permanent magnet close to that magnetic field, it will be repelled or attracted.

Motors take advantage of this magnetic field by mounting coils of wire on a shaft, so when the magnet repels the coils, the shaft begins to spin. In order to keep the shaft spinning, you need to keep flipping the magnetic field so the series of repel, attract, repel, and so on continues and the shaft keeps spinning. Different motors do this in different ways.

Project 6-1: DIY Motor with Magnet Wire

Let's make a simple motor to better understand how it generates mechanical energy.[1]

Shopping List:

- 10 ft length of magnet wire (RadioShack 278-1345; use the green spool)
- Ceramic disk magnet or other strong magnet (McMaster 5857K15)
- Two big paperclips
- Large eraser, piece of clay, or block of stiff foam

- Two alligator clips (like RadioShack 270-1540)
- 9V battery clip (like RadioShack 270-324)
- 9V battery

Recipe:

1. Measure and cut 10 ft of the green wire.
2. Wrap the green wire tightly around the magnet a bunch of times to form a tight coil. Leave about 1.5 in unwrapped on each end.
3. Remove the coil from around the magnet. Loop the ends inside the coil, then back out, to secure it from unraveling. Make sure the finished coil looks symmetrical.
4. Using a knife, remove the coating from the wire on one side of each end at a 45° angle (see Figure 6-1). Scrape each side of the wire such that when the coil hangs at 45°, the scraped part faces down or up. You should see the shiny copper now.

FIGURE 6-1 Wire coiled with end scraped

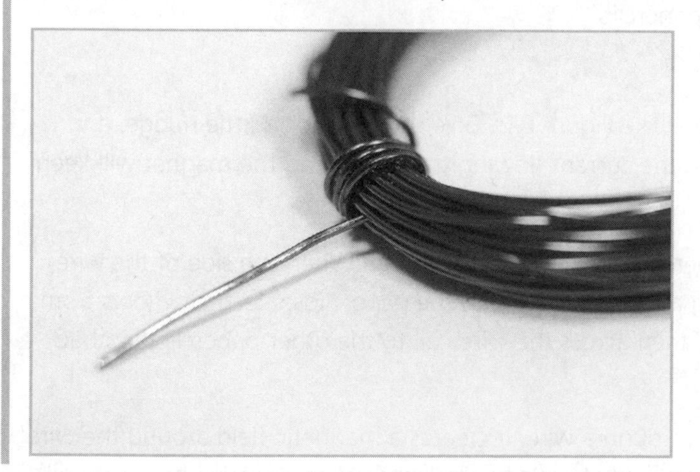

5. Using the paperclips and the eraser, make a cradle for the arms of the wire coil about 1.5 in apart (see Figure 6-2).

6. Place the wire coil in the paperclip cradle. Make sure it spins when you give it a nudge and doesn't get off center. If it does, adjust the wire coil until it looks symmetrical and it balances.

7. Place the magnet on top of the eraser, under the wire coil. Refine the spacing if necessary so the magnet doesn't touch the wire coil or the paperclips.

8. Attach one battery lead wire to the base of each paperclip with an alligator clip.

FIGURE 6-2 DIY motor setup

9. Your setup should look like Figure 6-2. Give the wire coil a little nudge, and the reaction between the current flowing through it and the magnet will keep the motor turning!

The wire coil sits on the paperclips, which are conductive. When the side of the wire coil with scraped-off coating makes contact with the paperclips, electricity flows from the battery to the paperclip, then across the wire coil to the other paperclip and back to the battery.

When electric current flows through a wire, it creates a magnetic field around the wire (see Figure 6-3). This magnetic field attracts the magnet sitting directly under the coil. Because the electricity is flowing in the opposite direction on the other side of the coil,

FIGURE 6-3 White arrows show direction of current flow and black arrows show the direction of the resulting magnetic field in the wire coil

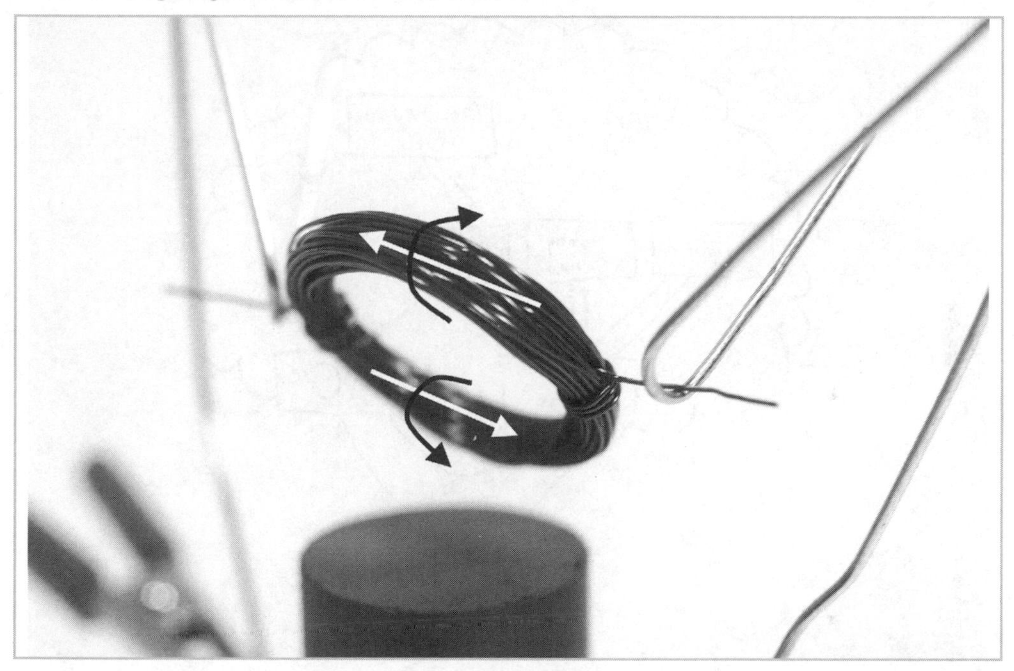

one side will repel the magnet, and the other side will attract it. In order to keep the wire spinning, we need to turn off this flow of current when one side of the coil is close to the magnet, or it will get stuck. So by scraping off the insulation on only one side of the wire, we are telling it to attract, turn off, attract, turn off, attract, and so on, and the momentum of the coil keeps it spinning!

Types of Rotary Actuators

All motors work under the same principles as our DIY motor, but different motors accomplish this in different ways. Each motor type in the motor family has pros and cons, is controlled in a different way, and is well suited to a different set of uses.

The most commonly used type of rotary actuator is the electric motor that spins and creates rotary, or circular, motion. Figure 6-4 shows the rotary motor family tree. There are some cousins I left off the tree, but these are all the motor types we're primarily concerned with in this book.

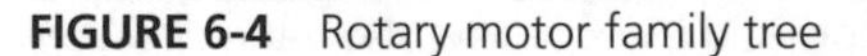

FIGURE 6-4 Rotary motor family tree

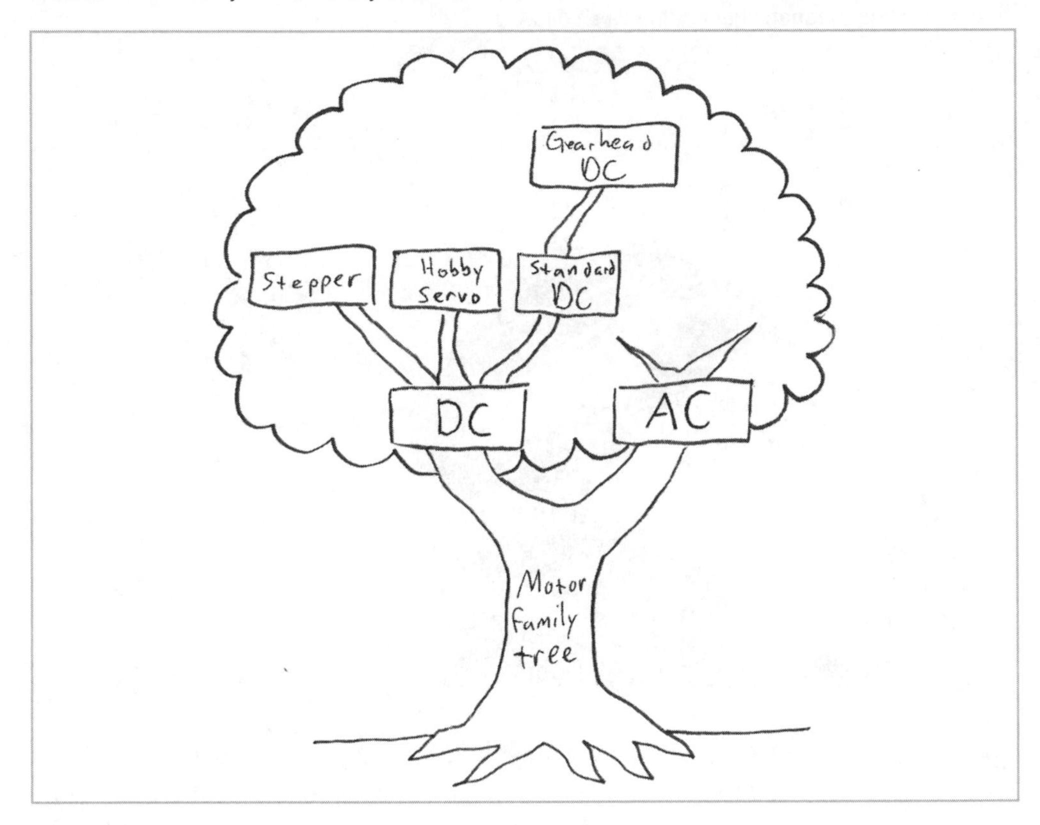

We'll explore each motor's personality by demystifying data sheets for each type.

DC Motors

As you learned in Chapter 5, DC is the kind of constant flow of electricity you get from batteries. Your cell phone also works on DC power, so your charging cable includes a bulky box that converts the AC power from a wall outlet into DC form the phone can use.

Figure 6-5 shows all the members of the DC branch of the motor family tree: DC motor, DC gearhead, hobby servo, and stepper motor.

FIGURE 6-5 DC motor family, clockwise from top left: DC motor, DC gearhead, hobby servo, and stepper motor (images used with permission from SparkFun Electronics and ServoCity)

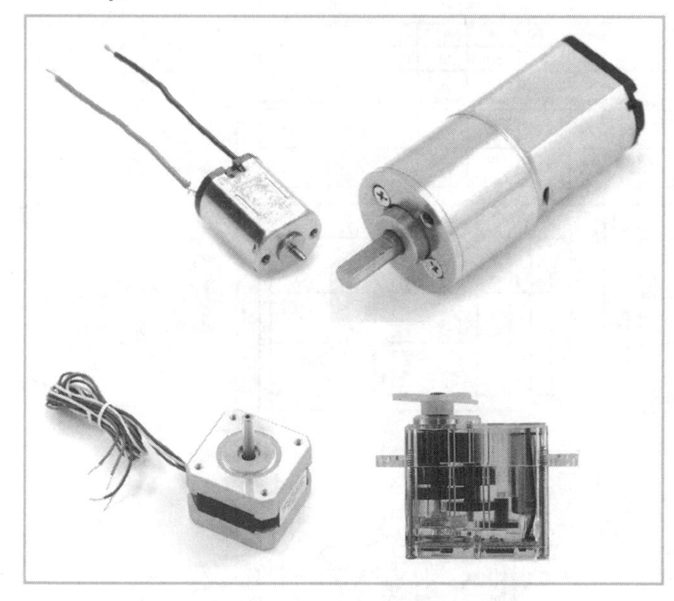

Standard DC Motors

The most basic motor you'll use is the standard DC motor, also called a DC toy motor. You'll find these in everything from toy cars to electric screwdrivers. The insides look like our DIY motor wrapped in a motor housing that resembles a can. Coils of wire are secured to the central shaft, and magnets are attached to the inside of the motor housing. There are also slightly more sophisticated versions of the DIY motor's scraped ends (a commutator) and paperclips (brushes) that enable the field to flip back and forth, as opposed to turning on and off. This makes even small motors more powerful than the DIY version in Project 6-1.

The motor has only two electrical connections, so all you need to do to make a 9V DC motor turn is hook it up to a 9V battery. To reverse the direction, reverse the connections to the battery. If you lower the voltage, it will still work over a certain range, but spin slower. If you raise the voltage, it will spin faster.[2]

DC toy motors usually need between 1.5V and 12V. They spin at speeds anywhere from 1,000 to 20,000 rpm or more. A good example is SparkFun's ROB-09608.

FIGURE 6-6 Data sheet for DC toy motor, ROB-09608 (image used with permission from SparkFun Electronics)

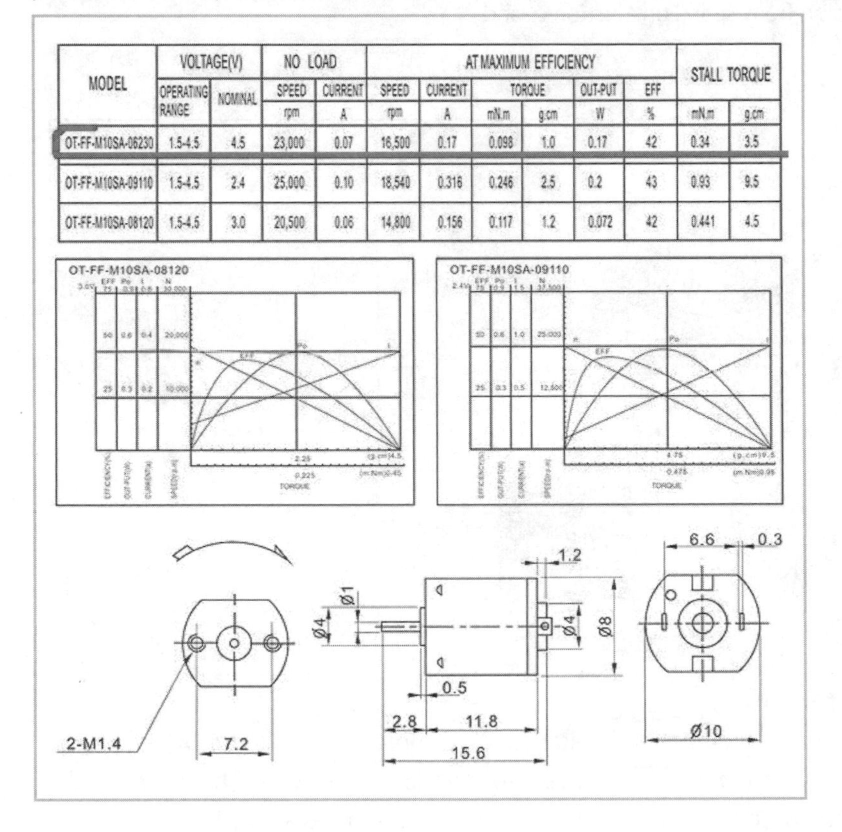

MODEL	VOLTAGE(V)		NO LOAD		AT MAXIMUM EFFICIENCY						STALL TORQUE	
	OPERATING RANGE	NOMINAL	SPEED	CURRENT	SPEED	CURRENT	TORQUE		OUT-PUT	EFF		
			rpm	A	rpm	A	mN.m	g.cm	W	%	mN.m	g.cm
OT-FF-M10SA-06230	1.5-4.5	4.5	23,000	0.07	16,500	0.17	0.098	1.0	0.17	42	0.34	3.5
OT-FF-M10SA-09110	1.5-4.5	2.4	25,000	0.10	18,540	0.316	0.246	2.5	0.2	43	0.93	9.5
OT-FF-M10SA-08120	1.5-4.5	3.0	20,500	0.06	14,800	0.156	0.117	1.2	0.072	42	0.441	4.5

The data sheet for this motor is shown in Figure 6-6 (www.sparkfun.com/datasheets/Robotics/ROB-09608.jpg).

Whoa, that's a lot of numbers! Let's step through this to make sense of what the data sheet is telling us and find the important parts.

- **VOLTAGE** The first column shows that the operating range is 1.5–4.5V, and nominal is 4.5V. This means that the motor will spin if you give it anywhere from 1.5V to 4.5V, but it really likes 4.5V the best. Your standard AA battery is 1.5V, so this motor will work with just one of those, but you could string three 1.5V AA batteries together in series to give the motor the 4.5V it prefers.

- **NO LOAD** The next column is split into speed and current. *No load* means this is what the motor is going to do when there is nothing attached to the shaft. Under the no load condition, this little motor is going to spin at 23,000 rpm! That's fast. And it's going to take only 0.07A to do it.

NOTE ***Sometimes you'll see motor speed in revolutions per second (rps) or radians/second (rad/s). There are 2π radians in one revolution, and 60 seconds in 1 minute. To convert from rad/s to rpm, multiply the rad/s by (60/2π) to get rpm. Or just go to www.onlineconversion.com/frequency so you don't need to remember the conversion.***

- **STALL TORQUE** Let's skip to the last column. This tells us that the motor will stall, or stop moving, when resisted with 0.34 millinewton-meters (mNm) of torque. Think of this as the maximum strength of the motor. This measurement of torque is in the familiar *force* × *distance* units, but if you can relate better to imperial units, go to www.onlinecoversion.com/torque to change it to something else. It turns out that 0.34 mNm equals about 0.05 oz-in. This is very weak, so this tiny motor could barely spin a 0.05 oz weight at the end of a 1 in stick glued to the motor shaft. You can feel how little torque this is by pinching the shaft with your fingers. It stops almost immediately. You can always stall DC toy motors with your fingers since the stall torque is so low.
- **AT MAXIMUM EFFICIENCY** This column contains a lot of numbers that are useful to review. Efficiency describes the relationship of mechanical power delivered to electrical power consumed. DC motors are most efficient at a fraction of the stall torque (in this particular case, maximum efficiency is around one-fourth of the stall torque). This torque corresponds with the EFF label on the bump on the graph of torque versus current in the middle of Figure 6-6. The motor uses power most efficiently at this torque. You can use the motor at a torque closer to its full stall torque, but it will be slower, and less of the electrical power will be converted to mechanical motion, which is particularly draining if you're running on batteries.

DC Gearhead Motors

The next step in motor complexity is the DC gearhead motor. This is just a standard DC motor with a *gearhead* on it. A gearhead is just a box of gears that takes the output shaft of the standard DC motor and "gears it up" to a second output shaft that has higher torque, but turns slower. How much slower depends on the gear ratio. This should sound familiar from earlier chapters that talked about mechanical advantage. Here, we're trading speed for torque: a 100:1 gearhead ratio will give us 100 times more torque than without the gears, but also will be 100 times slower. DC gearhead motors usually range from about 3V to 30V and run at speeds from less than 1 rpm to a few hundred rpm.

The GM14a from Solarbotics is an example of a tiny gearhead motor. You can even see the little gears. The data sheet, found under the Specs. tab on the Solarbotics website (www.solarbotics.com/products/gm14a/specs/), is shown in Figure 6-7. As shown here, on most DC gearhead motors, the gearhead is the end the shaft extends

FIGURE 6-7 Specs for DC gearhead motor GM14a from Solarbotics

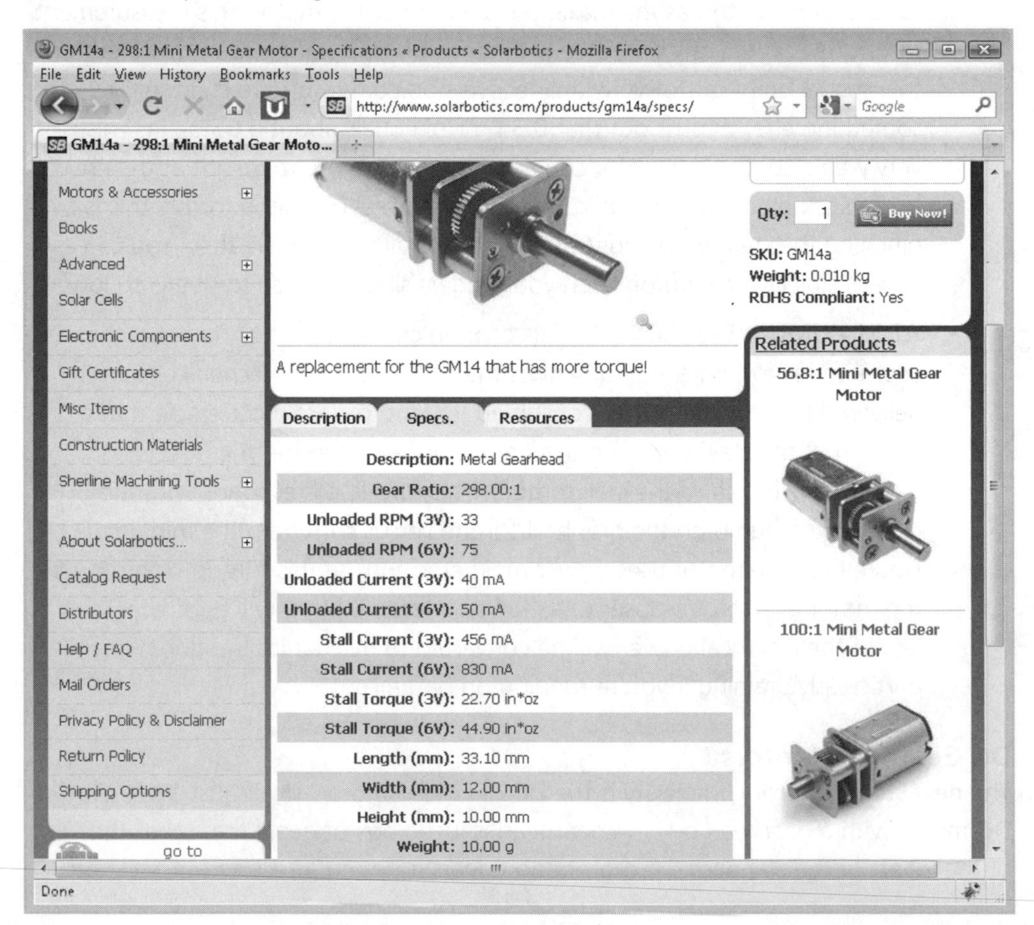

from (usually centered, but not always), and the motor is the end where the power is connected.

All motor data sheets look different, and the terminology can vary, but don't let that scare you. Let's look down the list in Figure 6-7.

- **Gear Ratio** This doesn't tell us anything yet, because even though we know it has 298 times more torque than the tiny motor did without the gearhead, we don't know anything about the tiny motor.
- **Unloaded RPM** This is the same as the no load speed in Figure 6-6. The speed here at 3V is only 33 rpm—*much* slower than the 23,000 rpm of the DC toy motor! The next line shows the unloaded RPM at 6V. The two values indicate that the motor will run on anything between 3V to 6V just fine, so the specs give you the speed for each extreme.
- **Unloaded Current** This is the same as the no load current spec in Figure 6-6. Milliamps are used here instead of amps. A rating of 40mA is 0.040A, which is even less than the 0.070A required by the DC toy motor.
- **Stall Current** This is the current the motor needs at the stall torque.
- **Stall Torque** At 6V, the stall torque here is 44.90 in-oz, which is about *900 times* more torque than the DC toy motor! I told you DC gearheads are stronger.

All DC motors have similar relationships among speed, power, efficiency, current, and torque. You've learned that maximum *efficiency* happens at about one-fourth of stall torque. As you can see in the data sheet for the DC toy motor and Figure 6-8, maximum *power* happens at one-half the stall torque.

Standard Hobby Servo Motors

There are two types of hobby servo motors: standard and continuous rotation. Standard servos are by far the more popular. They are usually found in radio-controlled models like planes and boats.

> **NOTE** ***In industry terminology,* servo *refers to any motor with built-in feedback of some sort.* Feedback *just means there is some way to know where the output shaft is. I'll call the ones covered in this book* hobby servos *to distinguish them from industrial servo motors.***

FIGURE 6-8 Relationships among speed, power, efficiency, current, and torque in DC motors

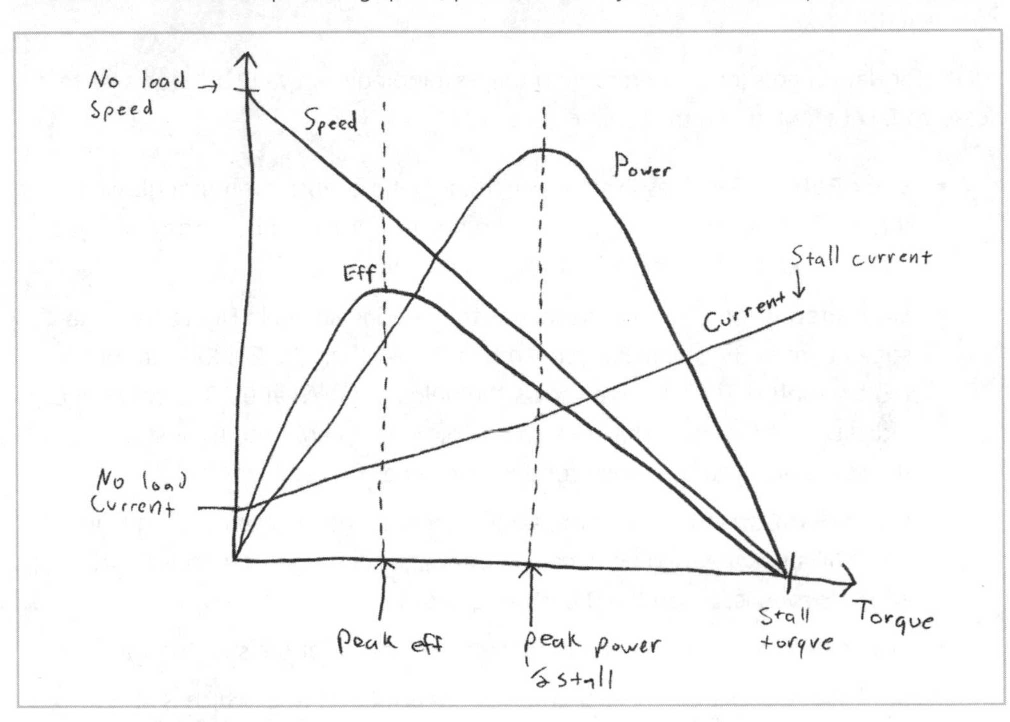

Standard hobby servo motors are just little DC gearhead motors with some smarts in them, as shown in Figure 6-9. When you give the smarts a certain kind of *pulse*—basically just turning the power on and off in a specific pattern—you're actually telling the servo motor where to point the shaft.

Instead of having just two wires you attach to a power source like the DC motors described earlier, these motors have three wires and are controlled by pulses. Standard servos have ranges between 60° and 270° (typically 180°), so they are most useful for pointing and positioning tasks. They also typically use 4.8V to 6V.

FIGURE 6-9 Anatomy of a hobby servo motor (image used with permission from ServoCity)

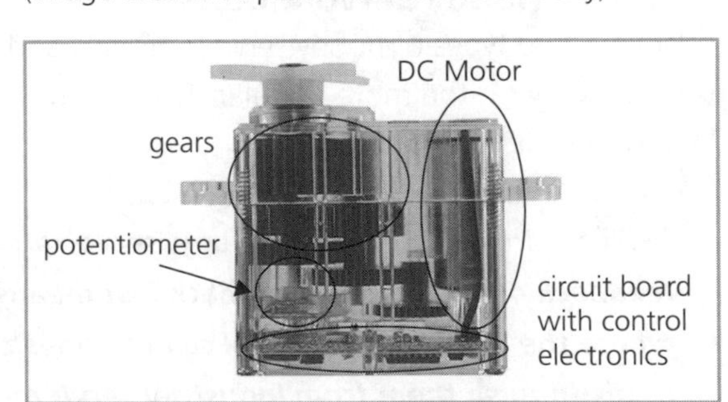

FIGURE 6-10 Detailed specifications for the Hitec HS-311 standard hobby servo motor

Hitec and Futaba are two popular brands of hobby servo motors that you can find at ServoCity and other sources. Figure 6-10 shows the detailed specifications for a Hitec HS-311.

Let's go through the items of interest in the specifications list, skipping the Control System and Required Pulse lines for now (we'll return to them when we talk about motor control later in the chapter.)

- **Operating Voltage** This means the motor will work if you give it anywhere from 4.8V to 6V.
- **Operating Temperature Range** This indicates the environment in which you can safely use the motor. The limits here are set by the sensitive electronic components on the printed circuit board inside the motor housing.

- **Operating Speed** This is equivalent to the no load speed on the DC motors, but worded a little differently. Because servos don't rotate all the way around, you won't see rpm. These specs tell us that when given 6V, the motor will move 60° (or one-sixth of a full rotation) in 0.15 second with no load on the shaft.
- **Current Drain** This is similar to no load current on the DC motors. For this servo, we see two numbers. At 6V, the servo will draw 7.7mA of current just doing nothing, and 180mA when moving with no load on the shaft.
- **Stall Torque** This is the same as on the DC toy and gearhead motors. The stall torque is the highest torque the motor can give, and happens when you stop it or stall it when it's trying to move. This motor shows 49 oz-in of torque at 6V when stalled. This is equivalent to about 3 in-lbs.
- **Operating Angle and Direction** These tell us the servo can move 45° in either direction, for a total range of 90°. On this particular motor model, you can pay $10 more to get a 180° range.
- **Gear Type** This is the most relevant line for us in the rest of the lines in the specification, which describe the motor parts. Lower torque servos will have plastic gears usually made out of nylon. Stronger servos with higher torque use metal gears.

Continuous Rotation Hobby Servos

A continuous rotation servo is a modification of the standard servo motor. Instead of determining position, the pulses tell the motor how fast to go. You give up knowing the position of the servo arm here, but you gain speed control and 360° movement. A continuous rotation servo is a great option if you have something that needs to spin continuously but you want an easy way to control the speed, such as for an electronic toy mouse to chase your cat around.

You can either buy servos that are already modified for continuous rotation, like the Hitec HSR-1425CR, or get a standard hobby servo and perform some surgery to modify it yourself. If you're wondering which servos can be modified for continuous rotation, check ServoCity's Rotation Modification Difficulty List (www.servocity.com/html/rotation_modification_difficul.html).

Stepper Motors

The stepper motor combines the precise positioning of standard hobby servos and the continuous rotation of DC toy and gearhead motors. The central shaft of a stepper has a series of magnets on it in the shape of a gear, and there are several wire coils surrounding this gear magnet on the inside of the motor housing. It is a bit like an inside-out version of the previously described DC motors, which have the coils on the shaft and the magnets on the housing.

Steppers work by moving in a bunch of little increments, or steps. If you step them fast enough, it looks like continuous motion. Each time one of the coils is energized, it pulls one of the teeth on the shaft toward it to complete one step. For example, a 200-step motor moves in a full 360° circle at 1.8° per step.

These motors have four to eight wires you need to use to control the pulses to make the shaft step continuously, so they're more complicated to control than the previously described motors. They are squatter looking than the rest of the DC motor family, and have less torque than you might expect for their size and weight. However, they're also the fastest way to integrate both speed and position control into a project. Printers and scanners use stepper motors to control the speed and location of the print head with the ink and rotate the paper through them. So if you see a discarded printer on the curb on garbage day, you just found yourself at least two free stepper motors.

A good example of a simple stepper motor is SparkFun's ROB-09238. Figure 6-11 shows the feature list from the website (www.sparkfun.com/commerce/product_info.php?products_id=9238).

Let's step through the list to see what we have here.

- **Step Angle** This is in degrees of 1.8. If you divide 360 by 1.8, you get 200 steps for one revolution. We'll talk about how to create these steps in the "Motor Control" section later in this chapter.
- **2 Phase** This stepper is bipolar (4 phase is unipolar). We'll also look at this characteristic in the "Motor Control" section.
- **Rated Voltage** This is 12V, which is just the voltage for which the stepper was designed. Give it more, and you're likely to burn out the motor. Give it less, and it might not turn at all.

FIGURE 6-11 Features of SparkFun's ROB-09238 stepper motor

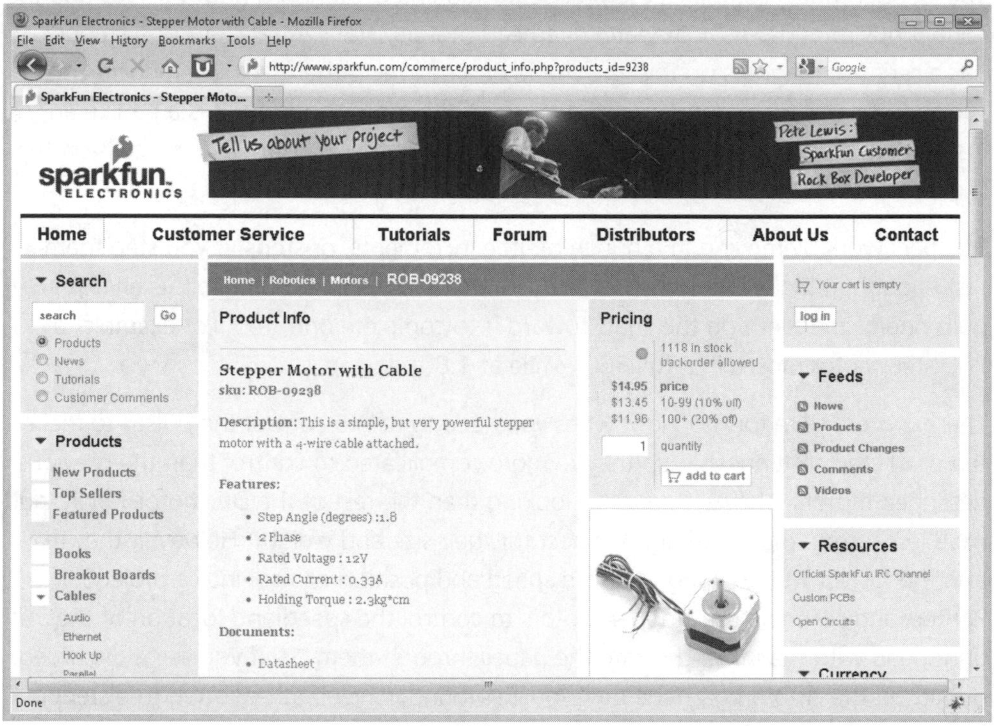

- **Rated Current** Shown as 0.33A, this is the current the winding will draw at the rated torque. It's a good idea to size your power supply for a maximum current higher than this limit, so your motor will hit its own current limit before hitting the limit of its power supply.
- **Holding Torque** This is similar to the stall torque in the motors described previously. The difference is that stepper motors effectively stall every step because there is a series of wire coils around the motor that are activated in a sequence. This motor has 2.3 kg-cm (about 2 in-lbs) of holding torque, which is the torque of the motor when it's powered and *holding* its position at one of the steps. If you try to drive something that needs more torque than the motor is rated for, it will probably slip and you'll lose the benefit of precise positioning. Stepper motors don't inherently know anything about their position. They just know to rotate the number of steps you tell them to rotate. If you exceed the holding torque and the motor slips, all bets are off.

Some other features of stepper motors you might see are *detent torque* and *dynamic torque*. Detent torque is the torque of the motor when it's *moving* from step to step, which is lower than the holding torque, since the shaft is between two holding positions. Dynamic torque is kind of an average of detent and holding torque, and is approximately 65% of the holding torque.[3] As a rule of thumb, don't expect a stepper motor to give you more than 65% of the rated holding torque while it's pulling or pushing something.

The information included on this feature list is enough to choose the motor, but once you have it in your hands, you'll need to know some more details before you can use it—like how to mount it and which wire is which. Luckily, on SparkFun's web page for the motor, there is a link to the data sheet shown in Figure 6-12.

Moving from left to right on the diagram, one of the first numbers you see is the *diameter of the shaft*. Diameter is denoted with the ø symbol, and in this case is 5mm. You also see two small numbers to the right of this, which represent the tolerance of the shaft. They indicate the range of actual dimensions for the 5mm shaft. The small *0* on top indicates the largest shaft size is 5mm + 0 = 5mm, and the bottom number –.013 means that the shaft can be as small as 5 – .013 = 4.987mm. This will be important in the next chapter when we talk about attaching things to motor shafts.

Farther along to the right, you see there are four M3 tapped holes that go 4.5mm deep. M3 is a standard metric screw, and you'll need four of them to mount this motor. There must be no more than 4.5mm of the end of the screws sticking into the motor, or they won't fit.

The wiring diagram on the right shows that the red and green wires control one phase of the motor, and the yellow and blue ones control the other. This will be important when we get to the "Motor Control" section. You've already learned about the important columns in the table in Figure 6-12. The additional information is nice to have, but doesn't matter to us, so feel free to ignore those values, and definitely don't let them confuse you.

AC Motors

You'll find AC motors in many household appliances, like blenders and fans, because they are continuously rotating and use the AC from the wall to drive them. They can be useful if you have a stationary project and just need a plug-and-play motor that

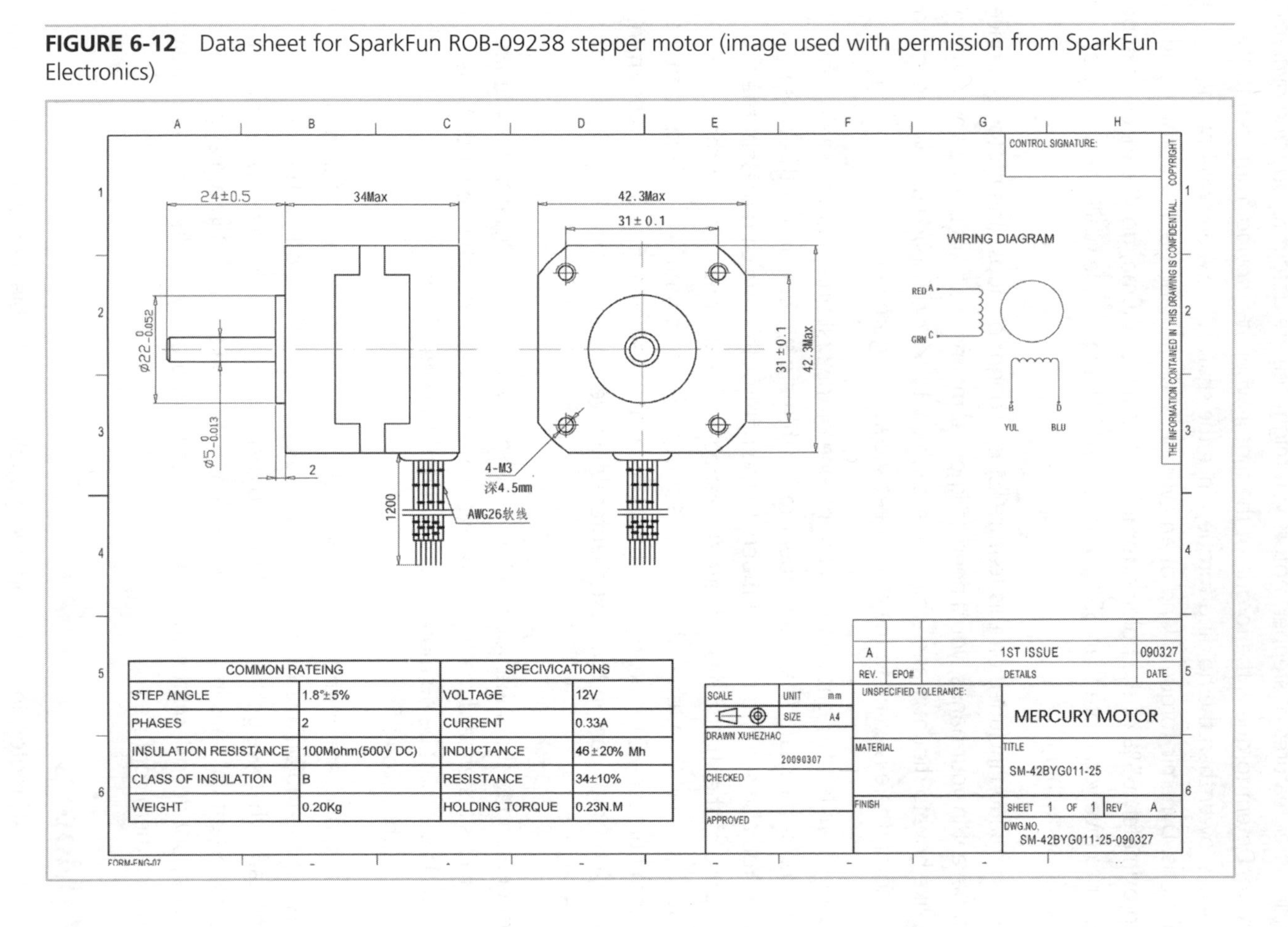

FIGURE 6-12 Data sheet for SparkFun ROB-09238 stepper motor (image used with permission from SparkFun Electronics)

turns all the time and is pretty powerful. However, attempting to control AC motors can be dangerous. You're playing with 120V from the wall, which is much higher than the voltages needed by the DC motors.

An AC motor can draw as much current as it wants from the wall supply, up to about 15A before it trips a breaker in your house. The combination of high voltage and high current is enough to seriously hurt you if something goes wrong. In addition, AC motors near logic circuits are likely to drive those circuits crazy (see the "Helpful Tips and Tricks for Motor Control" section later in this chapter).

I don't recommend using AC motors for general mechanism projects. However, if you can use them without modification or control, they can be helpful. SparkFun carries a PowerSwitch Tail (COM-09842), which isolates the lethal AC power but still allows you to control whatever plugs into it. If you want to do more with AC motor control, and have the time to study up on AC motors, you are encouraged to seek out other sources of information so you can work safely and effectively.

FIGURE 6-13 Using a rotary solenoid to launch ping-pong balls (CC-BY-NC-SA image used with permission from Greg Borenstein)

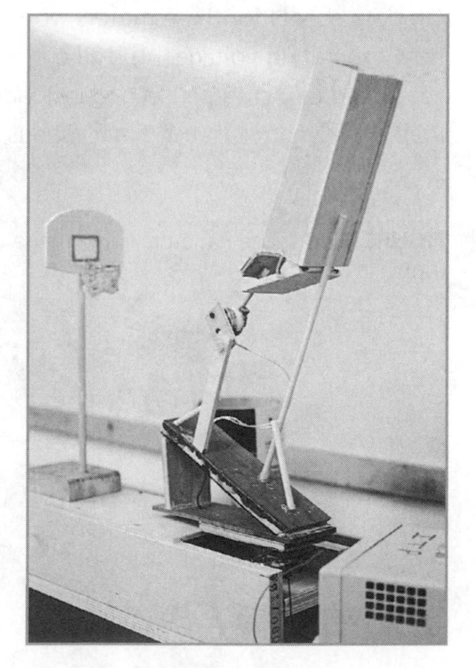

Rotary Solenoids

Rotary solenoids are good for quick rotary movements through a short range of motion. They are really just modified linear solenoids (see the upcoming "Solenoids" section) that force the plunger into a guide that makes it rotate.

Rotary solenoids are pretty expensive for their limited application, but are ideal for throwing ping-pong balls at mini basketball hoops (see Figure 6-13). Ledex (www.ledex.com) manufactures a wide selection of these.

Types of Linear Actuators

Linear motors are far less common than rotary motors. There are quite a few other ways to create linear output from rotary input (see Chapter 7 for more on this). However, linear motors can be handy when you have a specific need. Figure 6-14 shows the two main types of linear actuators: linear motors and solenoids.

Linear Motors

Linear motors, like the ones from ServoCity shown in Figure 6-14, are DC gearhead motors that interact with an Acme or ball screw assembly to push a plunger in and out. We'll talk more about these kinds of screws in the next chapter.

Linear motors can do a great deal of work, but you will pay the price for the convenient packaging (they start at around $130). A former student of mine used them to create lifting shoe mechanisms strong enough to hold and lift her weight (see Figure 6-15).

On data sheets for these motors, you'll see a lot of terms that should look familiar by now: operating voltage, no load current, and so on. Since the action is linear, you'll see speed in inches per second instead of rpm. You should also see ratings for static load and dynamic thrust. *Static load* is the weight of something you can put on the

FIGURE 6-14 Linear motors (left) (image used with permission from ServoCity) and solenoid (right)

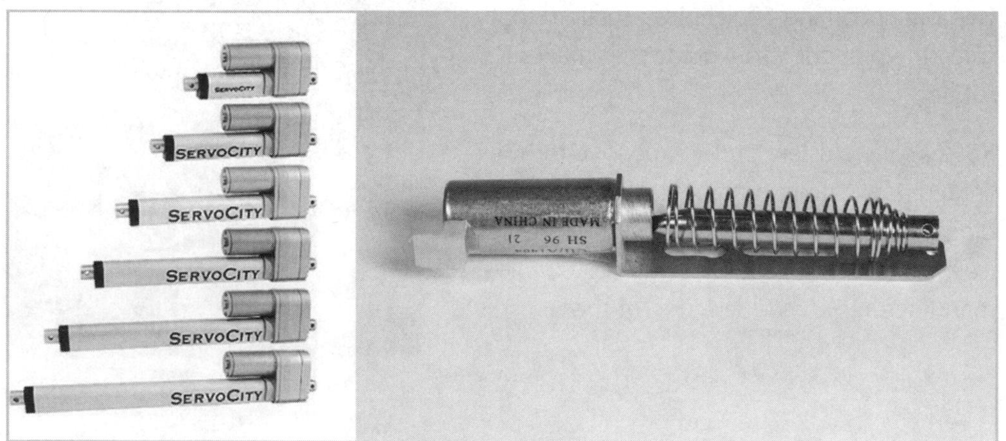

FIGURE 6-15 Linear motor controlling shoe lift (credit: Adi Marom)

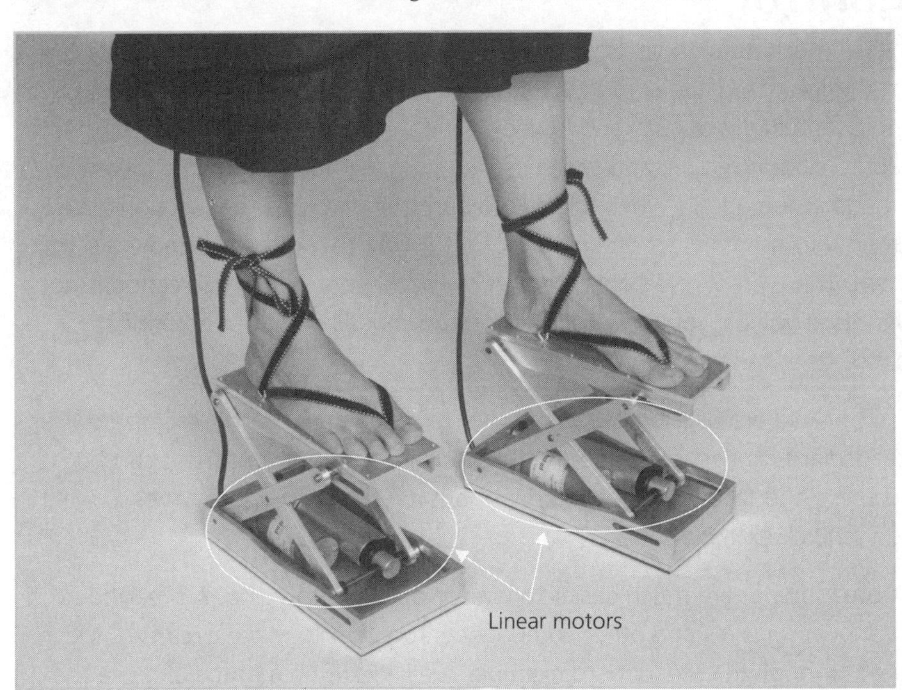

plunger and expect it to hold. *Dynamic thrust* is the maximum weight of something you can expect the motor to move. For example, the 25 lb actuator from ServoCity in Figure 6-14 (the smallest one) will not lift you up if you weigh 150 lbs, but it will hold your weight if fixed in one place.

Solenoids

Solenoids work like a motor that translates (moves in or out) instead of spinning. A solenoid consists of a housing, a plunger, and usually a spring that returns the plunger to a resting state once the power is off. There's a coil of wire around the plunger, and when electricity flows through that coil, it either attracts or repels the plunger to give you a short, linear stroke—good for pushing buttons and making robotic instruments.

If you have a doorbell, it most likely has a solenoid in it. When you press the button, it closes a circuit that makes a solenoid turn on, which moves the plunger and hits a chime.

Motor Control

Many times in mechanism projects, you want to do more than just turn a motor on and off. You might need it to spin a certain number of times, point a camera at a certain angle repeatedly, or raise and lower a window shade. You can also make your motor react to certain sensors and switches, like using a photocell to help lower your window shade automatically when it gets too bright. In the following sections, we'll talk about how to go from just getting a motor to work to more advanced ways to control them. There are whole books written on motor control, so this section is not exhaustive, but it will get you (and your motor) moving. I'll point out additional sources as we go along.

In the spirit of rapid prototyping, we'll try to minimize soldering and maximize our use of breadboards and ready-made modules to talk with our motors. It can be time-consuming and takes special equipment, but it is a handy skill to have, so we'll go through a quick example.

Solderless breadboards are much easier to use for quick prototyping. A *breadboard* is a way of connecting wires and other components together to make circuits quickly. Once you go through the breadboard example, we'll kick it up a notch and use the Arduino prototyping platform—a kind of mini-computer—to give our motors more complicated instructions. Finally, we'll integrate an off-the-shelf module and an Arduino to control a stepper motor. If this all sounds like Greek to you, don't worry. We'll go through each project step by step.

Basic DC Motor Control

All you need to do to get a DC toy or DC gearhead motor to spin is hook it up to a power source within its desired voltage range.

NOTE ***The examples here use red, black, and yellow wires for power, ground, and signal. These appear as gray, black, and white in the images.***

Project 6-2: DC Motor Control 101—The Simplest Circuit

If you hook up a 9V battery to a 9V DC motor, it will spin. Reverse the battery connections, and it will spin the other way. Let's take two components—a battery and a motor—and join them in a simple circuit.

Shopping List:

- DC toy motor
- Corresponding battery (9V used here)
- Battery snap or holder with wire leads (like RadioShack 270-324)

For example, you could use a small DC motor (SparkFun's ROB-09608 already has the wire leads on it) and just one AA battery, because the motor will run off 1.5V. The DC toy motor in Figure 6-16 shows a 9V battery and snap connector, and a 6V motor that will run on 3V to 9V. All Electronics (www.allelectronics.com/) is a great source of battery holders for just about any size.

FIGURE 6-16 A simple circuit with a DC motor and battery

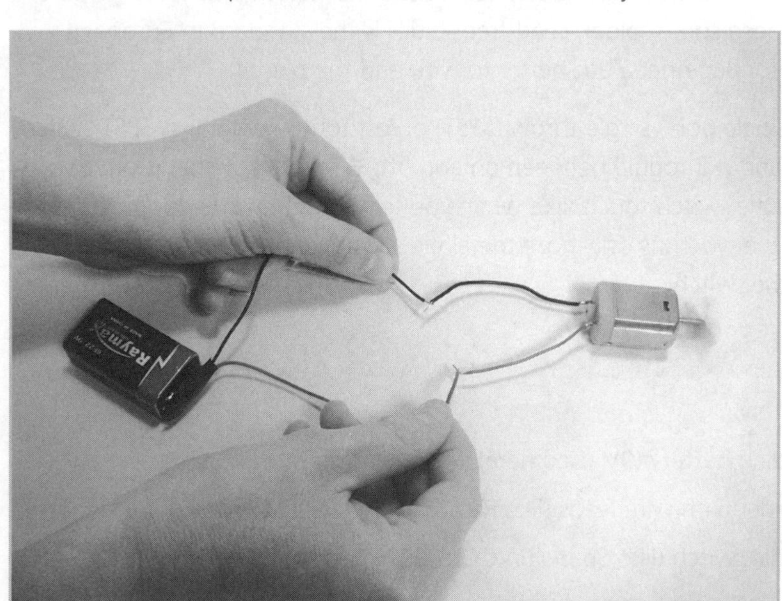

Recipe:

1. Touch the black wire from the motor to the black wire on the battery.
2. Touch the red wire from the motor to the red wire on the battery. Your motor should spin!
3. Reverse the black and red wires. The motor will spin the other way. The motor shown in Figure 6-16 has a small duct tape flag to make movement more obvious.

Project 6-3: Solder a Circuit

Solder is like conductive hot glue that lets you stick metal things together to conduct electricity. For this project, you'll need a soldering iron, which is like a pen with a hot tip that you plug in. A cheap one like RadioShack 64-2802 is fine for the amount of soldering we'll do in this book. This kit comes with a small stand and some solder, so you'll have three of the items you need for this project. If you plan to spend much time with electronics, you may want to spring for a nicer model with interchangeable tips and a temperature control dial like Jameco Electronics (www.jameco.com/) part 46595. You'll also need some solder. Lead-free solder is the standard in Europe. It's a little harder to use for beginners, but better for you and the planet.

You'll also need a single pole, single throw (SPST) on/off toggle switch. A SPST switch will have two legs and will toggle between on and off. When the switch is on, two metal pieces inside the switch touch, like when you touched the wires together in Project 6-2. When the switch is off, those metal pieces are pushed apart so no power can flow through the switch.

Shopping List:

- DC toy motor
- Corresponding battery (9V used here)
- Snap or holder with wire leads (like RadioShack 270-324)
- On/off toggle switch (like SparkFun COM-09276) or other SPST switch

- Soldering iron
- Solder (SparkFun TOL-09162)
- Stand, preferably with a sponge to wipe the tip on (like SparkFun TOL-09477)
- Helping hands (like SparkFun TOL-09317)
- Multitool with pliers or other pliers
- Rubbing alcohol (optional)
- Stiff brush (optional)

Recipe:

1. Plug your soldering iron in and set it on a stand.

CAUTION ***The soldering iron will get extremely hot!***

2. Clean the motor wires and switch terminals or wires with rubbing alcohol and a stiff brush. Although not strictly necessary, this step will make soldering a lot easier and make the connection better.

3. Loop the bare metal end of the red wire from your motor into one of the legs of the switch. Gently squish it with the pliers so it doesn't jiggle around too much. Turn the switch to the off position.

4. Position the switch in one of the clips of the helping hands so you don't need to hold it.

5. Unroll a little solder and touch it to the soldering iron tip. If the soldering iron is at the right temperature, the solder will melt instantly and stick to the tip. This is called *tinning* the tip, and makes your job easier.

6. Position your soldering iron on one side of the motor wire/switch connection. After a few seconds, the wire from the motor and the switch leg will heat up.

7. Touch the solder to the other side of the motor wire/switch connection, as shown in Figure 6-17. If you touch the solder directly to the soldering iron, it will melt quickly, but will not usually form a strong joint. The idea is to heat up the stuff you are joining, and let that stuff heat the solder. If you do it correctly, you'll see a shiny blob of solder melt into and around the motor wire/switch connection. Don't worry if it isn't pretty.

8. Do the same thing with the red wire from the battery clip or holder. The battery should *not* be attached yet. Use the pliers and helping hands as needed to hold the wires still while you work. Hold the solder in one hand and the soldering iron in the other hand, and hold anything else with the helping hands or whatever you can to secure the parts in place while you solder (duct tape, clamps, cable ties, and so on).

9. Solder the black wire from the battery pack to the black wire of your motor. You may want to twist the two bare ends together with the pliers first and use the helping hands so the wires stay put while you solder. Your circuit should look like Figure 6-18 (without the battery).

FIGURE 6-17 Soldering the switch

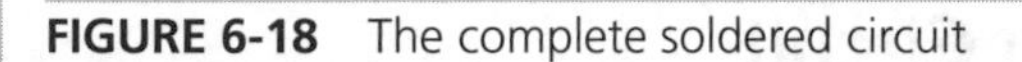
FIGURE 6-18 The complete soldered circuit

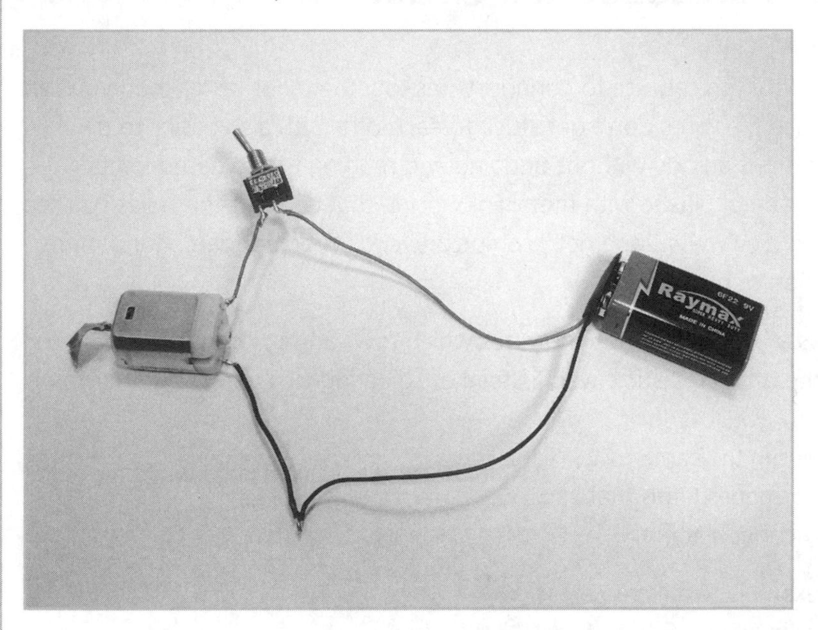

10. It's a good idea at this point to relieve any strain on your wires so the soldered joints don't break—a practice commonly called *strain relief*. Ideally, your soldered joints should not be used as mechanical joints. You also don't want to leave conductive parts exposed where you might short them against a wrench on your desk. Use heat-shrink tubing, hot glue, electrical tape, or cable ties wherever necessary.

11. Attach the battery, turn your switch on, and watch the motor move! Turning the switch on allows power from the batteries to flow through your soldered joints to the motor.

Project 6-4: Breadboard a Circuit

A breadboard is a tool you can use to connect wires together a lot faster than you can with soldering. Since the wires don't get stuck to each other, it's also easier to try different configurations quickly without undoing and redoing the soldered joints. Breadboards are made of plastic with metal links inside that connect the holes you see in the plastic cover. (See www.tigoe.net/pcomp/code/circuits/breadboards for a more complete description of breadboards.)

Figure 6-19 shows a breadboard and indicates which rows and columns are connected underneath the plastic cover. Instead of soldering two wires together to create a connection, you just stick each wire into holes in the same row, and the metal strips underneath that row automatically connect them.

FIGURE 6-19 Breadboard indicating some of the connected rows and columns

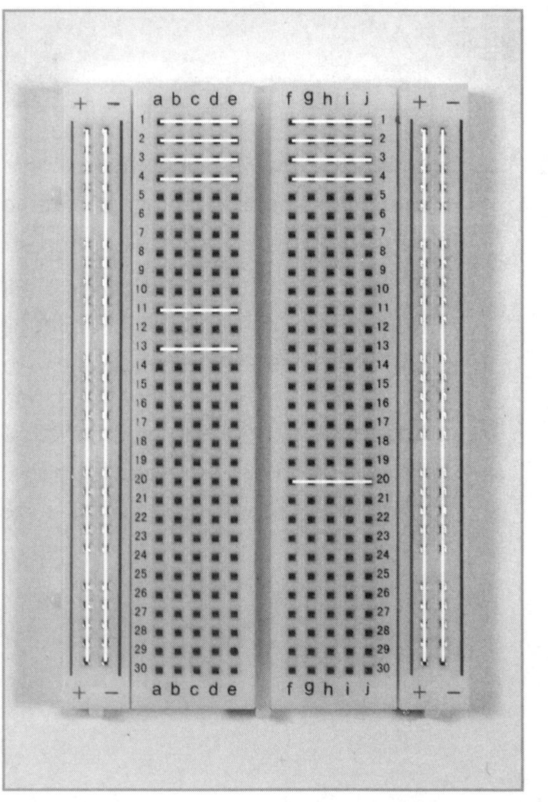

In the following example, we'll create the same circuit as in Project 6-3, but use the breadboard to hold the wires instead of soldering them together. You'll need some jumper wires to create circuits on your breadboard. You can make your own jumper wires by cutting short lengths off red and black insulated solid wire spools (also called *hook-up wire*, like SparkFun PRT-08023 and PRT-08022). All you need is a pair of wire strippers (like SparkFun TOL-08696) to get started.

Wire comes in two flavors: solid and stranded. *Stranded wire* is made up of a bunch of tiny wires as thin as your hair that are twisted together and covered with plastic insulation.

Solid wire is just that—a long, solid piece of wire that is covered with plastic insulation. Stranded wire is more flexible, but solid wire is stiffer, so it's easier to plug into breadboards. Wire size is measured in gauges. The wire spools specified above are 22 gauge, which works well in breadboards (the higher the gauge, the thinner the wire).

To strip the plastic insulation off the ends of a piece of wire from your spool, you need to find the groove at the end of your wire strippers that corresponds to the wire gauge number. Place the piece of wire in this groove with about 1/4 in of wire sticking out of one side. Then squeeze the wire strippers together on and off while you rotate the wire. This will cut the insulation but not the wire itself. Once you see a cut all the way around the wire, pull the insulation off with your fingers. Follow the same steps for the other end of your wire piece, and you have your own jumper wire!

Shopping List:

- DC toy motor
- Corresponding battery (9V used here) and snap or holder with wire leads (like RadioShack 270-324)
- On/off toggle switch (like SparkFun COM-09276) or other SPST switch
- Breadboard (like All Electronics PB-400)
- Jumper wires (like SparkFun PRT-00124) or hook-up wire to make your own (as just described)

Recipe:

1. Solder jumper wires to the legs of the switch and the terminals of the motor (if they don't already have wire leads). On DC motors, it doesn't matter which terminal is which, so you can just pick one.

2. Plug one leg of the switch into one of the breadboard columns marked with a plus (+) sign and the other to a row of your choice on the breadboard. Turn the switch to the off position.

3. Plug the red wire of your motor into that same row, and the black wire to one of the breadboard columns marked with a minus (–) sign.

4. To get power to your breadboard, plug the red wire of the battery into the breadboard column marked with a + sign. Now plug the black one into the column with a – sign. This makes the whole + column power and the whole – column ground, so it completes your circuit. (Refer to the appendix for other ways to power a breadboard than directly from batteries.) Your circuit should look similar to Figure 6-20.

NOTE ***By convention, with breadboards (and electronics in general), red is power/on/+ and black (sometimes blue or green) is ground/off/–. If you use the marked side columns for power (+) and ground (–), it will be easier to follow your circuits. Red shows up as gray in the black-and-white photos here.***

5. Now flip the switch to on, and your motor should spin!

FIGURE 6-20 The same circuit as in Project 6-3 on a solderless breadboard

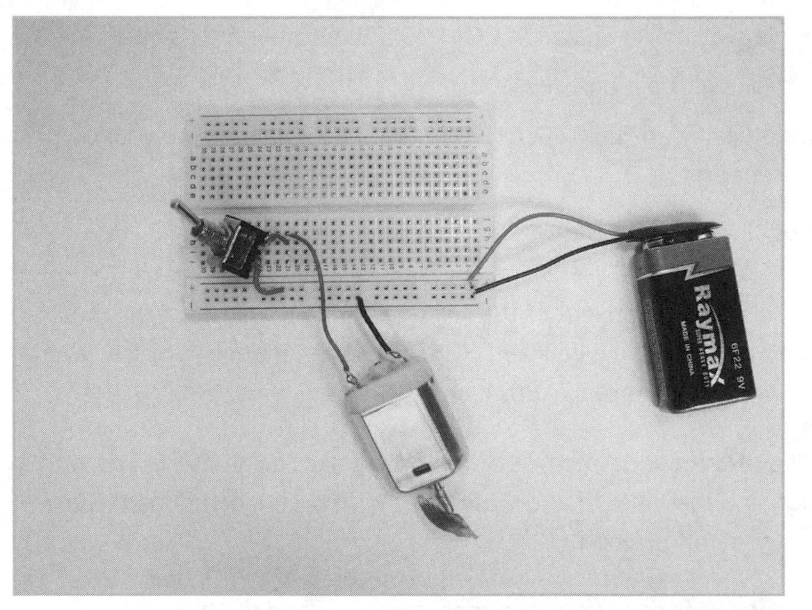

Project 6-5: Motor About-Face

In Project 6-2, we changed the direction of the motor by switching the red and black wires manually. It's great that all we need to do to switch the direction of a DC motor is change the direction of current flow, but how do we do that without disconnecting and reconnecting wires all the time?

The simplest way to switch direction is by using an integrated circuit (IC) chip called an H-bridge and a three-legged switch called an SPDT (for single pole, double throw) switch. Instead of just two positions (on and off), an SPDT switch has three positions (on, off, and on). When the switch is at either extreme on position, two metal pieces inside the switch touch. But unlike the SPST switch from the previous example, these metal pieces are independent, so this switch can control separate circuits. When the switch is in the middle (off) position, those metal pieces are pushed apart so no power can flow through the switch.

Inside the H-bridge chip is a series of electronic gates. In this example, when your switch is at one extreme, the gates in the chip will let current flow through the motor in only one direction. When you toggle the switch to the other extreme, the gates in the chip will reverse and make current flow through the motor in the opposite direction.[4]

Shopping List:

- DC toy motor with wire leads
- Corresponding battery (9V used here) and snap or holder with wire leads (like RadioShack 270-324)
- Breadboard (like All Electronics PB-400)
- Jumper wires (like SparkFun PRT-00124) or hook-up wire to make your own (see Project 6-4)
- On-off-on toggle switch (like SparkFun COM-09609) or other SPDT switch
- H-bridge motor driver chip (SparkFun COM-00315)
- Four AA batteries and holder (like SparkFun PRT-00552)

Recipe:

1. Place the H-bridge in the middle of the breadboard with the small notch facing up, as shown in the bottom image of Figure 6-21.

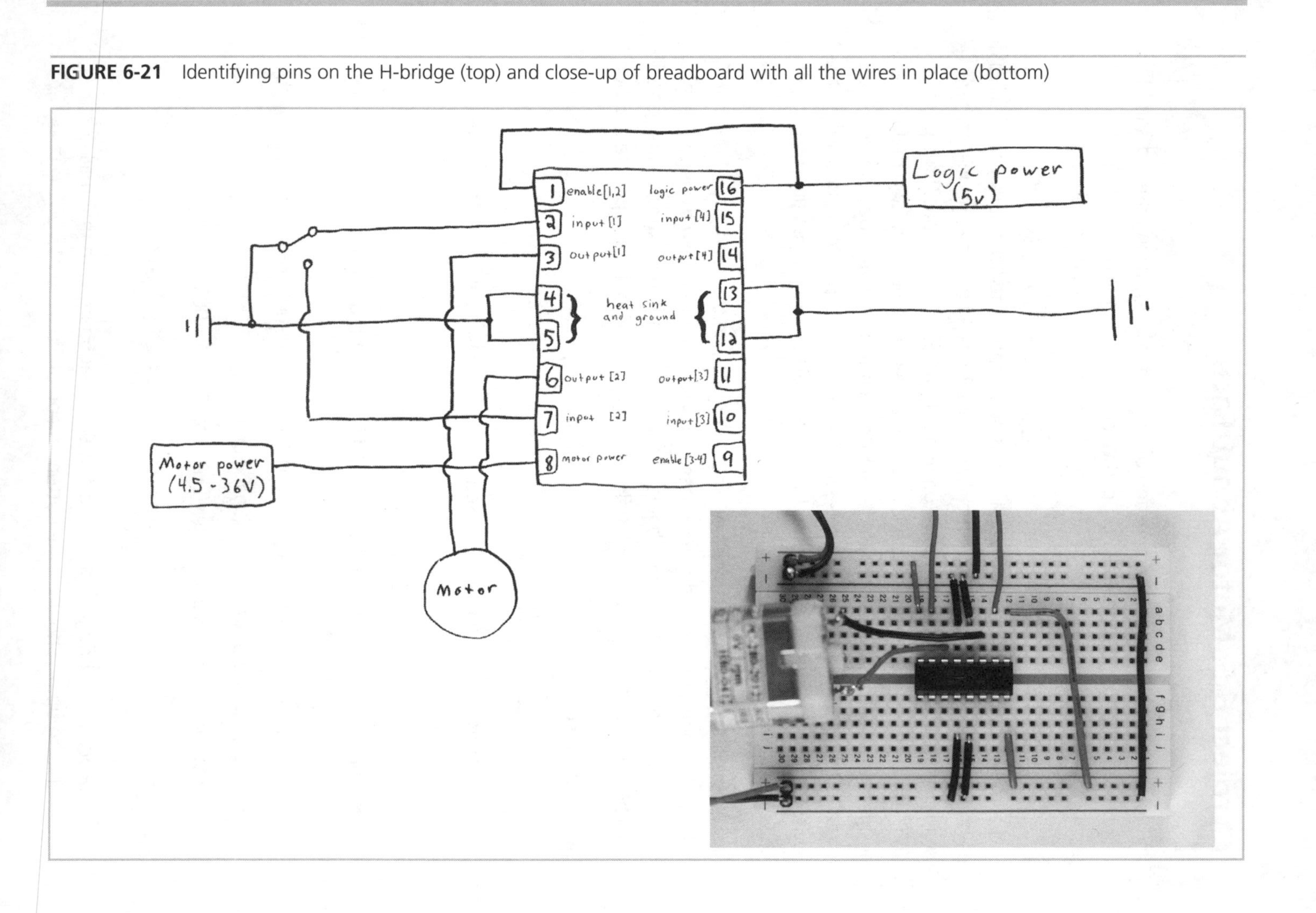

FIGURE 6-21 Identifying pins on the H-bridge (top) and close-up of breadboard with all the wires in place (bottom)

2. Plug the DC toy motor wire leads into the rows next to pins 3 and 6 on the H-bridge. Refer to the top image in Figure 6-21 to see which pin is which.

3. Solder jumper wires onto your SPDT switch. The example uses red for each side and black for the center (ground) leg.

4. Plug the two red wires of the SPDT switch into the rows next to pins 2 and 7 on the H-bridge. Connect the black wire from the middle leg to the ground column on the side of the breadboard marked with a – sign (ground). Refer to Figure 6-22 for the full setup. Make sure the switch is in the center (off) position.

5. Use jumper wires to connect pins 4, 5, 12, and 13 of the H-bridge to the ground columns.

6. The H-bridge chip needs about 5V to work. If you use four alkaline AA batteries as shown, that will add up to about 6V, which will work just fine (four rechargeable batteries will equal about 4.8V, which will also work). Plug the black wire from the battery holder into the ground column on the right side of the board and the red wire into the power column on the right side.

FIGURE 6-22 DC motor direction control with an SPDT switch and H-bridge chip

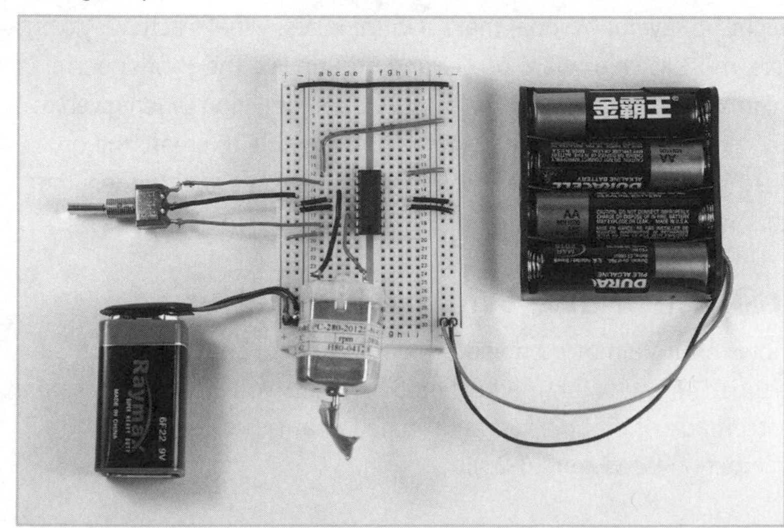

7. Connect pin 1 of the H-bridge to this 5V power column. This is the enable pin, which means it needs to be powered to tell the H-bridge chip you're ready to go. Connect pin 16 to this power column to give the circuit inside the chip power.

8. It's always a good idea to *use separate power supplies for the motor and the control logic parts of circuits*. Chips like the H-bridge we're using always want 5V, but motors usually want something different. Connect your motor battery (a 9V in this example) to the power and ground columns on the left side of the breadboard. This H-bridge chip will allow you to run motors that need up to 1A of current, which should be good for most of the motors we'll talk about in this book.

9. Even though motor power and control power come from different places, the ground columns should still be linked so they share a common zero point. You can do this on your breadboard by using a long jumper to link the two ground columns across the top of the board.

10. Connect pin 8 to the motor power column on the left side of the board.

11. Try to flip the switch from on to off to on, and see how the motor spins. It should spin clockwise at one extreme, stop in the middle, and then spin counterclockwise at the other extreme.

At this point, you might be saying, "Whoa, that's a lot of wires. What's actually going on here?" For starters, most IC chips want power and ground like the H-bridge. An H-bridge will allow current to flow through the motor in one direction when given a digital on (high or 5V) signal. We're using a switch in this example to create the on signal. When you flip the switch to the other extreme, another on signal triggers the H-bridge to allow current to flow through the motor in the opposite direction.

Speed Control with Pulse-Width Modulation

By now you know how to turn a motor on and off with a switch, but what if you want to control the speed? Pulse-width modulation (PWM) lets you do this by creating a *duty cycle*—the percentage of on time versus off time—that is between 0% and 100% of a given time period (see Figure 6-23).

FIGURE 6-23 A PWM signal

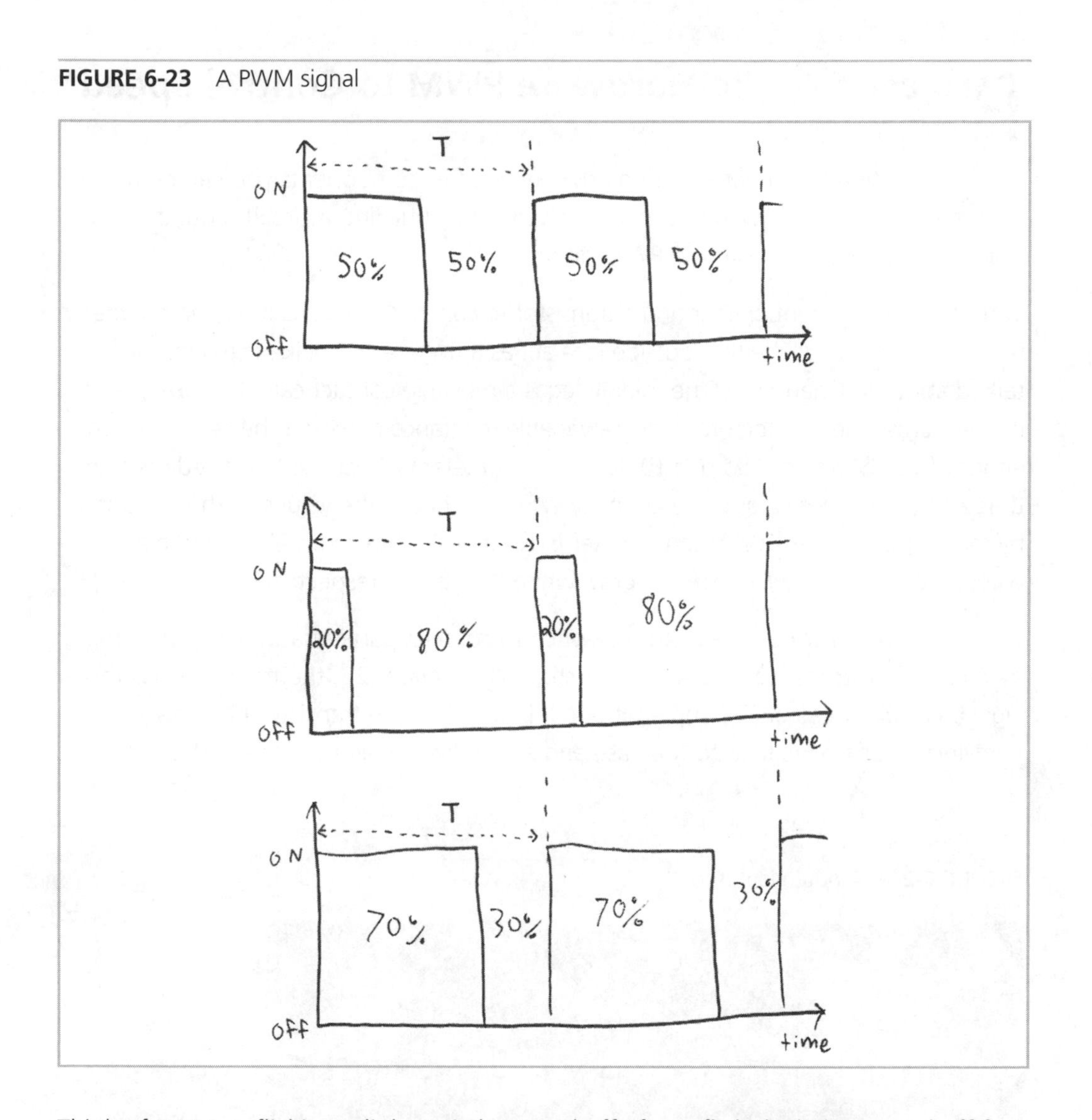

Think of PWM as flicking a light switch on and off. If you flick the light on and off fast enough, the average of the dark and light makes it look like the light is on, but just dimmer than if you leave it on. The same goes for a PWM signal to control a motor. Instead of giving the motor its full-rated voltage, you flick the power on and off fast enough that the average voltage is below what your power source gives you. For example, if you have a 9V power source trying to drive a DC gearhead motor that wants 3V to 6V, you could give it a pulse width at 50% of the time interval to equal a voltage of 4.5V, and make the motor happy.

Project 6-6: Use Hardware PWM to Control Speed

You can create a PWM signal with hardware—that is, components you can hold—or in software. We'll start with the hardware version by building a circuit around a chip called a 555 timer to create the PWM signal.[5,6]

You will need a potentiometer and a transistor to complete this circuit. A *potentiometer* is a variable resistor. The two outside legs act as a fixed resistor (like the ones we talked about in Chapter 5). The middle leg is a movable contact called a *wiper*, which moves across the resistor, producing a variable resistance between the center leg and either of the two sides.[7] So our 100KΩ potentiometer will act like two fixed resistors that add up to 100KΩ, and the knob allows us to choose the values of those resistors by moving the wiper. The potentiometer in Figure 6-24 has red, yellow, and black wires soldered to it (which appear gray, white, and black, respectively, in the figure).

We'll use a *transistor* as an electronic switch to connect parts of a circuit, just as the mechanical switches we've used do. As shown in Figure 6-25, the transistor has three legs: base (B), collector (C), and emitter (E). In an NPN type transistor (like this), applying a positive voltage to the base and a negative voltage to the emitter allows

FIGURE 6-24 A potentiometer

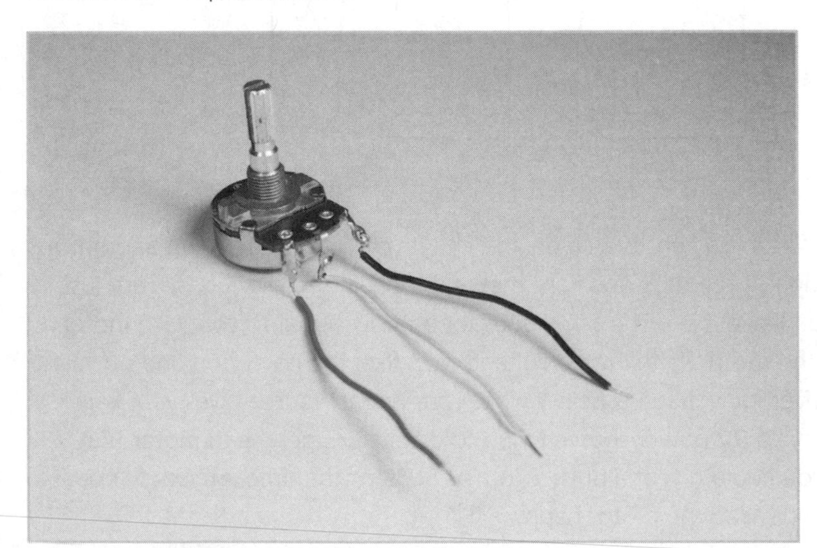

FIGURE 6-25 A transistor

current to flow from collector to emitter. We need a transistor here because even though we can send timing signals to the motor directly from the 555 timer chip, we can't actually send the motor power *through* it. That would fry the chip (the chip can handle only up to 200mA, and most motors use more than that). So we use the transistor like an electronic switch to allow power to flow to our motor only when the 555 timer says it's okay.

Shopping List:

- DC toy motor with wire leads
- Corresponding battery (9V used here) and snap or holder with wire leads (like RadioShack 270-324)
- Breadboard (like All Electronics PB-400)
- Jumper wires (like SparkFun PRT-00124) or hook-up wire to make your own (see Project 6-4)
- On-off-on toggle switch (like SparkFun COM-09609) or other SPDT switch
- Four AA batteries and holder (like SparkFun PRT-00552)
- 555 timer chip (SparkFun COM-09273)
- TIP120 Darlington transistor (Digi-Key TIP120-ND or Jameco 32993)

- 100KΩ potentiometer (Jameco 29103)
- Two 0.1 μF capacitors (electrolytic shown here, but you can also use ceramic, like SparkFun COM-08375)

Recipe:

1. Connect the 5V power and ground (from the AA battery pack) to the power and ground columns on one side of the breadboard.
2. On the other side of the breadboard, connect the motor power (9V battery here) to power and ground. Use a long jumper to link both ground columns on the breadboard.
3. Plug the 555 chip into the breadboard with the small dimple on the top left (see Figure 6-26).

FIGURE 6-26 Layout of 555 timer chip circuit

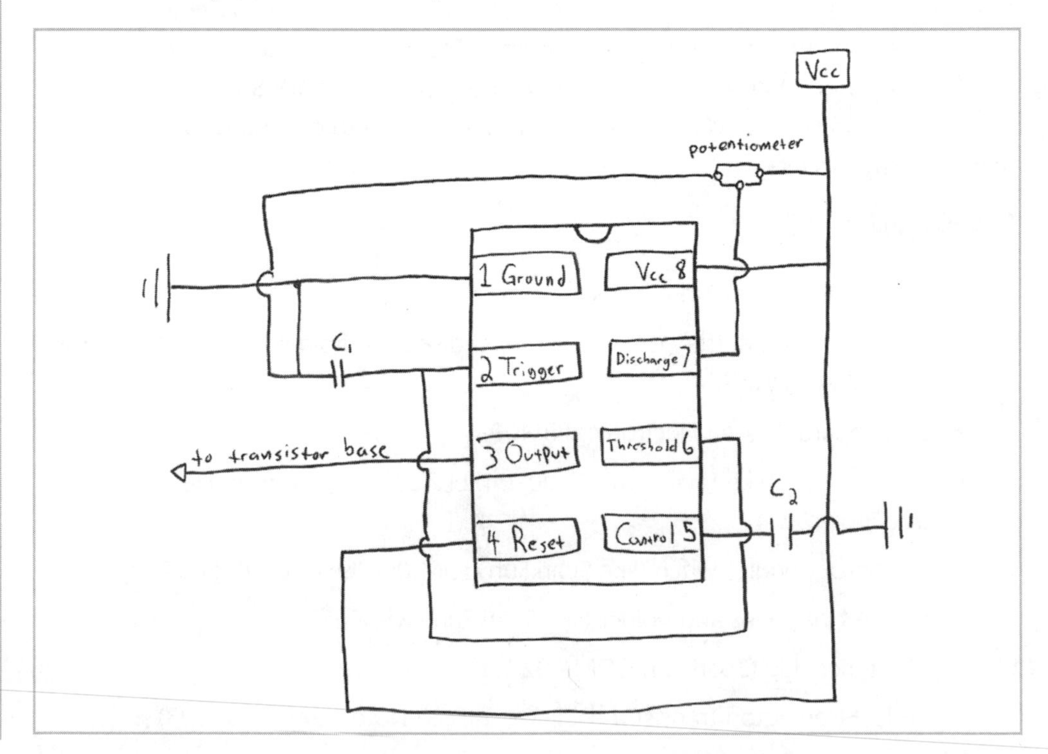

4. Connect pin 1 to ground.
5. Connect pin 2 and pin 6 of the 555 timer with a short jumper.
6. Also connect pin 2 to one of the outside legs of a potentiometer.
7. Also connect pin 2 through a 0.1 μF capacitor to ground.
8. Connect pins 4 and 8 with a small jumper wire. Connect pin 8 to the 5V battery power column and to the other outside leg of the potentiometer. Refer to Figure 6-27 for a close-up of the completed breadboard circuit.
9. Connect pin 5 to ground through a 0.1 μF capacitor.
10. Connect pin 7 to the middle leg of the potentiometer.
11. Plug the transistor into the breadboard as shown so each of the three legs has its own row (see Figure 6-27).
12. Connect the output pin 3 on the 555 chip to the base of the transistor.

FIGURE 6-27 Using a 555 timer to PWM a motor, detailed view

13. Connect the emitter leg of the transistor to ground.

14. Connect one leg of the motor to the collector (middle leg) of the transistor. Connect the other leg of the motor to the 9V battery power column. Your circuit should look something like Figure 6-28.

15. Now your motor should turn on. If it doesn't, turn the knob of the potentiometer clockwise or counterclockwise until you see the motor spin. Watch how the motor speeds up or slows down when the potentiometer is turned. In this circuit, we have the 555 timer wired as a pulse generator, where the length of the pulse depends on the ratio of resistor values from the potentiometer.

FIGURE 6-28 Using a 555 timer to PWM a motor, full-circuit view

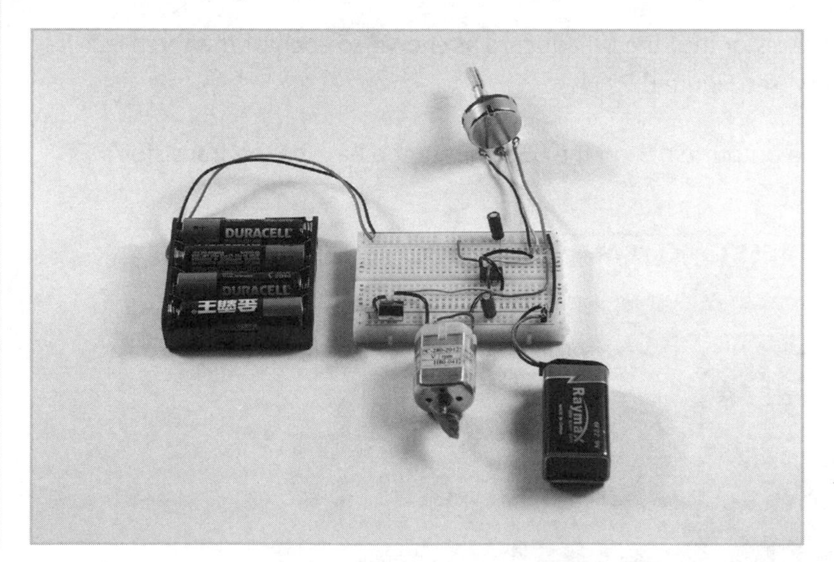

Advanced Control of DC Motors

The next step up from using the breadboard, chips, and switches is using a microcontroller, such as the one on the Arduino prototyping platform, to talk to your motor. This is like giving your project a brain. The Arduino can do just about everything we've done with hardware in the preceding example with just a few lines of code.

Project 6-7: Use Software PWM to Control Speed

We will re-create the hardware PWM project in software so you can get a feel for what the Arduino can do. This example assumes you've downloaded the software, installed the drivers for your PC or Mac, and identified the port your Arduino is plugged into. Refer to the "Arduino Primer" section in the appendix for how to set up the Arduino to communicate with your computer.

Unfortunately, you usually can't plug motors directly into the Arduino. The Arduino can source up to only 40mA on each of the input/output pins, and up to 500mA through the power pins when connected through USB. A lot of motors you'll use need more current than this. We get around this issue by using the Arduino to give the motor instructions through a transistor, and giving the motor a separate power supply. This is similar to Project 6-6, except we'll replace the 555 timer with an Arduino.

In this project, we'll create a sketch that listens for input from the stuff you have plugged in (switches, sensors, and so on) and then talks to components you want to control (such as motors). We'll build the circuit first, and then go over how to turn a motor on and off through a transistor with code from the Arduino, and finally use PWM for speed control through the Arduino. (You can find plenty of well-documented example sketches of using the Arduino to talk to motors and other components. For example, see http://arduino.cc/en/Tutorial/HomePage for basic sketches. In most cases. you can just start with these examples and modify them as you see fit.)

Shopping List:

- Arduino with USB cable
- DC toy motor with wire leads
- Corresponding battery (9V used here) and snap or holder with wire leads (like RadioShack 270-324)
- Breadboard (like All Electronics PB-400)
- Jumper wires (like SparkFun PRT-00124) or hook-up wire to make your own (see Project 6-4)

- On-off toggle switch (like SparkFun COM-09276) or other SPST switch
- TIP120 Darlington transistor (Digi-Key TIP120-ND or Jameco 32993)
- 220KΩ resistor (Jameco 30470)
- Diode (SparkFun COM-08589)

Recipe:

1. Connect the 5V power and ground pins on the Arduino to power and ground on one side of the breadboard with jumper wires (see Figure 6-29). On the other side of the breadboard, connect a 9V battery to power and ground. Make sure that the ground is linked between the ground columns on the breadboard.

2. Plug the transistor into the breadboard as shown in Figure 6-30, so each of the three legs has its own row. Connect the emitter pin of the transistor to ground on the breadboard.

3. Connect pin 9 on the Arduino to the base pin of the transistor.

FIGURE 6-29 Powering a breadboard with the Arduino

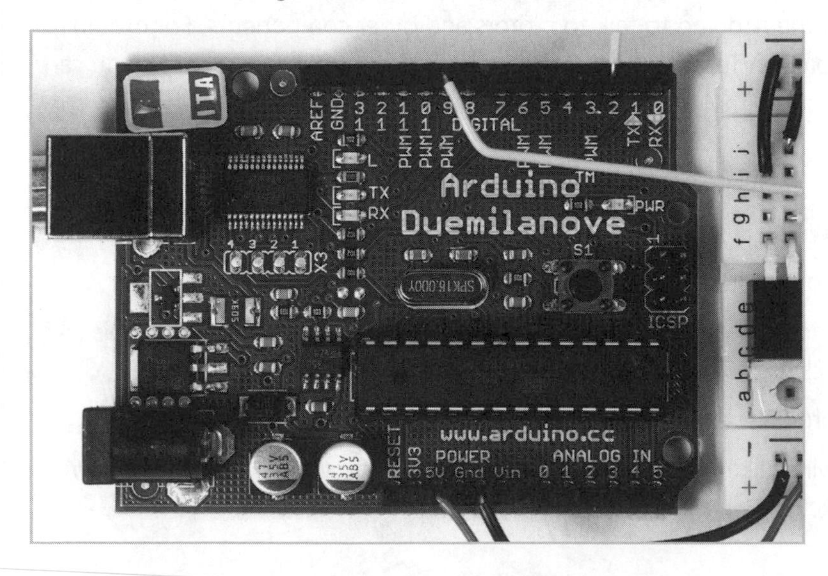

FIGURE 6-30 Close-up of the transistor wiring

4. Connect the collector of the transistor to ground through the diode. Make sure it's pointing in the right direction, with the stripe mark closest to the middle of the board.

NOTE ***Diodes allow power to flow in one direction and block it in the other. Although not strictly necessary, this is good practice to make sure current is flowing only in the direction you want it to (in this case, into the collector from 9V power). The diode protects the TIP120 transistor from back voltage (power flowing the wrong direction through the motor) generated when the motor turns off.***

5. Connect one leg of the motor to the collector of the transistor on the breadboard. Connect the other leg to the column with the 9V battery power.

6. Place the toggle switch (in the off position) on the other side of the breadboard and connect one leg with a signal wire to pin 2 on the Arduino. Also connect this leg to ground through a 220KΩ resistor. Connect the other leg of the switch directly into the 5V power column on the breadboard fed from the Arduino. Your circuit should look like Figure 6-31.

FIGURE 6-31 The Arduino setup to drive a motor through a transistor

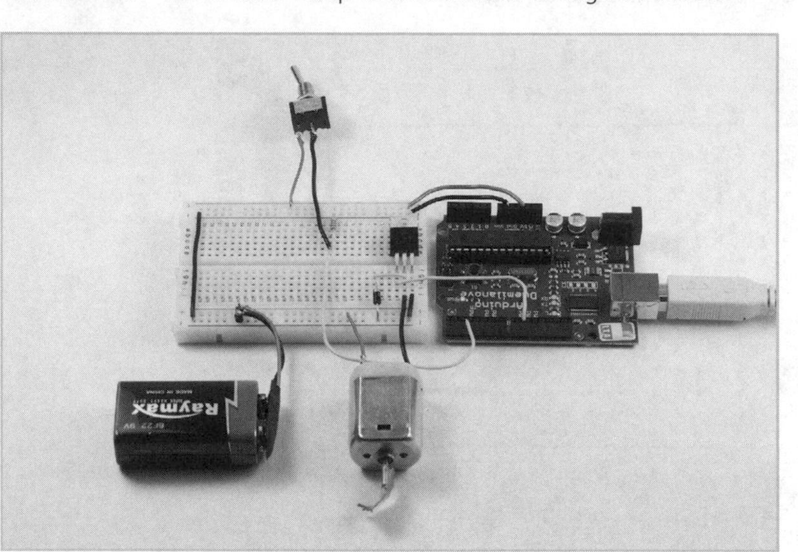

7. Open the Arduino application on your computer and start a new sketch. Type in the following code, verify it, and then upload it to the Arduino.

```
/*
Using Arduino to turn on a motor with input from a switch

Created June 2010
By Stina Marie Hasse Jorgensen, Sam Galison, and Dustyn Roberts
Adapted from code at
http://itp.nyu.edu/physcomp/Tutorials/HighCurrentLoads
*/

const int transistorPin = 9;    // connected to base of transistor
const int switchPin = 2;    // connected to switch

void setup()
{
pinMode(switchPin, INPUT); // set the switch pin as input:
pinMode(transistorPin, OUTPUT);  // set the transistor pin as output:
}
```

```
void loop()
{
if (digitalRead(switchPin) == HIGH) // if switch is on (HIGH)...
  {
  digitalWrite(transistorPin, HIGH);   // turn motor on (HIGH)
  }

else if (digitalRead(switchPin) == LOW)  // if switch is off (LOW)...
  {
  digitalWrite(transistorPin, LOW);   // turn motor off (LOW)
  }
}
```

8. Flip the switch from off to on and see how the motor turns on. When you flip the switch off, the motor should stop. The signal to turn on or off goes from the switch, to the Arduino, and then to the base of the transistor, and allows motor power to flow from 9V power through the transistor to the motor.

9. Now that your motor will turn on and off through a transistor, we'll introduce speed control. You may have noticed a few of the digital input pins on the Arduino board have "PWM" written next to them. These are specifically set to recognize PWM directions from the Arduino code language using the `analogWrite` command. To test this function, open a new sketch and type the following code, verify it, and then upload it to the Arduino.

```
/*
Using Arduino's built in PWM code (analogWrite) for motor speed control
to turn on a motor with input from a switch

for more on PWM with Arduino, see http://arduino.cc/en/Tutorial/PWM

Created June 2010
By Stina Marie Hasse Jorgensen, Sam Galison, and Dustyn Roberts
Adapted from code at
http://itp.nyu.edu/physcomp/Tutorials/HighCurrentLoads
*/

const int transistorPin = 9;     // connected to base of transistor
const int switchPin = 2;     // connected to switch

void setup()
{
pinMode(switchPin, INPUT); // set the switch pin as input:
pinMode(transistorPin, OUTPUT); // set the transistor pin as output:
}
```

```
void loop()
{
if (digitalRead(switchPin) == HIGH)  //if switch is on (HIGH)...
  {
  for (int i=0; i <= 255; i++)  //ramp up speed slowly
    {
    analogWrite(transistorPin, i);  //send value of i to transistorPin
    delay(10);
    }

  delay(500); //wait half a second

  for (int j = 255; j >= 1; j--)  //ramp down speed slowly
    {
    analogWrite(transistorPin, j);
    delay(10);
    }

  delay(500); //wait half a second
  }  //end if

else if (digitalRead(switchPin) == LOW)  // if switch is off (LOW)...
  {
  digitalWrite(transistorPin, LOW);   // turn motor off (LOW)
  }

}  //end loop
```

10. When the switch is turned on, the motor should start spinning slowly, speed up, and then slow back down. This cycle will repeat until you turn the switch off.

Arduino Extensions

If you want robust speed and/or direction control, you might want to check out ready-made modules that interface with your Arduino and do the hard work for you. These modules can make your life easy by incorporating many of the things in the "Helpful Tips and Tricks for Motor Control" section later in this chapter. You'll pay for this convenience, but sometimes it's worth it. For example, SparkFun's ROB-09670 is a motor driver that has an H-bridge already in it, along with other conveniences like direction-indicating LEDs. SparkFun also sells a Digital PWM Motor Speed Controller (ROB-09668), which can control the speed of your motor with PWM without sacrificing torque. Adafruit Industries (www.adafruit.com) sells a Motor/Stepper/Servo Shield for Arduino that can make things even easier. All you do is plug the shield in on top of your existing Arduino, attach the motor wires in the right spots, download the library, and copy a few lines of code. The kits come unassembled, and you need to do a fair amount of soldering to get started. However, there are excellent tutorials linked right from the site.

Hobby Servo Control

All hobby servo motors have circuits inside them that respond to pulses. Each servo has three wires: power, signal, and ground. You need to plug one wire into ground (usually the black one), one wire into a power source in the motor's working range (usually the red one), and one wire into something that can give it pulses (usually the yellow one). This signal is known as *pulse-proportional modulation* (PPM)[8] (also known as pulse-position modulation) and is similar to PWM. This is what the Control System and Required Pulse lines at the top of the hobby servo motor specifications shown earlier in Figure 6-10 tell us.

The smarts inside a servo motor expect a pulse every 20 milliseconds (ms), or 50 times a second. Different servos vary, but most servos use a pulse width between 0.5 ms and 2.5 ms out of this 20 ms to send a signal to the servo motor (see Figure 6-32). This signal, or pulse, is similar to repeatedly turning the light on for 0.5 to 2.5 ms, then turning it off until a total on/off time of 20 ms passes, and then repeating the cycle (see Figure 6-32). As with a PWM signal, you can create this pulse in hardware or with software.

NOTE ***There are 1,000 microseconds (µs) in 1 ms. Because the letter*** **u** ***looks like the Greek letter*** **µ** ***but is easier to type, you will often see servo data sheets that state the servo range as 500 to 2,500*** **usec.**

Standard Hobby Servo Control

Standard hobby servos are controlled by pulses that tell them which direction to point. The specs shown earlier in Figure 6-10 indicate that the servo's range is 600 to 2,400 µs, with 1,500 µs neutral. The circuit inside the servo knows that a 600 µs pulse width out of 20 ms means point to one extreme (0°), and a 2,400 µs pulse width means point to the other extreme (180°). Any pulse width between 600 and 2,400 µs moves the motor to a position proportionally between 0° and 180°. If you want the motor to stay put, you just keep sending the same pulse width.

FIGURE 6-32 A PPM signal

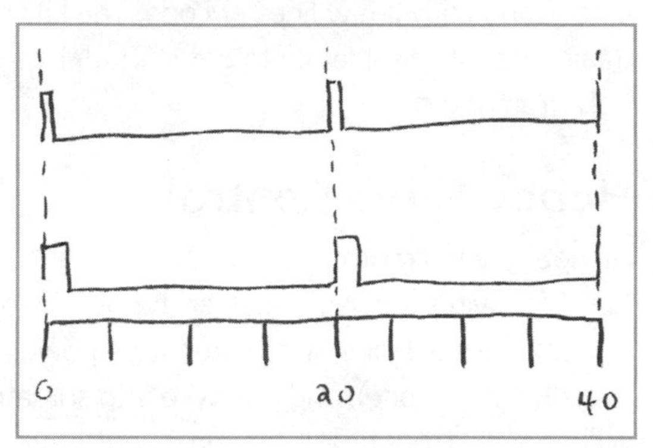

A potentiometer meshes with the gears in the servo to tell you exactly where the shaft is at all times. This is called closed-loop *feedback*. You can generate this pulse in hardware or software, or use a radio-controlled (RC) transmitter (like the ones found in model airplane kits) to send the signal to a receiver that talks to the motor.

Project 6-8: Control a Standard Hobby Servo

We'll use the Arduino in this example to generate the pulse, as we did in Project 6-7. However, instead of using the pulse to control the speed of a DC toy motor, it will control the pointing direction of a servo motor.[9] This time, we'll use a code library (which is just a bunch of code that's already written for you).

NOTE ***You can also take the long way and not use the servo code library. It's more involved but also gives you more control. For details, see Section 4.1, "Using the pulse method," at http://itp.nyu.edu/physcomp/Labs/Servo.***

Shopping List:

- Arduino Duemilanove with USB cable
- Servo motor (Hobbico CS-60 used here)
- Breadboard (like All Electronics PB-400)
- Jumper wires (like SparkFun PRT-00124) or hook-up wire to make your own (see Project 6-4)
- Male header pins (SparkFun PRT-00116)

- Diagonal cutters (like SparkFun TOL-08794)
- Photocell (10KΩ – 100KΩ, Digi-Key PDV-P9007-ND used here) and resistor (10KΩ, like SparkFun COM-08374 used here)

NOTE ***You can also use a 1KΩ – 10KΩ photocell (SparkFun SEN-09088). In that case, you should use a 1KΩ resistor (SparkFun COM-08980) to get the best response.***

Recipe:

1. Connect 5V power and ground from the Arduino to the power and ground columns on one side of the breadboard. Use jumper wires to jump connect these to power and ground on the other side of the board.

2. Clip off a set of three header pins from one side of the long row with the diagonal cutters or just snap them off by hand. Choose three rows on the top of the breadboard to hold the pins, and then plug the servo motor connector onto the header (see Figure 6-33).

FIGURE 6-33 Standard hobby servo control circuit

3. Connect the red wire to 5V power, the black wire to ground, and the yellow wire to digital pin 2 on the Arduino. Sometimes this third wire is green or orange, but it will always be different from the red (power) and black (ground) wires.

4. Connect one leg of the photocell directly to power.

5. Connect the other leg to a row on the breadboard of your choice. Then connect this row to ground through the resistor. Also connect this row to analog pin 0 on the Arduino. Your circuit should now look like Figure 6-33.

NOTE ***We need to use the analog pins with a photocell because it's not just an on-off type of input like the switches we've used until now. The photocell will actually indicate a value between 0 and 1023 (as will any analog sensor), depending on how much light is hitting it.***

6. Open a new sketch in Arduino and type the following code. Then verify and upload the code to the Arduino.

```
/*
Servo control from an analog input using the Arduino Servo library

Created June 2010
By Stina Marie Hasse Jorgensen and Dustyn Roberts
Adapted from code at http://itp.nyu.edu/physcomp/Labs/Servo
*/

#include <Servo.h>       // include the servo library

Servo servoMotor;     // creates an instance of the servo object
                      // to control a servo
int analogPin = 0;       // the analog pin that the sensor is on
int servoPin = 2;        // the digital pin for the yellow servo
                         // motor wire
int analogValue = 0;     // the value returned from the photocell

void setup()
{
servoMotor.attach(servoPin);  // attaches the servo on Arduino pin
                              // 2 to the servo object
}
```

```
void loop()
{
// read the analog input from the photocell (value between 0 and
// 1023)
analogValue = analogRead(analogPin);

// map the analog value from the photocell (0 - 1023) to the angle
// of the servo (0 - 179)
analogValue = map(analogValue, 0, 1023, 0, 179);

// write the new mapped analog value to set the position of the
// servo
// servoMotor.write(analogValue);

delay(15);   // waits for the servo to get there before getting
             // another photocell reading
}
```

7. Try to make the servo motor move by using your finger to block the photocell, and then moving it away and letting light hit it. It should move back and forth, but depending on the light in the room, you probably won't get the full range of the servo by doing this.

Continuous Rotation Servo Control

If you have a continuous rotation servo (that you bought or modified from a standard servo), you no longer have control over position. Instead, the signal you give the servo controls the speed.

The maximum speed you can expect with no load is already stated for you in the data sheet. For example, for the Hitec HS-311 servo motor, that speed is 60° in 0.15 seconds. Since there are 360° in one revolution, that means a modified HS-311 servo could finish one full revolution in 0.15 × 6 = 0.9 seconds. Because there are 60 seconds in a minute, dividing 60 seconds by 0.9 seconds gives a speed of 67 rpm.

Stepper Motor Control

There are two main types of stepper motors: unipolar and bipolar. They have between four and eight wires coming from the housing, and there are no standards as to what the wire colors mean. You use all these wires to give power to different parts of the motor in a specific sequence. The specifics of the sequence determine the stepping behavior (forward, backward, one-half step at a time, and so on). Both unipolar and bipolar steppers can be controlled by the same stepping sequence, but are wired differently.

Unipolar Stepper Motors

Unipolar, or four-phase, steppers have five, six, or sometimes (but rarely) eight wires. They have four sets of wire coils alternating around the outside of the motor housing (hence the term *four-phase*). Unipolar steppers energize the coils all at the same *polarity*, or direction of current flow (hence the term *unipolar*).

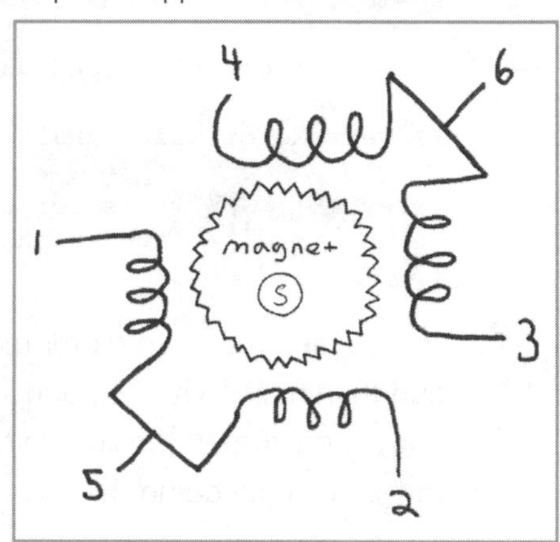

FIGURE 6-34 Schematic of a six-wire unipolar stepper motor

A five-wire stepper is the same as a six-wire stepper with the center connections (wires 5 and 6 in Figure 6-34) joined. The six-wire configuration shown in Figure 6-34 is the most popular and probably what you'll find when you pull a stepper motor out of a printer. If you come across an eight-wire, or universal stepper motor, it actually has four independent coils with two connections to each. These can be wired as a unipolar or bipolar stepper.

NOTE ***A six-wire unipolar stepper is just like a bipolar stepper motor but with center connections on each coil. It can also function as a bipolar stepper motor if the manufacturer has designed it that way.***

There are many options for controlling your stepper motor. To minimize time spent with breadboards and programming, it's best to consider ready-made modules that can handle all the hard work of feeding current to the correct wires the right way. Here are some suggestions:

- SparkFun's EasyDriver (ROB-09402) will work with unipolar stepper motors with six or eight wires that are wired as bipolar steppers. This module will work with anything that can generate a 0 to 5V pulse (your Arduino comes in handy here).
- You can use the Arduino to drive the motor directly, but there is more programming involved and you need some extra chips and a breadboard. Luckily, this is mostly done for you. Check out the Arduino stepper tutorial and library at www.arduino.cc/en/Tutorial/Stepper. The code works the same for unipolar and bipolar stepper motors.

- Another option is the Adafruit Industries (www.adafruit.com) Motor/Stepper/ Servo Shield for Arduino. All you need to do is plug the shield in on top of your existing Arduino, attach the stepper motor wires in the correct spots, download the library, and copy a few lines of code. The shield works for five- and six-wire unipolar steppers as well as bipolar steppers.

Bipolar Stepper Motors

Bipolar, or two-phase, stepper motors have four wires (see Figure 6-35). They have two independent sets of wire coils in alternating positions around the housing (hence the term *two-phase*). Bipolar steppers move by energizing the coils first in one direction and then reversing the direction as the shaft is turned (hence the term *bipolar*). A bipolar stepper motor will always be stronger than a unipolar motor of the same size. The same options are available to control the motor as with the unipolar type: SparkFun's EasyDriver, the Arduino with the stepper library and some breadboard work, the Adafruit motor shield, and plenty of others.

FIGURE 6-35 Schematic of a bipolar stepper motor

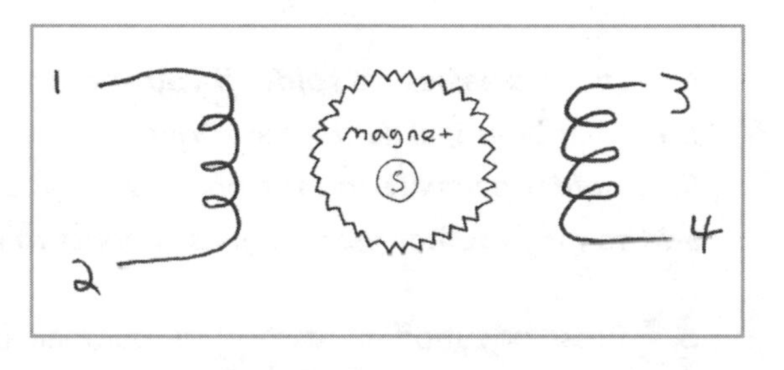

Project 6-9: Control a Bipolar Stepper Motor

In this example, we'll use a bipolar stepper motor and control it with SparkFun's EasyDriver.

Shopping List:

- Arduino with USB cable
- Breadboard (like All Electronics PB-400)
- Jumper wires (like SparkFun PRT-00124) or hook-up wire to make your own (see Project 6-4)

- Stepper motor (SparkFun ROB-09238)
- EasyDriver (SparkFun ROB-09402)
- Male header pins (SparkFun PRT-00116)
- Diagonal cutters (like SparkFun TOL-08794)

Recipe:

1. Break or cut off one set of four, one set of three, and one set of two male headers.

2. Solder male headers onto the EasyDriver (see Figure 6-36). The set of four lines up with the four motor holes, the set of three lines up with the GND/STEP/DIR holes, and the set of two lines up with the GND/M+ holes.

NOTE ***It's easiest to solder if you stick the long ends of the headers into the breadboard, slide the EasyDriver on the short ends, and then heat up the little solder pads around the holes while you add solder. Be careful not to add so much solder that the pins connect to each other!***

3. Break or cut off another set of four male headers and solder them onto the end of the stepper motor wires (see Figure 6-37). Make sure red and green are next to each other on one side, and blue and yellow on the other.

4. Plug the stepper motor header into the breadboard in line with the motor pins on the EasyDriver. The red and green wires should be next to A on the EasyDriver, and the blue and yellow wires next to B.

5. Jump a ground pin from the Arduino to the GND pin on the EasyDriver.

6. Connect Arduino pin 8 to DIR.

7. Connect Arduino pin 9 to STEP.

FIGURE 6-36 Soldering headers onto the EasyDriver board before (top) and after soldering a few header pins (bottom)

FIGURE 6-37 Soldering male headers onto the stepper motor wires before (top) and after strain relieving with a cable tie and insulating with hot glue (bottom)

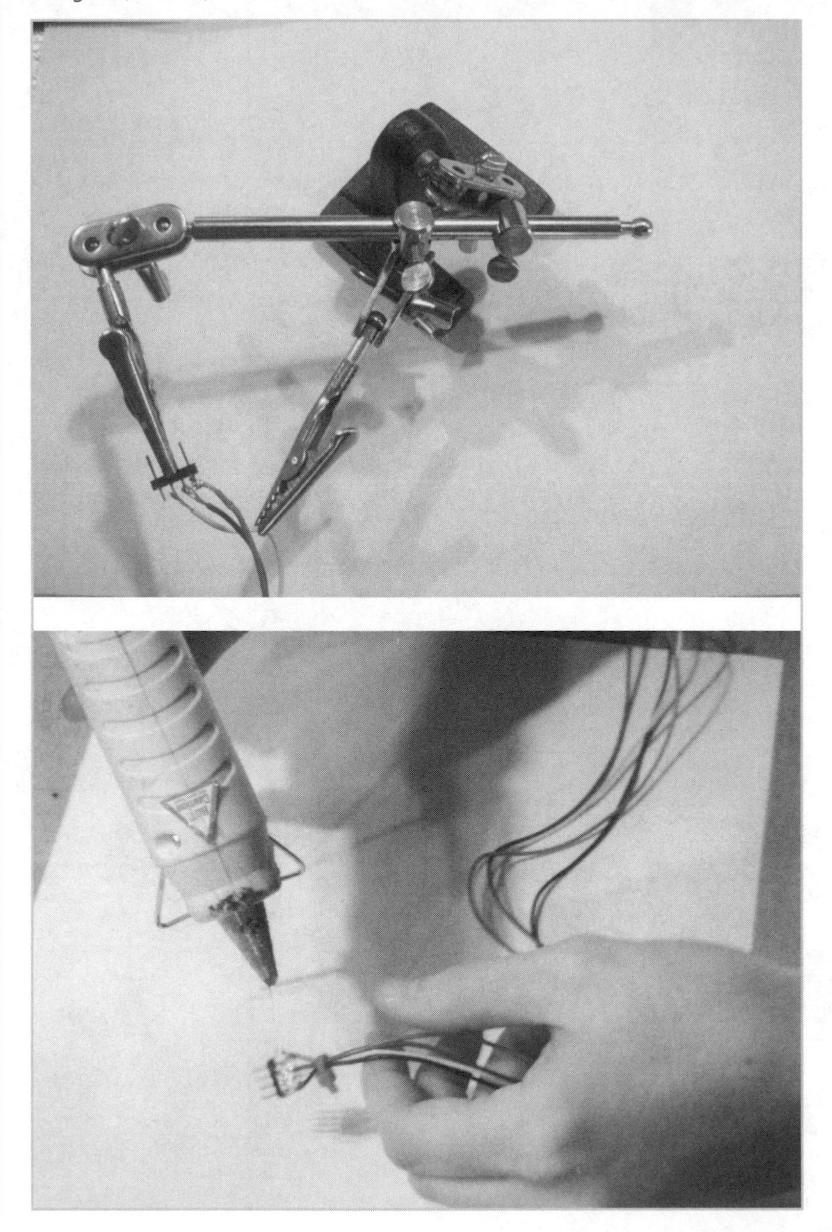

FIGURE 6-38 Stepper motor circuit all wired up

8. Connect a 12V power supply to the M+ and GND pins on the EasyDriver. The example uses a benchtop supply set to 12V. (Refer to the appendix for other ways to get power to your breadboard.) Your circuit should now resemble Figure 6-38.

9. Type the following code, verify it, and upload it to your Arduino.

```
/*
This code is to get one stepper motor moving through
SparkFun's EasyDriver board.
```

```
Created 2010.06.01
By Ben Leduc-Mills and Dustyn Roberts
*/

#include <Stepper.h>  //import stepper library

#define STEPS 200  //this should equal the number of steps our
                   //motor is rated for
Stepper stepper(STEPS, 8, 9); //goes to Arduino digital pins 8
                              //(DIR) and 9 (STEP)

void setup()
{
stepper.setSpeed(200);  //set speed of stepper in RPM
}

void loop()
{
  stepper.step(6400);  //turn one full rotation
  delay(100);  //wait 1/10th of a second
  stepper.step(-6400);  //turn one full rotation the other way
  delay(100);  //wait 1/10th of a second
}
```

10. Your motor should now rotate back and forth! Try putting a little flag of tape on the motor to help you see what's going on.

Linear Motor Control

Linear motors are essentially DC motors that interact with a power screw assembly to make a plunger go in and out, so you can control them in the same way that you control DC motors. Just apply a voltage across the two wires within the stated operating range, and you're set.

The total distance the plunger travels from out to in (or in to out) is called the *stroke distance*. You may want to use the Arduino and/or switches to limit the stroke distance to create whatever movement you want. Most linear motors come with integrated switches and/or a potentiometer to help you control the speed and position.

Solenoid Control

A solenoid is controlled like a DC toy or gearhead motor, in that all you need to do is apply the correct voltage with enough current across the two connections, and it will move. Pull-type solenoids pull the plunger into the housing, and push-type solenoids push the plunger out.

Solenoids are either continuous duty or intermittent duty. *Continuous duty* means that you can turn the solenoid on, and the plunger will either push or pull, and then stay there as long as it's powered. *Intermittent duty* means that when you turn on the solenoid, it will either push or pull for only a set amount of time (sometimes called *maximum on time*). When you remove power from a solenoid, the plunger does not return to its original position on its own. Usually, there will be a spring to return it after it has been pushed or pulled.

Helpful Tips and Tricks for Motor Control

Whenever you turn a motor on or switch directions, you create mechanical stress on the motor, as well as electrical stress on the attached cables, circuits, and batteries. Current can flow backward through the motor, something called *blowback* or *back voltage*, causing back electromotive force (EMF), and that's all bad.[2] Mechanical and electrical stress decrease the life span of the motor and can wreak havoc on any connected control electronics. Many of the ready-made modules mentioned earlier in the chapter limit this stress and prolong motor life with tactics like slowly ramping up speed, regulating voltage, smoothing the current flow, and using some of the following helpful tips and tricks.

Diodes Are Your Friends

When you reverse the direction on a DC motor or turn on a solenoid, you create a power spike that can sometimes be harmful to other components in your circuit. When this happens, electrical energy can flow in directions you didn't intend. Diodes are little electronic components that let current flow through them in only one direction. They help protect your circuits by ensuring electricity can flow only the way you want it to flow. To use a diode like the one shown in Figure 6-39, all you need to do is put it in line with the intended direction of current flow and make sure it's facing the right direction.

FIGURE 6-39 How to use a diode and LED in a circuit

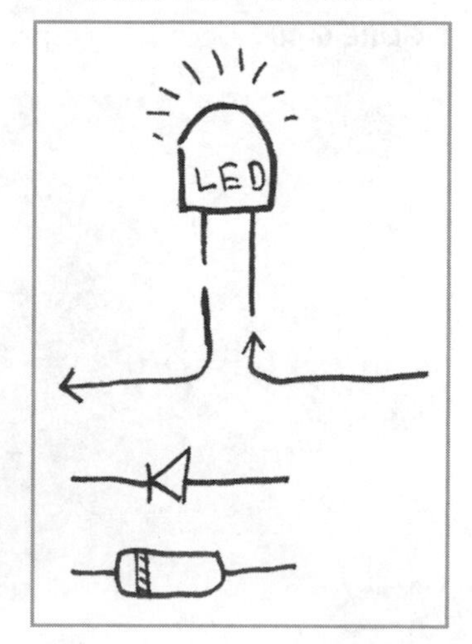

Light-emitting diodes—LEDs for short—emit light when current runs through them. Like all diodes, these need to be put in a circuit in the correct orientation, or else they will act as a wall and not an open gate. Luckily, most LEDs come with a short leg (ground) and a long leg (power).

The side of the LED over the shorter ground leg is usually flat, so you can still tell which side is ground, even if you clip the legs shorter. Install the LED in the circuit so that current flows through it from the long leg to the short leg.

LEDs don't limit voltage on their own, so you need to put a resistor in series with your LED before hooking it up to most power sources, or else you'll fry it. For example, if you're using a 5V power supply, a 220KΩ resistor will work well with most LEDs. At higher voltages, you'll need proportionally higher resistors to protect your LED.

Decoupling Capacitors

You can use capacitors to your advantage in circuits with motors to smooth out the energy spikes that happen when motors are turned on or change directions. Capacitors used for this purpose are called *decoupling capacitors*.

A popular approach is to solder a capacitor across the two leads of a DC motor before ever using it, just to be safe. You may also see capacitors soldered from the motor leads to the motor housing (see Figure 6-40). Make sure to use ceramic (not electrolytic) capacitors for these applications, since they don't care which way the current flows into them. This will allow you to run the motor clockwise or counterclockwise with no worries.

FIGURE 6-40 Decoupling capacitors soldered to DC motor leads

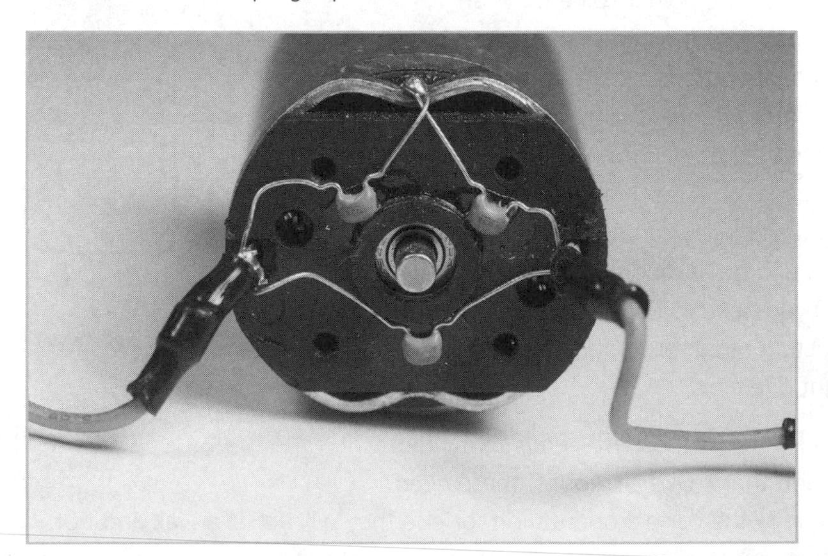

Using decoupling capacitors is a quick-and-dirty method to keep energy spikes in your mechanism's circuit under control, rather than using some of the ready-made smart modules we talked about earlier that do this kind of thing for you. A 0.1 μF capacitor generally works well for bridging the connections on a DC motor and smoothing energy spikes.

Separating Logic and Motor Power Supplies

It's a good idea to separate the power supplies for your motor and the logic—the breadboard circuit or Arduino—that is controlling it. There are a few good reasons for this:

- Most controllers (like the Arduino) and chips (like the 555 timer and H-bridge) take power at 5V. Your motor will most likely want something different. If you try to power your 5V Arduino and your 12V motor from the same battery pack, one of those paths is going to waste a lot of energy or not work at all. Isolating the power supplies means you can choose the right supply for each job.
- If your motor needs more current and voltage than you can safely run through an Arduino (anything over 500mA), you absolutely need a separate power supply.
- Even if your circuit or Arduino can supply the voltage and current your small motor needs, you will still see noise in the system from turning motors on and off and switching directions. Diodes and decoupling capacitors can help this situation, but it's still a good idea to keep the power supplies separate and avoid the problem altogether.

NOTE ***Even if you separate the power supplies for your controller and motor, you must connect the ground wires together. It's good practice to keep the circuit and the power supply grounds at the same low energy level in order for the logic to talk to the motor effectively.***

Relays and Transistors

Transistors and relays are like electronic switches. Mechanical switches are switched on and off with your finger. Transistors and relays are switched by an electrical signal. You need them when working with motors, because most of the time, the amount of current the motor needs and the amount of current allowed to flow through your circuit or Arduino (500mA for the Duemilanove model) are different. By using a transistor or relay, and a separate power source for your motor, you can just tell the transistor or relay to open when you want your motor to get power.

The TIP120 is a common transistor to use for this purpose, and SparkFun's COM-00100 is an easy-to-use relay that will plug directly into your breadboard. The relay will allow you to turn on a motor that needs as much as 5A at 12V with only a 12mA and 5V signal, which an Arduino can easily send.

Motorless Motion

Although motors are the most common actuators, there are a few other options worth mentioning. These include fluid pressure and artificial muscles.

Fluid Pressure

We talked about fluids in the alternative energy section of Chapter 5, so you know a fluid is anything that flows—air, water, or maple syrup. Fluids always take the shape of the container they're in, so they exert pressure in all directions in that container. This pressure depends on the depth and weight of the fluid:

$$Pressure = Depth \times Density \times Gravity$$

Viscosity is a measure of the thickness of a liquid. Water has a low viscosity, maple syrup has a medium viscosity, and silly putty has a high viscosity.

Both hydraulic fluid and compressed nitrogen are used in the open-assist gas springs (such as McMaster 9416K14) common on lids, windows, and car trunks. Close-assist gas springs are common on screen doors to avoid slamming. These allow smooth motion in one direction and resist motion in the other, providing help in the stroke direction where help is needed. The compressed gas does the work, and the hydraulic fluid stops the plunger from slamming at the end of stroke.

Hydraulics

Hydraulics are concerned with liquid-driven mechanisms. Liquids are incompressible, so when you try to squish them, they push back. Hydraulic cylinders are normally operated at high pressures (about 1,000 psi or more) and used in backhoes and industrial machinery.

Using hydraulics is a bit like working with AC power. You want to make sure you know what you're doing before you try, or the consequences could be lethal. The equipment also tends to be very expensive. The high pressures and forces that hydraulic systems can create aren't usually necessary for the kinds of projects this book encourages.

Pneumatics

The field of pneumatics deals with gas-driven mechanisms. Gases *are* compressible, and they can store the energy it takes to compress them for later use. Pneumatics are normally used at much lower pressures (around 100 psi) than hydraulics, which makes them much safer. Pneumatic actuators are used where electric motors are dangerous (as in underground mines) or impractical (as in common dentistry tools). Pneumatic drills and nail guns are commonly used for DIY construction projects.

Cheap resources for compressed air include air brush kits, bike tire pumps, car tire pumps, and portable tabletop compressors. Compressors need electricity to squish the air before you can use it as a source of power. Black & Decker (among others) sells a small, portable inflator (Model ASI300) you might find useful for projects in this area.

Air muscles are another technology that uses compressed air. Think of them as a sealed version of those mesh-woven Chinese finger traps you played with as a kid. They work by inflating the mesh-woven tube so the overall result is contraction. The ones from Images (www.imagesco.com/catalog/airmuscle/AirMuscle.html) can contract to 75% of their relaxed length. You will need an air pump that can reach at least 50 psi, so a small compressor or even a bike tire pump would work fine.

Artificial Muscles

There are two flavors of materials emerging that contract when you feed them electrical energy: electroactive polymer actuators and nitinol. Since they mimic human muscle motion, both technologies are commonly referred to as *artificial muscles* or *muscle wire*. They are attractive options for engineers and designers because they could potentially take up much less space and be lighter than motors, leading to mechanisms with more human-like actuators and motion. However, the technologies are immature, require high current, and can be hard to apply.

Shape Memory Alloy

Wire made of nitinol (a nickel-titanium mix) is one example of a material that will shrink when heated past a certain point, and then return to its original length at room temperature. This effect is called *shape memory*, since the wire "remembers" what it's supposed to look like. The metal mix is known as shape memory alloy (SMA).

Dynalloy (www.dynalloy.com) is the main manufacturer of an SMA called Flexinol, which is designed to be durable enough to create movement in mechanisms. SparkFun carries some actuators from Miga motors, like the Miga NanoMuscle (ROB-08782), which can be used to make small linear movements.

The recurring complaint about muscle wire is that it contracts only about 3% to 5% of the original wire length, which limits its practical applications, but you can probably make an inchworm robot with a few LEGO pieces and a short SMA wire. For project ideas, check out the *Muscle Wires Project Book* by Roger Gilbertson (Mondo-Tronics, 2000).

Electroactive Polymer Actuators

Electroactive polymer actuators (EAPs) are similar to SMA, but based in plastic instead of metal (although some metal-plastic composites are emerging). They have the ability to contract up to 380%, which is extreme in comparison to nitinol. Researchers have tried using it to make arm-wrestling robots and even control a fish-shaped inflatable blimp, but again, the technology is immature and has not gone mainstream yet, Figure 6-41 shows a four-fingered gripper actuated by EAPs. See the article at www.empa.ch/plugin/template/empa/*/74071) for more information about EAPs.

FIGURE 6-41 Gripper actuated by EAPs (courtesy of Yoseph Bar-Cohen, Jet Propulsion Laboratory/Caltech/NASA)

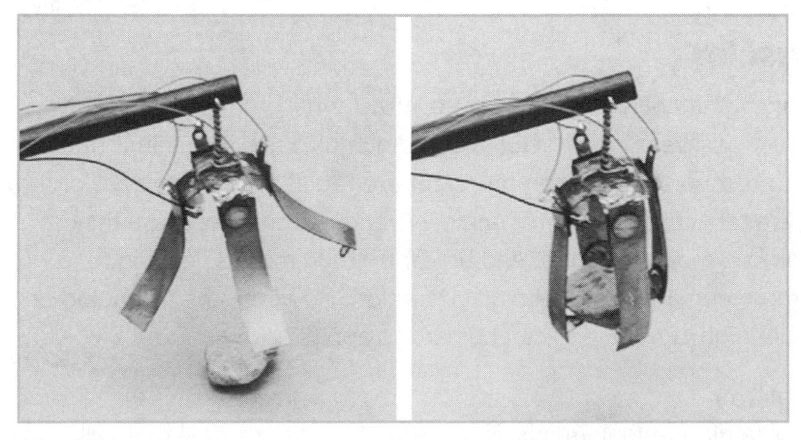

References

1 Mike Passaretti, Honeybee Robotics, introduced me to this project.

2 Dan O'Sullivan and Tom Igoe, *Physical Computing: Sensing and Controlling the Physical World with Computers* (Boston: Thomson, 2004).

3 Dennis Clark and Michael Owings, *Building Robot Drive Trains* (New York: McGraw-Hill, 2003).

4 Tom Igoe, "DC Motor Control Using an H-Bridge" (http://itp.nyu.edu/physcomp/Labs/DCMotorControl).

5 Tom Igoe, "Using a Transistor to Control High Current Loads with an Arduino" (http://itp.nyu.edu/physcomp/Tutorials/HighCurrentLoads).

6 Gordon McComb, *The Robot Builder's Bonanza*, ed. Michael Predko (New York: McGraw-Hill, 2006).

7 Tom Igoe, "Components" (http://itp.nyu.edu/physcomp/Labs/Components#toc4).

8 Dennis Clark and Michael Owings, *Building Robot Drive Trains* (New York: McGraw-Hill, 2002).

9 Tom Igoe, "Servo Motor Control with an Arduino" (http://itp.nyu.edu/physcomp/Labs/Servo).

References

7 The Guts: Bearings, Couplers, Gears, Screws, and Springs

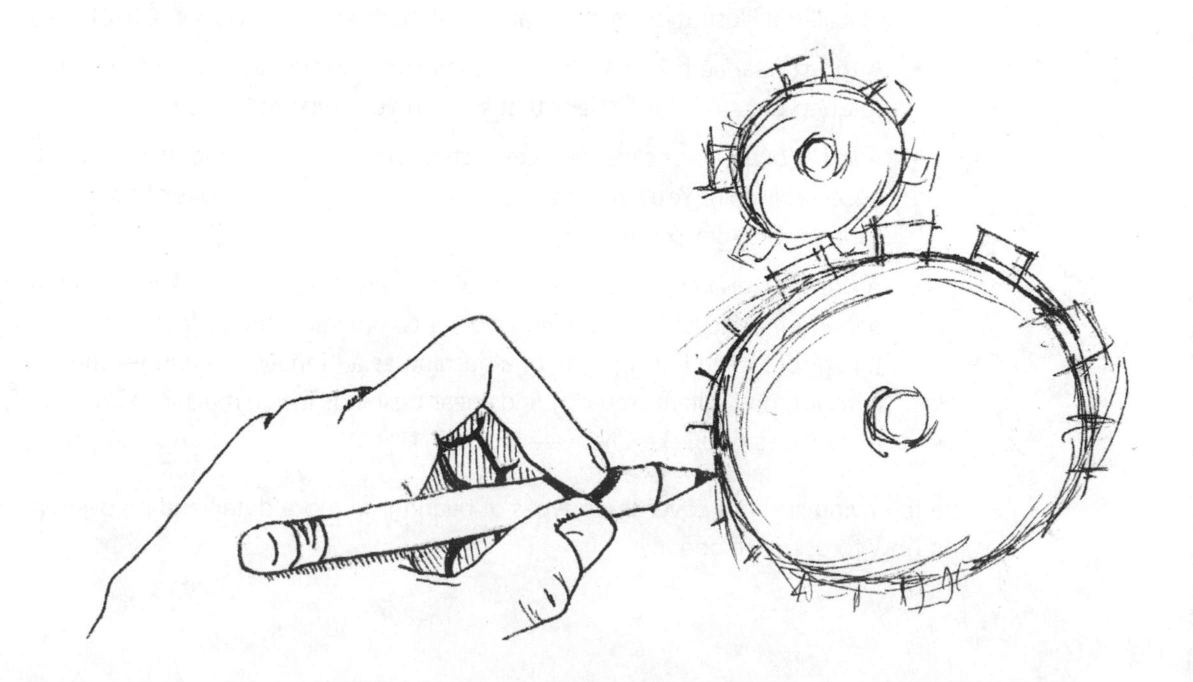

The guts of a mechanism are everything that happens between the input and output. The input is your energy source, which can range from a hand crank to an electric motor. The output is what you want to happen—does your mechanism crawl, spin, point, or shake? Maybe you need to attach a gear to your motor shaft or figure out how to make something spin with lower friction.

The components we'll cover in this chapter are integral to being able to work through your ideas and make them into reality. The majority of them can be found through a quick search on McMaster and other suppliers I'll point out along the way.

Bearings and Bushings

Bearings are components that are used between moving parts and stationary parts for support and reduction of friction. A bearing can be as simple as a drilled hole in a block of wood, or it can be an actual steel ball bearing, as in inline skates or skateboard wheels. You can also find bearings inside motors, where they help to support the motor shaft and keep it running smoothly.

Bearings are categorized by the kind of load they support:

- A *radial bearing*, like the type in your inline skates, supports radial loads. (Recall the illustration of radial and axial loads in Figure 1-26 in Chapter 1.)
- A *thrust bearing* handles the axial loads. You can find this kind of bearing in rotating bar stools and chairs that support your weight but still spin.
- A *linear bearing*, or *slide*, reduces friction in sliding components that don't necessarily spin. You can find this type of bearing on the sides of filing cabinets and dresser drawers.
- A *bushing* (also known as a *sleeve*, *plain*, *plane*, or *journal bearing*) is a type of bearing that doesn't have rolling elements, but still reduces friction for radial, thrust, or sliding loads. Think of a bushing as a "female" bearing—one without, um, rollers. You can find linear bushings inside machines like MakerBot's CupCake CNC (see Figure 7-1).

The following sections cover these types of bearings in more detail and go over when and how to use each one.

FIGURE 7-1 Linear bushings on MakerBot's CupCake CNC (image used with permission from MakerBot Industries)

Radial Bearings

The purpose of any radial bearing is to support a spinning shaft or rod and keep it running smoothly, even if things like gears and pulleys create radial loads on the supported shaft. Some radial bearings have rolling elements that reduce friction. These are called *ball bearings* when the rolling element is a ball, or *roller bearings* when the rolling element is more like a long cylinder or needle (see Figure 7-2).

FIGURE 7-2 Radial ball and roller bearings (credit: McMaster-Carr)

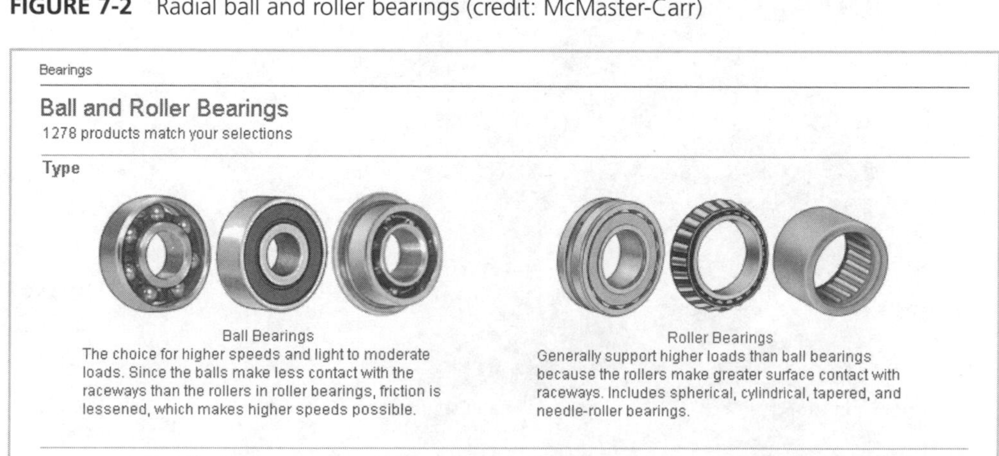

FIGURE 7-3 Radial bushings (credit: McMaster-Carr)

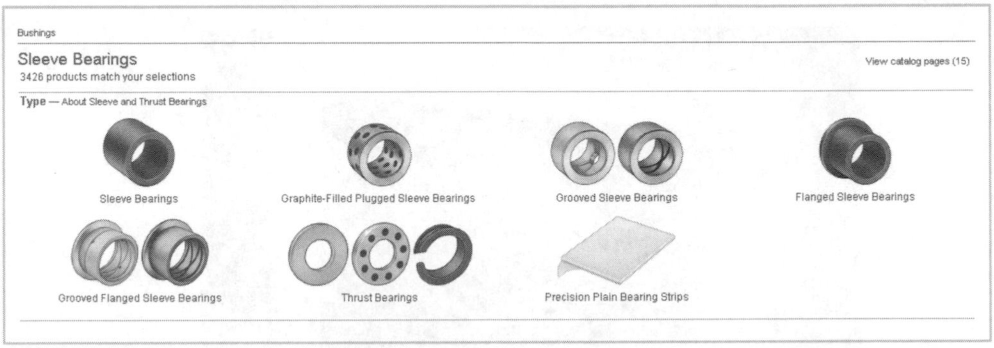

Bushings

Sleeve Bearings

3426 products match your selections

View catalog pages (15)

Type — About Sleeve and Thrust Bearings

Sleeve Bearings | Graphite-Filled Plugged Sleeve Bearings | Grooved Sleeve Bearings | Flanged Sleeve Bearings

Grooved Flanged Sleeve Bearings | Thrust Bearings | Precision Plain Bearing Strips

You can also find plastic and metal sleeves that have less friction than a drilled hole in a block of wood does, but aren't quite as frictionless as ball bearings. We'll call these *radial bushings* (see Figure 7-3).

Radial Ball and Roller Bearings

Your basic roller-skate bearing is a radial ball bearing and is by far the most popular type. It's easier to understand and find bearings for your projects if we go over a bit of bearing anatomy and vocabulary first (see Figure 7-4).

FIGURE 7-4 Bearing anatomy

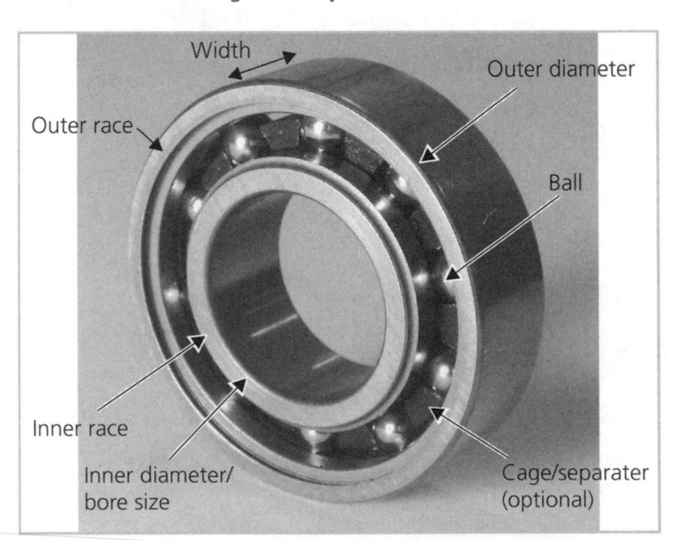

- **Outer diameter** The outer dimension of the bearing.
- **Outer race/ring** The short cylindrical part outside the rolling elements.
- **Inner race/ring** The shaft you use should fit snugly into the inner diameter of the bearing, so the shaft and inner race rotate together.
- **Inner diameter** Also called *bore size*, *bore diameter*, or just *for shaft size* in reference to the size shaft it is designed to fit over.
- **Ball/roller** The spherical or cylindrical rolling elements, usually made of hardened steel.
- **Width** The thickness of the bearing.
- **Cage/separator/spacer/retainer (optional)** This helps keep balls separate so they don't run into each other. Bearings without cages where the balls can roll around without constraint are called *full-complement* bearings.
- **Seal or shield (optional, not shown)** Some bearings are open so you can see all the rolling elements, and some have one or more seals or shields to stop gunk from getting into the bearing.

Here are a few more useful bearing vocabulary terms:

- **ABEC rating** Sometimes bearings are rated with an ABEC number. ABEC stands for Annular Bearing Engineers Committee. The ABEC rating ranges from 1 to 9 (in odd numbers) and is a measure of precision. The higher the ABEC number, the more precise the bearing, and of course, the more expensive it is. More precision generally leads to longer life from less friction and wear, faster spinning, and more reliable performance. For reference, skateboard and Inline skate wheels are normally equivalent to ABEC-3.
- **Revolutions per minute (rpm)** This is how fast you expect your bearing to be spinning. If you can estimate this, you can use the number to narrow down your options on sites like McMaster that ask for an rpm range. Their ranges are generally *really* high— maybe 15,000 rpm—so will rarely make or break your design. You should always buy bearings that are rated for many more rpms than you need.
- **Static load and dynamic load** You might see options for static load, dynamic load, and dynamic radial load capacity ranges on sites like McMaster and Stock Drive Products (www.sdp-si.com/estore) when you look for bearings. Static load

is how much the bearing can handle while not moving, like the bearings in your inline skates if you're just standing still. This radial load acts perpendicular to the axis on which the bearing rotates. Dynamic load is how much the bearing can handle while moving. For example, you wouldn't use bearings on your inline skates with a dynamic load rating of 10 lbs if you weigh 200 lbs. Dynamic load ratings are usually more than twice the static load ratings.

NOTE ***Bearings can handle more load when they're spinning because more of the rolling elements are sharing the load. When a bearing is not moving, all the load is concentrated on just a few rolling elements, so is more likely to cause wear and dimples in the bearing material.***

Ball bearings are the best choice when you have high speeds and light to moderate loads, as in skateboards and inline skates. Each ball only contacts each race (inner and outer) at one point, so there is very little rolling friction. Roller bearings can handle heavier loads, since the weight spreads out over a line along a cylinder and not just a point on a ball, but friction is slightly higher than in ball bearings because of this extra contact. Needle roller bearings have rolling elements that are longer and thinner than cylindrical bearings. They are useful when radial space is limited.[1]

To use a bearing properly, you want one race of the bearing to stay still while the other one moves. Generally, you install bearings on smooth shafts, but it's possible to install a bearing on a snug-fitting threaded rod as well. Although unconventional, this does secure the inner race to the rod so they rotate as one. Figure 7-5 shows an

FIGURE 7-5 Installing a bearing on a threaded rod (images used with permission from MakerBot Industries)

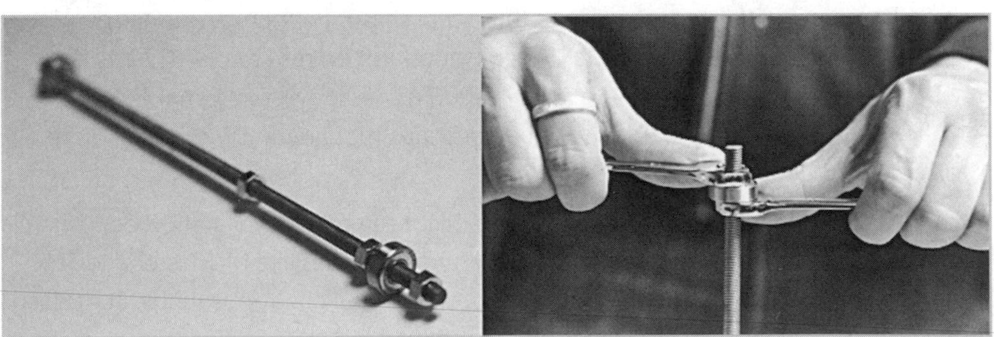

example of a bearing installation where the outer race will be held stationary while the inner race spins with the threaded rod.

You can also install a bearing so the inner race stays still and the outer race moves. This is how inline skates and skateboard wheels are mounted. The inner races are squished together, while the outer races fit snugly into a plastic wheel, so the wheel and outer race of the bearing rotate together (see Figure 7-6).

FIGURE 7-6 Bearings in inline skates mounted so inner races are clamped while outer races are free to spin

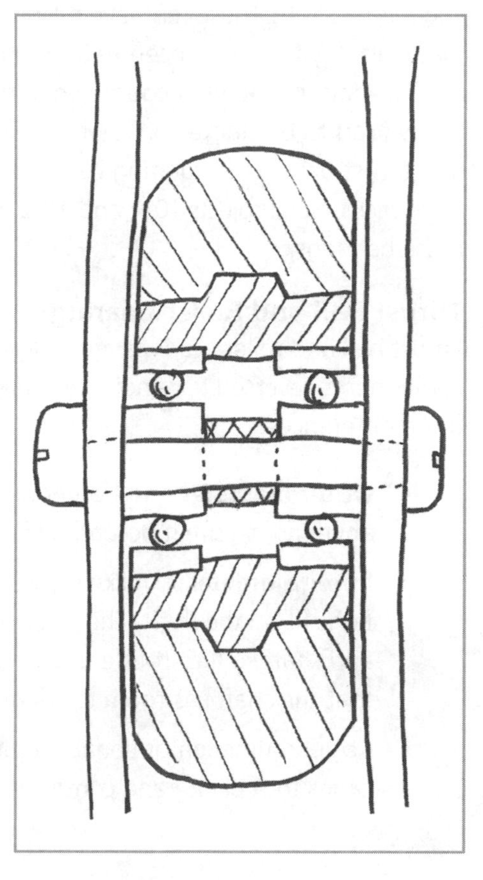

Radial Bushings

Radial bushings are a better choice for low speeds, light loads, or when precision frictionless movement just isn't necessary (or in your budget). A radial bushing looks just like a section of a small pipe or straw. These bushings usually come in a variety of plastics, bronze, and sometimes aluminum or steel with a low-friction coating on the inside like Teflon or Frelon. Oilite bushings are a special kind of bronze construction that allows many tiny open pores to be filled with oil and create a very slick surface.

NOTE ***Teflon is DuPont's brand name for a slippery plastic with the molecular name of polytetraflourethylene, abbreviated PTFE. So if you see a PTFE on McMaster, it's the same thing as Teflon.***

Three measurements that define radial bushings are outer diameter, inner diameter, and length. Before installing a bushing, make sure it fits on your chosen shaft and spins without being too tight or too loose. To install a bushing, just press or hammer it into a hole the size of its outer diameter. If you have access to an arbor press (like McMaster's 2444A61), that's even better. Using some kind of lubricant (like WD-40, 3-IN-ONE, or certain greases) is always a good idea and will decrease friction even more.

Thrust Bearings

Thrust bearings (see Figure 7-7) support axial loads, which are parallel to and ideally in line with a shaft. These can have rolling elements or just be washers made of slippery materials. If you've ever been to a restaurant with a rotating center turntable, known as a lazy Susan, you've encountered a thrust bearing. This turntable allows a lot of heavy food to be stacked on it while still allowing you to spin it easily. You can also find thrust bearings in rotating bar stools, chairs, and on a smaller scale, in rotating spice racks. See Projects 10-1 and 10-2 in Chapter 10 for examples of how to use these bearings.

Thrust Ball and Roller Bearings

Thrust ball and roller bearings are similar to radial ball and roller bearings with the components reversed to handle axial loads. The vocabulary is mostly the same, with these differences:

- **Outer diameter** The outer diameter on thrust bearings shouldn't touch anything, so size it accordingly.
- **Inner diameter** Unlike radial bearings, the inner diameter *should not* be a tight fit on the shaft. There should be clearance between the inner diameter and shaft so they rotate freely relative to each other, but not so much slop that the shaft has room to wiggle around.
- **Cage** Although optional in radial bearings, thrust bearings always have cages to separate and contain the rolling elements.

FIGURE 7-7 Thrust bearings and bushings (credit: McMaster-Carr)

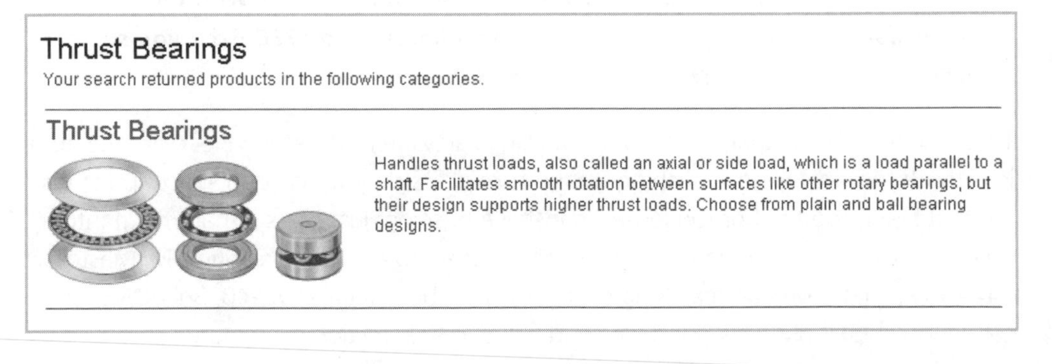

- **Shield (optional)** Some thrust bearings come assembled with thrust washers in a ready-to-use unit with a shielded cover. Shields help stop gunk from getting into the rolling elements.
- **Thrust load or axial load** This load rating describes how much weight the bearing can handle while spinning.

Thrust Washers and Bushings

Thrust bearings with no rolling elements are called *thrust bushings* or *thrust washers*. They look just like your average washer, except that they're made from slippery material and have a higher quality flat surface to support rotating things. They often come in sets with thrust ball and roller bearings to make sure the rolling elements have nice smooth, hard surfaces to interact with. You can use a thrust washer by itself as a thrust bushing to decrease friction if rolling elements aren't necessary.

Linear Bearings and Slides

Linear bearings allow motion in a straight line, often along a shaft. There are a variety of types with rolling elements in them. The most common are meant to ride on shafts, as shown in Figure 7-8. The cylindrical sleeve has a kind of cage that holds steel balls, as in other bearings, but these allow the bearing to roll along a shaft instead of spin around it. Linear bearings are designed to carry heavy loads on precision, hardened steel shafts, so the system components can get expensive pretty quickly.

Another type of linear bearing is a drawer slide or track roller. You've probably seen these on the sides of dresser, kitchen, filing cabinet, or shop drawers. They allow you to pull a drawer out while supporting the weight of the contents in a smooth, relatively frictionless motion. These can be repurposed for many different projects that need smooth, linear motion.

Linear bushings offer an economical alternative when you have light loads and a small bit of friction is okay. Linear bushings, also called linear plain bearings, look a lot like radial bushings. In their simplest form, they are just small, hollow cylinders of a slippery material like plastic or bronze. These are the type used in MakerBot's CupCake CNC (see Figure 7-1). Higher-end linear bushings have Teflon or other slippery linings on the inside surface. They perform better than linear ball bearings when dirt, water, and vibrations are involved, but have slightly higher friction. Some have grooves that allow dirt and debris to slide right through them.

FIGURE 7-8 Linear ball bearing configurations (credit: McMaster-Carr)

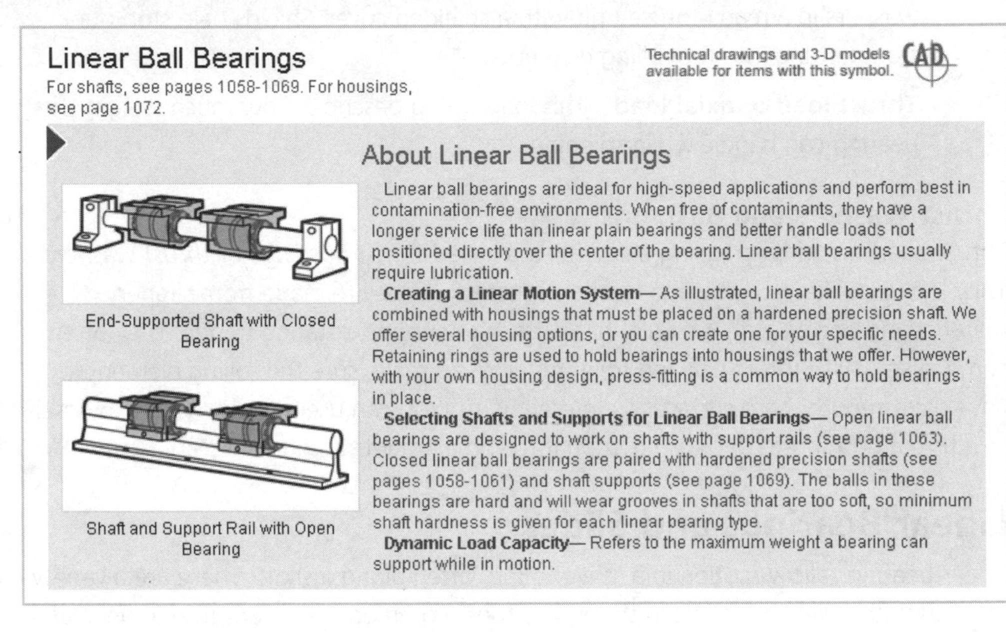

Linear Ball Bearings

For shafts, see pages 1058-1069. For housings, see page 1072.

Technical drawings and 3-D models available for items with this symbol. CAD

About Linear Ball Bearings

End-Supported Shaft with Closed Bearing

Shaft and Support Rail with Open Bearing

Linear ball bearings are ideal for high-speed applications and perform best in contamination-free environments. When free of contaminants, they have a longer service life than linear plain bearings and better handle loads not positioned directly over the center of the bearing. Linear ball bearings usually require lubrication.

Creating a Linear Motion System— As illustrated, linear ball bearings are combined with housings that must be placed on a hardened precision shaft. We offer several housing options, or you can create one for your specific needs. Retaining rings are used to hold bearings into housings that we offer. However, with your own housing design, press-fitting is a common way to hold bearings in place.

Selecting Shafts and Supports for Linear Ball Bearings— Open linear ball bearings are designed to work on shafts with support rails (see page 1063). Closed linear ball bearings are paired with hardened precision shafts (see pages 1058-1061) and shaft supports (see page 1069). The balls in these bearings are hard and will wear grooves in shafts that are too soft, so minimum shaft hardness is given for each linear bearing type.

Dynamic Load Capacity— Refers to the maximum weight a bearing can support while in motion.

Combination and Specialty Bearings

General-purpose radial ball and roller bearings are not designed to handle axial loads or torques, and thrust bearings are not designed to handle radial loads or torques (see Figure 7-9).

Ball bearings also tend to take up a lot of radial space, so they may not be feasible for use in smaller projects. It can also be hard to align everything in your system perfectly so the bearing functions as intended. Here are a few common bearing alternatives that address these problems:

- **Angular contact bearings** If you try to put an axial load on a radial bearing, it probably won't work well, and the inner or outer race will likely get damaged. However, in the real world, you rarely have pure axial or radial loads. Angular contact bearings have angled races, so they can handle radial loads as well as axial loads in one direction. Figure 7-10 shows a cross section and the direction of the applied load.

FIGURE 7-9 Right and wrong ways to load radial and thrust bearings

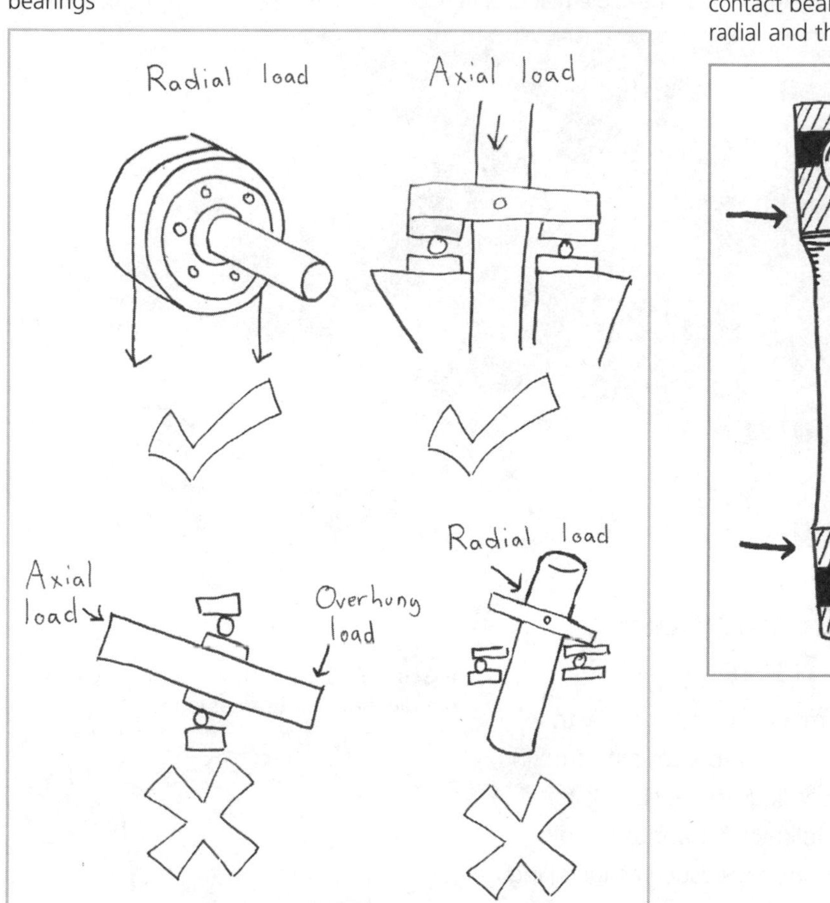

FIGURE 7-10 Angular contact bearings can handle radial and thrust loads.

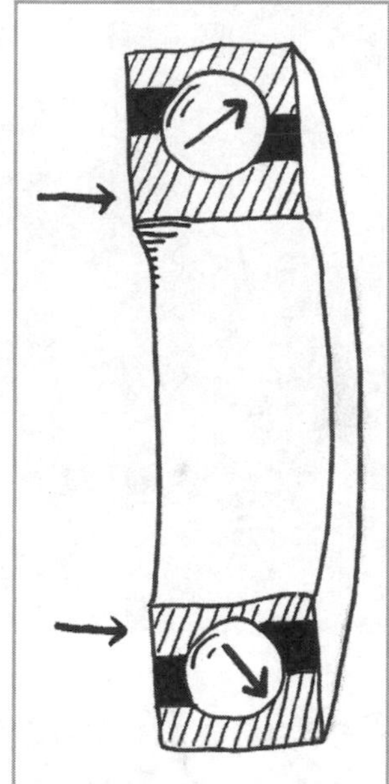

- **Spherical bearings** Spherical bearings have a spherical-shaped outer race that increases surface contact with the housing and boosts load capacity while accommodating misalignment. They handle radial and thrust loads, so are often used when these loads are present in combination with some misalignment, as on the gym equipment in Figure 7-11.
- **Combination bushings** Flanged bushings (see Figure 7-12) handle both radial and axial loads. You can install these just like radial bushings, and then use the flange as a thrust washer and/or spacer.

FIGURE 7-11 Spherical bearings are used on gym equipment to accommodate misalignment in the cable pull direction.

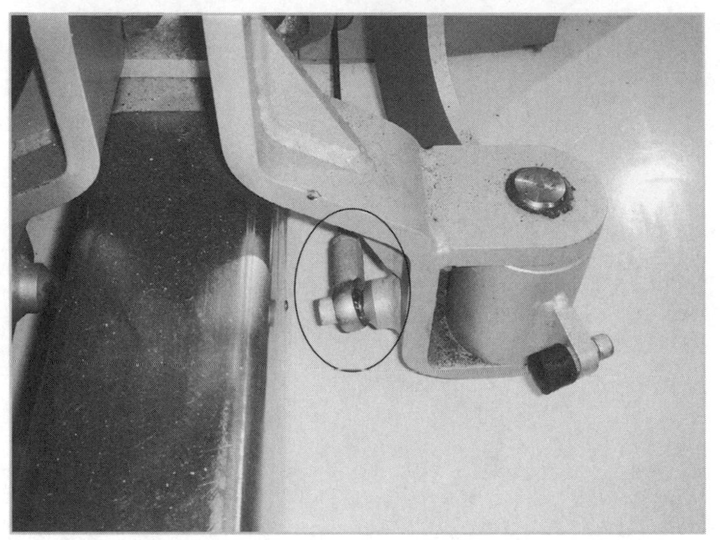

Bearing Installation Tips and Tricks

There are two main ways to work with bearings: build them into your structure or use mounted bearings (pillow blocks). We've already talked a bit about building them into your structure, such as installing bushings by pressing them into a hole just bigger than their outer diameter. For radial bearings or bushings, if you're working with wood, you should drill a small hole first, and then drill progressively larger holes until you get the right fit. You can also use counterbore bits (also known as Forstner bits), as shown in Figure 7-13, to create a recess for a bearing to sit in.

FIGURE 7-12 A flanged bushing can handle radial and thrust loads.

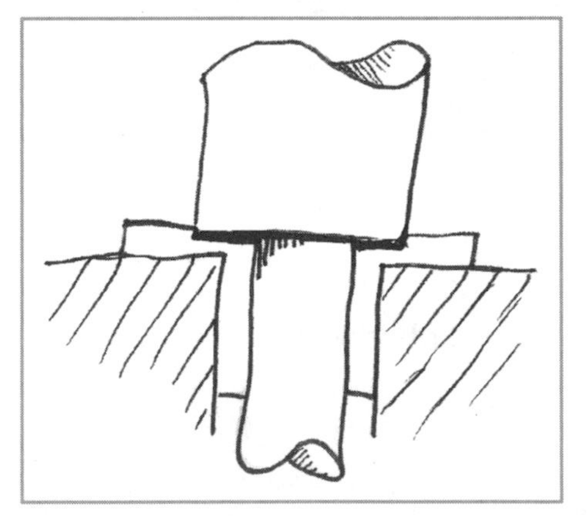

FIGURE 7-13 Counterbore bits can be used to create recesses for bearings (credit: McMaster-Carr).

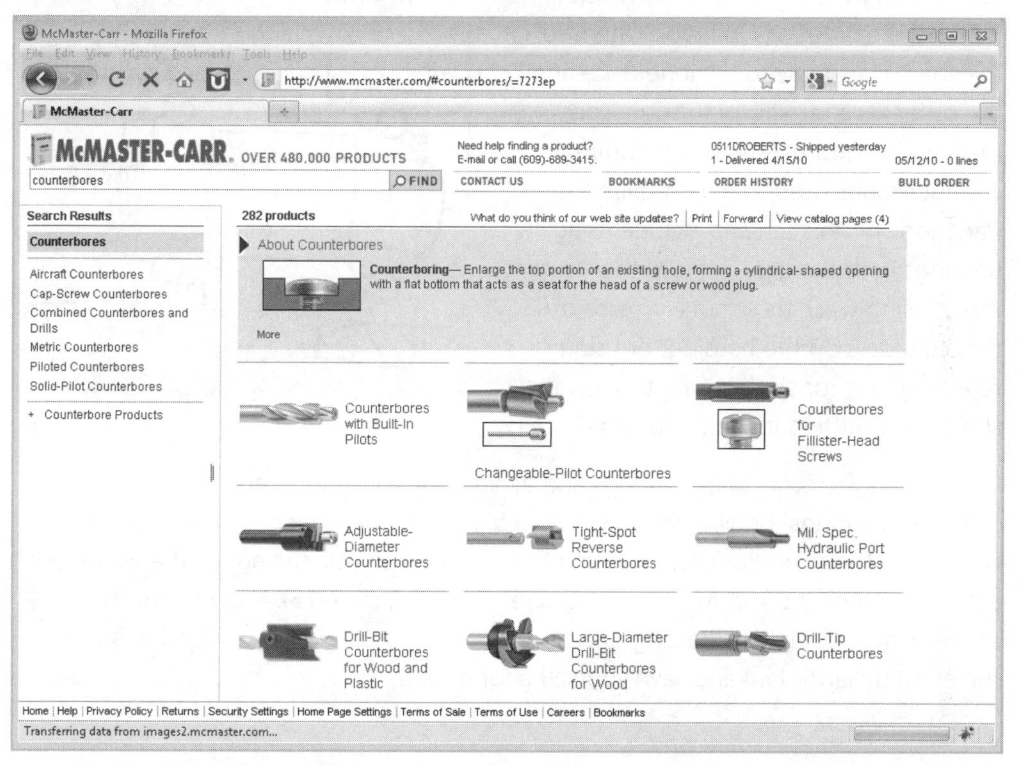

If you're working with metal like aluminum, you can use the same method of drilling progressively larger holes until you reach the correct outer diameter to hold your bearing snugly. You can also create counterbores the same way as in wood, but most of the counterbore bits that can handle metal are not designed to work with portable handheld tools.

The best way to create holes for bearings in metal is to use a drill press or a milling machine. A drill press is basically what you get when you mount a portable drill on a stable structure with a base. A milling machine is a fancier version of a drill press that allows the base to move in the x, y, and z axes so you can do more than just drill straight down (see Figure 9-4 in Chapter 9). You can use a counterbore drill bit in a drill press. The best tool to create a counterbore on a milling machine is called an *endmill*. An endmill looks like a drill bit with the tip cut off, so it can create holes with flat bottoms.

There are many ways to mount bearings and bushings. The important considerations with ball bearings are to mount them so the shaft fits snugly to the inner race and the outer race fits snugly to some sort of housing. Figure 7-14 shows common configurations used to mount a long shaft. Variations on this scheme include using bearings with flanges built into their outer races, using washers or nuts in place of the shoulders on the housing, or using retaining rings or shaft collars to keep the shaft from shifting inside the bearing.

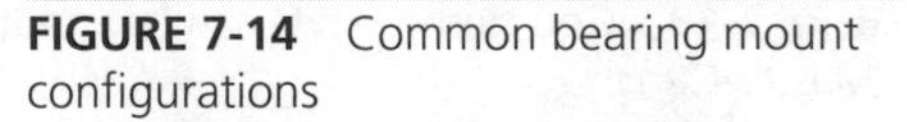

FIGURE 7-14 Common bearing mount configurations

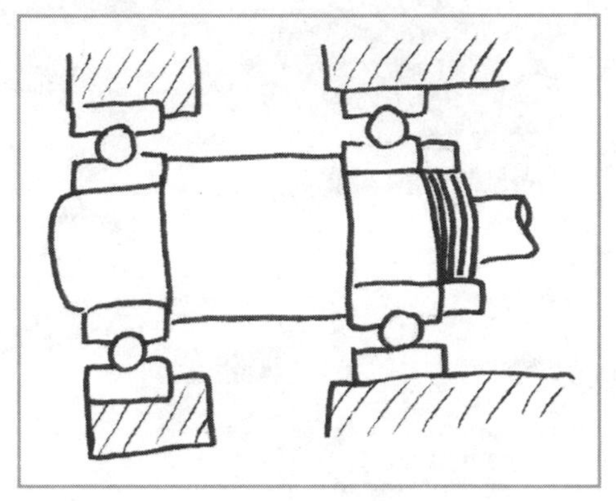

Another way to work with bearings and avoid using fancy tools and bits is to use mounted bearings, or pillow blocks. *Pillow blocks* are just bearings mounted in their own case. The case provides mounting holes or slots so you can adjust the alignment before tightening down the mounting screws (see Figure 7-15). You pay for the convenience, but with a starting price of around $3, you might be willing to spend the extra dollar or two and save yourself a lot of time.

FIGURE 7-15 Pillow blocks allow you to easily mount bearings to support rotating shafts (credit: McMaster-Carr).

Couplers

A *coupler*, or *coupling*, is anything that joins two rotating things to transfer torque from one to the other. Attaching, or coupling, something to your motor's shaft can be the first and biggest challenge you face when building your mechanism. Information about how do to this is rarely rounded up in the same reference. The methods depend on the motor type and shaft shape. The following sections summarize recommendations from different sources and years of experience, so you can easily see your options for extending a motor shaft, attaching it to an existing shaft, connecting it to a gear, and so on.

Working with Hobby Servos

Hobby servos make connecting anything to them very easy because they come with a *spline* (a little gear-shaped thing) already fixed to the motor shaft (see Figure 7-16). All the hardware designed to interface with servos has the female indent of the spline already in the part, so it just slides on, and then is usually secured with a screw into the motor shaft itself.

FIGURE 7-16 Servo motor with spline on motor shaft and extra attachments

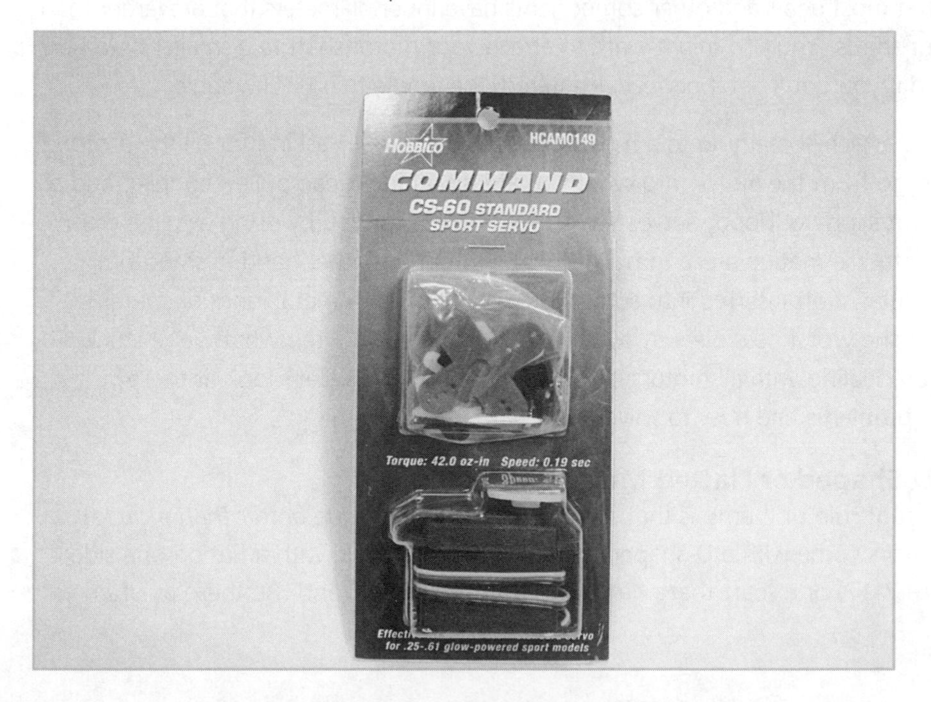

In the Servos & Accessories section, ServoCity (www.servocity.com/html/servo_shafts___couplers.html) has at least six different couplers you can use to extend the shaft, attach other parts, or attach another shaft. ServoCity also offers all kinds of arms, pulley wheels, gears, and sprockets (for chains) that attach directly to the spline on the shaft. If you can use hardware ready-made to interface with a servo, definitely do it.

Even if you don't use one of the ready-made servo accessories, you can still take advantage of the fact that the motor shaft is threaded and use a screw to attach something to it. You can also glue the flimsy plastic servo arms that come with most hobby servos to something more durable. Many hobby servo suppliers will sell you a small amount of screws that work with your motor, so you don't need to figure out what size they are and buy a box of 100 from McMaster. Screw size 4-40 (screw sizes are covered in Chapter 3) is common to standard servos, but it's worth checking to confirm before you try a random screw and ruin the threads.

Working with Other Types of Motors

Common shaft sizes for other motors you might work with (DC, DC gearhead, and stepper) range from as small as 1/16 in to around 3/8 in for larger motors. The problem here is that most gears and other components have inner diameters that are larger than the motor shafts. You also might want to attach your motor shaft to a smaller or larger shaft, and if you don't get it perfectly centered, the whole thing will wobble.

When you attach something to a motor shaft, you are really asking for all the motor torque to go from the motor into what you are attaching (gear, pulley, coupler, and so on) without slipping. Hobby servos solve this problem for you by using a spline that can bite into the mating piece to transfer torque. On the other hand, a smooth, metal, circular shaft inserted into something with a smooth metal inner diameter is just about the worst possible way to transfer torque, yet is often what we're stuck with when dealing with all motors other than hobby servos. Let's look at these common problems and how to solve them.

Using D-Shaped or Flatted Motor Shafts

An important rule of thumb is that *any shape transfers torque better than a circle*! Many motors come with a D-shaped or flatted shaft (a circle with a flat on one side; see Figure 7-17) or a shaft that's flatted on both sides. Find and use these as often as possible.

FIGURE 7-17 A gearhead motor with a flatted shaft makes it easier to attach components with set screw hubs (credit: ServoCity).

If the motor you need does not have a flat section to the shaft, you can always file in a small patch with a metal file. Others have grooves (called *keyways*) cut out of the motor shaft and mate with a component with a matching key cutout. Some motors even come with a tapped hole in the shaft so you can screw components directly into them. At the very least, you should take a file or sandpaper to a circular motor shaft to give it some texture to better enable components to hold onto it.

Attaching Components to Motor Shafts

If you're really lucky, you can find a motor that has a wheel or other component that matches the shape of your motor's output shaft. For example, Solarbotics sells a great little DC gearhead motor kit (www.solarbotics.com/products/gmpw_deal/) that includes a motor with a shaft that's flatted on both sides, a wheel with a matching profile, and a mounting screw to keep the wheel in place.

FIGURE 7-18 Components with set screw hubs, like this gear from ServoCity, are easy to attach to motors (credit: ServoCity).

If you're not that lucky, make your life easier by searching for components (gears, pulleys, sprockets, and so on) that come with a hub. A hub slides onto your motor shaft and is secured with a set screw or clamp (see Figure 7-18). Some components come with hubs but without a set screw or clamp to secure them. The term for this is *plain bore*. If the fit is too loose, you can always drill and tap your own hole for a set screw. (See Chapter 3 for details on how to drill and tap holes.)

If you're not lucky enough to find a component with a convenient hub, you can always *press fit* a component to your motor shaft. This is when the hole in your component is so close to the size of your motor shaft that you need to push it really hard to slide it on, and it will hold that position because of the stress of the fit. Figure 7-19 shows a gear that ServoCity has designed to press onto the shaft of small DC motors.

CAUTION ***A press fit is one of the weaker methods we've talked about for attaching components to a motor and is tricky to get just right. The act of pressing on the gear or other component can also damage the radial bearings in some motors because you are putting an axial load on the shaft when you press something onto it. You should use this method only after you've run out of other options.***

Another way to attach components is by using a clamp hub, also called a flanged coupling or mounting flange, like the one shown in Figure 7-20. This attachment allows you to grip onto circular motor shafts with the clamp and then use the mounting holes for gears, pulleys, wheels, or whatever you want. For larger diameter motor shafts, McMaster sells a mounting flange (9684T1) that does the same job.

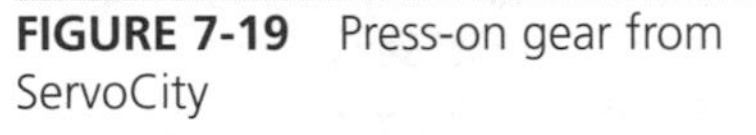

FIGURE 7-19 Press-on gear from ServoCity

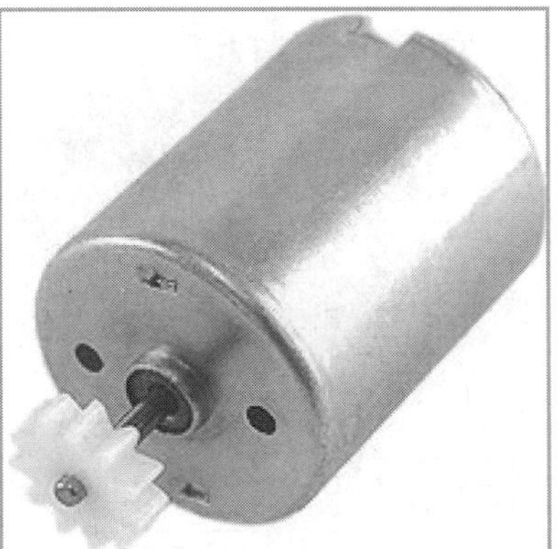

FIGURE 7-20 Clamp-style hubs offer strength and flexibility.

WM Berg (www.wmberg.com) is another good place to find hubs, shaft adapters, and other components in convenient sizes. (The product search on the WM Berg site doesn't have pictures at the time of this printing, so it may be easier to order the company's free print catalog.)

Increasing Shaft Size

Since motor shafts are often smaller than the components you need to attach to them, there are a few handy tricks you can use to fill the gap. One way is to use a shim. *Shim* is a general term for a thin thing that fills a gap, and can be made of wood, metal, or plastic. If you've ever tried wrapping duct tape around a shaft to make it fit tightly to a component, you were *shimming* the shaft. You can certainly stick with duct tape if that works, but a more professional approach is to get a roll of shim stock—basically thick tin foil—and cut a piece that wraps around your motor shaft.

You can find shim stock sheets and rolls at most hardware stores (and McMaster, of course). Soda and beer cans are also readily available sources of metal shims if you have some tin snips to cut off the ends. Shim stock comes as thin as 0.001 in, so if your component fits on your motor just a little too loosely, you can jam some layers of shim stock in that gap until you get a snug fit.

You can also use aluminum or brass tubes as a kind of shim to create a new uniform surface on the motor shaft. These are available from McMaster and most hardware and craft stores, and they come in many diameters. The walls can be almost as thin as plastic drinking straws, so you can layer one size on top of another if necessary.

Attaching the Motor Shaft to Another Shaft

Sometimes you need to extend a motor shaft to reach a wheel or rotate a long shaft. For example, if you are trying to automate your window shades, you might want to connect your motor to a rod that runs the width of your window and rolls up the window shade when you turn on the motor. There are three main options here, depending on the relative sizes of the shafts:

- **Insert a smaller shaft into a bigger shaft** Make a hole in the bigger shaft, stick the smaller one into it, and secure it with a set screw if necessary. See Project 9-2 in Chapter 9 for an example of how to drill a hole in the center of something without a lathe.

- **Use a rigid shaft coupling** Some types of couplers can join shafts of different sizes (see Figure 7-21). The inner diameter of the coupling is a tight fit to the shaft, and the set screws bite into the shaft a little to help transfer torque. Rigid shaft couplers come in a variety of styles, including clamped hubs (see Figure 7-22). Clamped hubs give you a tighter grip on both shafts, so they transfer torque better, but are not well suited to high-speed applications since the weight of the clamp hub is off center and can make the system wobbly.

NOTE ***As you can probably tell from the pictures in Figures 7-21 and 7-22, these set screw shaft couplers are relatively easy to make yourself in a pinch. Just take a short length of aluminum or plastic rod, drill a hole through the center the size of your motor shaft (it doesn't need to be perfectly centered), drill and tap two holes for whatever size screw you have lying around (see Chapter 3), and you're done. It's best to use a small vise, like McMasters 5312A2, to hold the material while you drill. If one shaft is bigger than the other, use a bigger drill bit to drill back through half of the coupling. The bigger drill bit will naturally center itself on the existing hole.***

- **Use a flexible shaft coupling** Flexible couplings compensate for a certain amount of misalignment of the shafts (parallel, angular, or axial) by giving a little if they aren't perfectly aligned. These are highly recommended because the coupling takes the stresses induced by poor alignment instead of making the motor work harder to turn something that's not on center.

If you go with flexible shaft coupling, rubber tubing is by far the simplest (but weakest) option. If you're lucky enough to find rubber tubing that has an inner

FIGURE 7-21 Rigid shaft couplers, set screw style

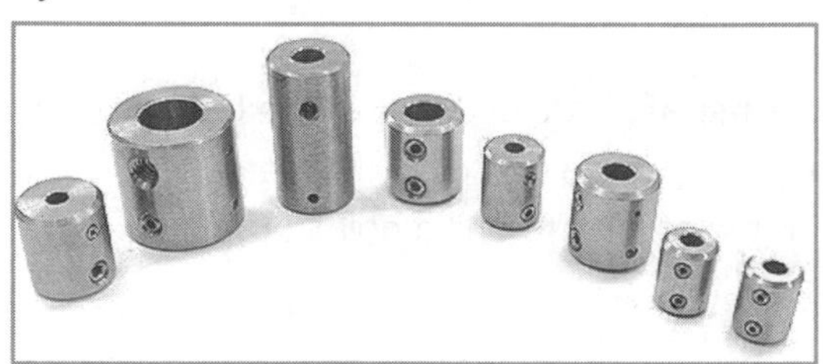

FIGURE 7-22 Rigid shaft couplers, clamp style

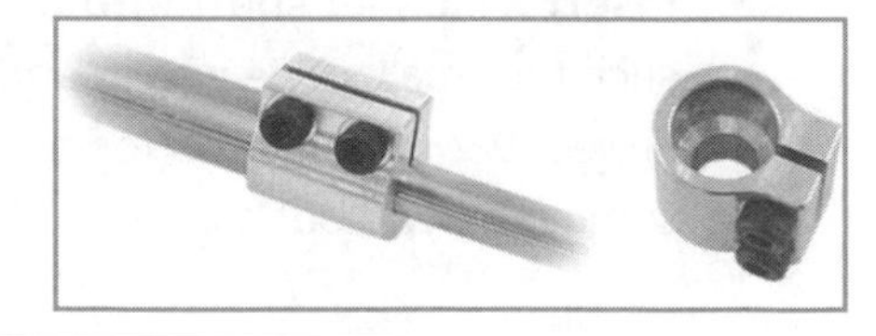

diameter that fits the motor and shaft you want to join, all you need to do is cut a short piece of it, and then push your motor shaft into one side and the shaft you want to connect into the other. If you want a tighter fit, you can put small hose clamps (like McMaster 5388K14) on each end of the tube to secure your coupler.[2] Search for "tubing" on McMaster for a dizzying array of options in every material and dimension you can think of.

For flexible coupling of two shafts in different planes, you need to use a universal joint (U-joint) (see Figure 7-23). These come in many different sizes and shapes, and sweep through a variety of angles. They can also be used to join shafts of different sizes. Many other flexible coupling options are available. Just search for flexible shaft couplings on McMaster (or any other components supplier website), and you'll find a wide array of options with funny names like spider, Oldham, and bellows couplings that accommodate different kinds of misalignment.

FIGURE 7-23 U-joints (credit: ServoCity)

Attaching Gears and Other Components to Shafts

The options for attaching components to shafts are basically the same as for attaching components to a motor shaft, with a few additions:

- **Press it** It's easier to press fit components at the ends of shafts. If you need to locate a component in the middle of a shaft, this is probably not the way to go.
- **Glue it** If your component will slide onto the shaft, you can try using a strong super glue or epoxy to hold it on. If both components are wood, wood glue is a good choice.
- **Pin it** If your component is wide enough to drill a small hole through its side or hub, you can match drill the component and shaft, and use a nail or dowel pin (wooden or metal) to hold them together (see Figure 7-24, left and middle images). *Match drilling* refers to lining up two things and drilling through them both at once to make sure they are perfectly aligned.

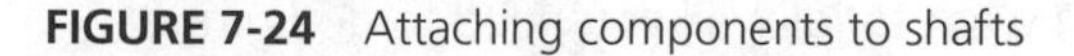

FIGURE 7-24 Attaching components to shafts

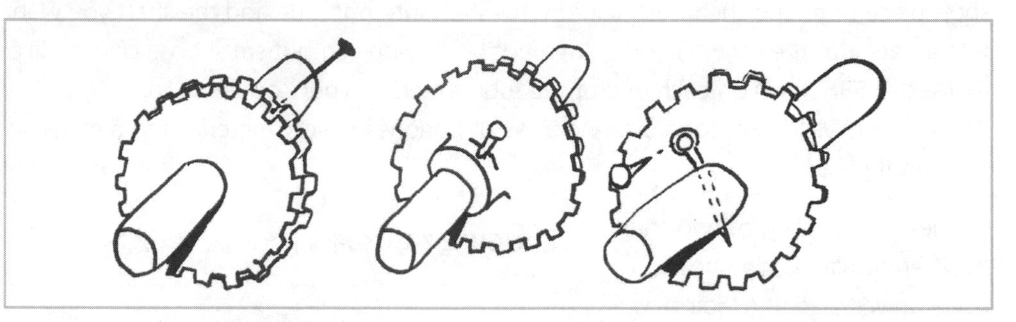

- **Screw it** If your component has a set screw hub, this is easy. If it has a plain bore with a hub, you can drill and tap a hole for a set screw. If it has no hub at all, you can make your own, or use an off-the-shelf mounting hub like the one shown earlier in Figure 7-20.
- **Screw and pin it** If you can drill a hole radially through your shaft, and you can drill a hole anywhere on the face of your component, you can probably connect them with a stiff wire or pin. This is displayed in the right image of Figure 7-24.
- **Pinch-clamp it** Use a shaft collar on either side of a flat gear or other component to hold it in place. If you squish the shaft collar together while you tighten their clamps or set screws, you can pinch the component as well. This method is used in Project 10-2 in Chapter 10 to secure the wind turbine parts that hold the blades.
- **Hold-and-stick it** Use epoxy putty to glue a shaft collar onto a flat component (and/or the shaft) to create your own hub. This method is used twice in Project 10-2 to secure the laser-cut gears to the wind turbine shaft and motor shaft.

Using Clutches

A clutch is a special type of coupling designed to connect or disconnect the driven part (shaft) from the driving part (motor), usually as a safety mechanism or to allow motion in only one direction. Some clutches, like part MSCB-4 from SmallParts (www.smallparts.com/), let you set the limit between where they slip or grip (see Figure 7-25).

FIGURE 7-25 Common spiral claw/rachet type clutch (credit: SmallParts.com)

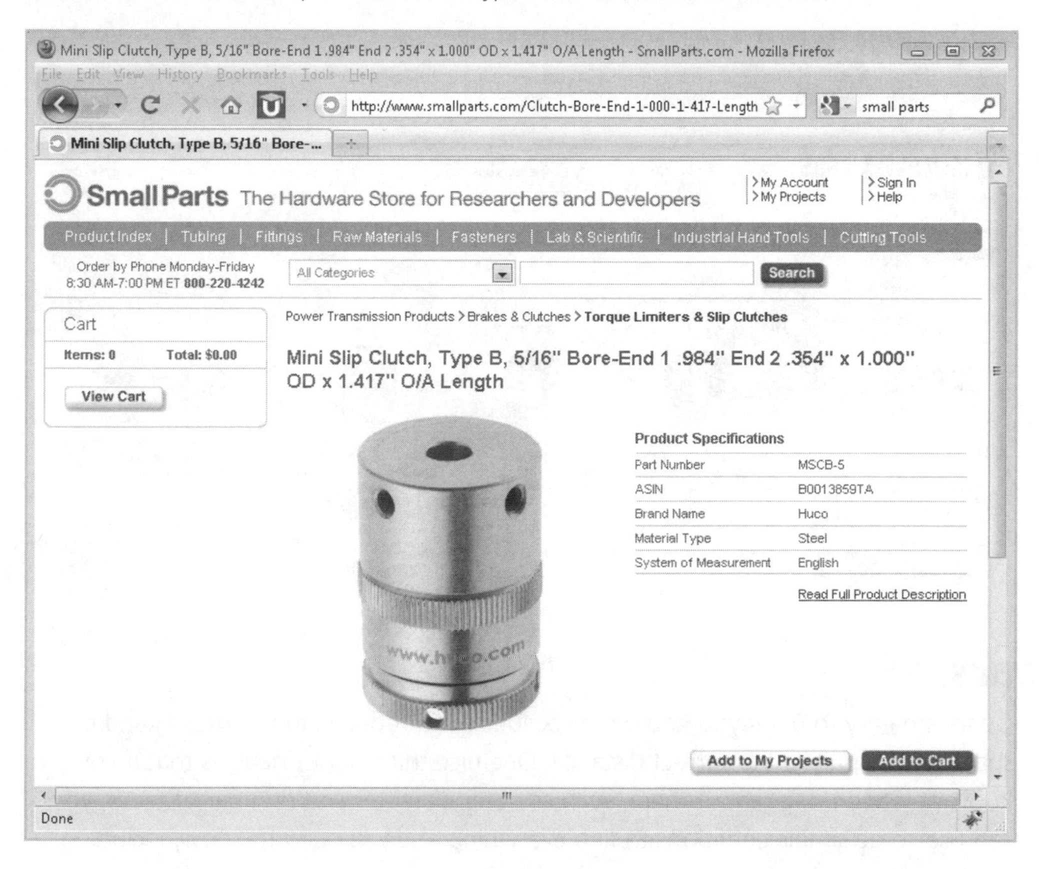

An example of a ratchet type clutch is seen in bicycles.[3] It engages the rear sprocket with the rear wheel when the pedals are moving forward, and lets the rear wheel move freely when the pedals are stopped or moving backward.

Shaft Collars

Shaft collars, also called lock collars, are like one half of a rigid shaft coupling. You can use them as mechanical stops or to limit movement on a shaft. They can also be used as spacers between gears or other components.

As shown in Figure 7-26, shaft collars come in set screw and clamp types, and can be made of metal or plastic.

FIGURE 7-26 Shaft collars (credit: McMaster-Carr)

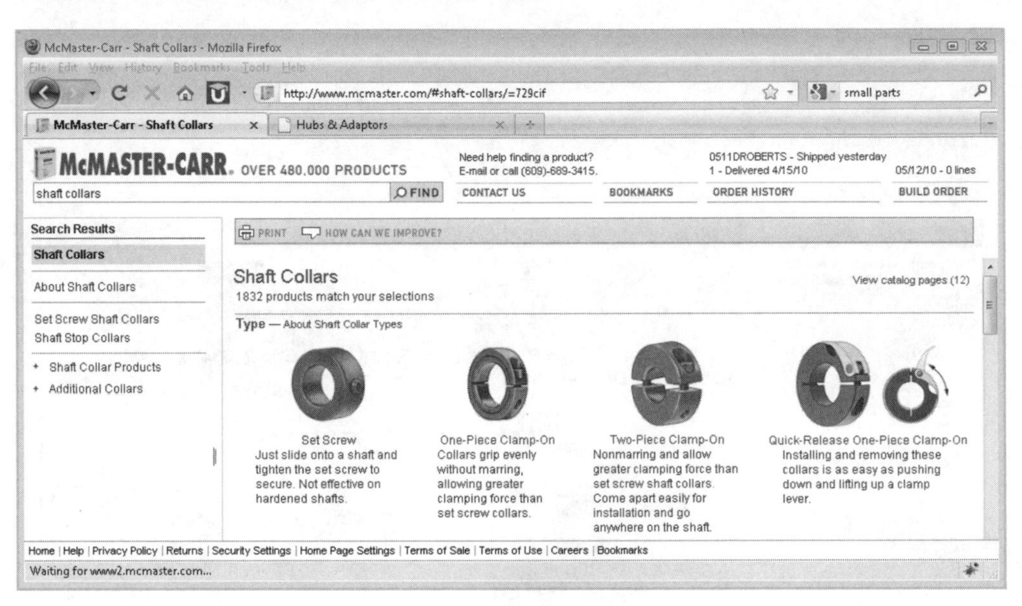

Gears

Gears are easy to use if you know the vocabulary (introduced in Chapter 1) and can space them apart at the correct distance. One nice thing about gears is that if you know any two things about them, such as outer diameter and number of teeth, you can use some simple equations to find everything else you need to know, including the correct center distance between them.

Before we talk about the types of gears, let's review the anatomy of a spur gear drive train in Figure 7-27 and the related vocabulary.

- **Number of teeth (N)** The total number of teeth around the outside of the gear.
- **Pitch diameter (D)** The circle on which two gears effectively mesh, about halfway through the tooth. The pitch diameters of two gears will be tangent when the centers are spaced correctly.
- **Diametral pitch (P)** The number of teeth per inch of the circumference of the pitch diameter. Think of it as the density of teeth—the higher the number, the smaller and more closely spaced the teeth. Common diametral pitches for hobby size projects are 24, 32, and 48.

FIGURE 7-27 Spur gear anatomy

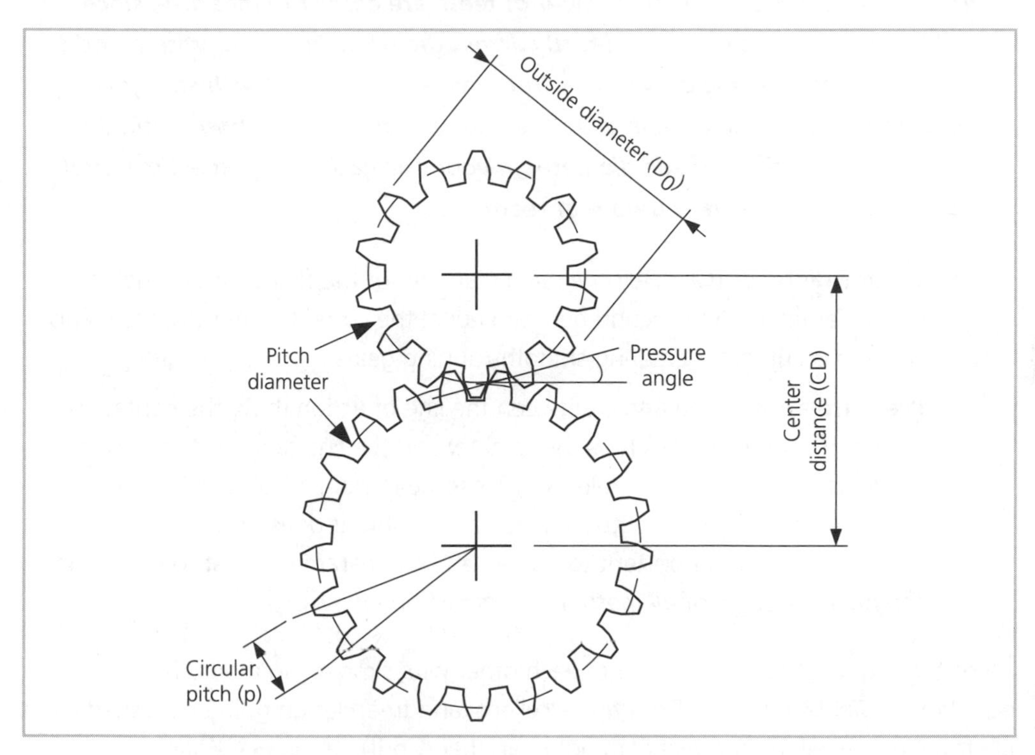

NOTE ***Remember that the diametral pitch and circular pitch of all meshing gears must be the same.***

- **Circular pitch (p) = π / P** The length of the arc between the center of one tooth and the center of a tooth next to it. This is just pi (π = 3.1416) divided by the diametral pitch (P). Although rarely used to identify off-the-shelf gears, you may need this parameter when modeling gears in 2D and 3D software (see Project 7-1).
- **Outside diameter (D_o)** The biggest circle that touches the edges of the gear teeth. You can measure this using a caliper like SparkFun's TOL-00067.

***NOTE* Gears with an even number of teeth are easiest to measure, since each tooth has another tooth directly across the gear. On a gear with an odd number of teeth, if you draw a line from the center of one tooth straight through the center across the gear, the line will fall between two teeth. So, just be careful using outside diameter in your calculations if you estimated it from a gear with an odd number of teeth.**

- **Center distance (C)** Half the pitch diameter of the first gear plus half the pitch diameter of the second gear will equal the correct center distance. This spacing is critical for creating smooth-running gears.
- **Pressure angle** The angle between the line of action (how the contact point between gear teeth travels as they rotate) and the line tangent to the pitch circle. Standard pressure angles are, for some reason, 14.5° and 20°. A pressure angle of 20° is better for small gears, but it doesn't make much difference. It's not important to understand this parameter, just to know that *the pressure angle of all meshing gears must be the same*.

All of these gear parameters relate to each other with simple equations. The equations in Table 7-1 come from the excellent (and free) design guide published by Boston Gear (www.bostongear.com/pdf/gear_theory.pdf).

TABLE 7-1 Gear Equations

TO GET	IF YOU HAVE	USE THIS EQUATION
Diametral pitch (P)	Circular pitch (p)	$P = \pi/p$
	Number of teeth (N) and pitch diameter (D)	$P = N/D$
	Number of teeth (N) and outside diameter (D_0)	$P = (N+2)/D_0$ (approx.)
Circular pitch (p)	Diametral pitch (P)	$p = \pi/P$
Pitch diameter (D)	Number of teeth (N) and diametral pitch (P)	$D=N/P$
	Outside diameter (D_0) and diametral pitch (P)	$D = D_0 - 2/P$
Number of teeth (N)	Diametral pitch (P) and pitch diameter (D)	$N = P \times D$
Center distance (CD)	Pitch diameter (D)	$CD = (D_1 + D_2)/2$
	Number of teeth (N) and diametral pitch (P)	$CD = (N_1 + N_2)/2P$

Project 7-1: Make Your Own Gears

In this project, we'll design and fabricate spur gears using free software and an online store, Ponoko, that does custom laser cutting at affordable prices.[4] If you have access to a laser cutter at a local school or hacker space, even better! You can also print out the template and fix it to cardboard or wood to cut the gears by hand.

We'll use Inkscape, a free, open source vector-based drawing program similar to Adobe Illustrator. It plays well with most modern Windows, Mac, and Linux operating systems (check the Inkscape FAQ at http://wiki.inkscape.org/wiki/index.php/FAQ for details). In Inkscape, you can draw gears with a built-in tool. One glitch is that the circular pitch is given in pixels, not inches, as in the equations in Table 7-1. You can get different gear ratios by just choosing a circular pitch that looks good and varying the teeth number, but if you want to make gears that interface with off-the-shelf gears, you need to pay a little more attention.

In Inkscape, there are 90 pixels (px) in 1 in by default. So if you set circular pitch to 24px in the Gear tool, that rounds to 0.267 in (24/90 = 0.2666…). Since diametral pitch (P) = π / circular pitch (p), the diametral pitch (P) in inches is = π / 0.267 = 11.781. You will not find any off-the-shelf gears with a diametral pitch of 11.781. As mentioned earlier, common diametral pitches are 24, 32, and 48. So if you plan to make gears to play nice with off-the-shelf gears, start with the diametral pitch of your off-the-shelf gear and use the equations in Table 7-1 to work backward to what your circular pitch should be in pixels in Inkscape.

Shopping List:

- 1/4 in wooden dowel
- Hobby knife

Recipe:

1. Download and install Inkscape from www.inkscape.org.
2. Download the Inkscape starter kit from www.ponoko.com/make-and-sell/downloads. This will give you a making guide (a PDF file) and three templates that relate to the sizes of materials Ponoko stocks. Unzip the file and save it to somewhere you'll remember.

3. Open a new file in Inkscape. Choose File | Document Properties from the menu bar to open the Document Properties window. Change the default units in the upper-right corner to inches. Back in the main window, change the rulers from pixels to inches in the toolbar. Your screen should look like Figure 7-28. Close the Document Properties window.

4. Choose Extensions | Render | Gear from the menu bar. You'll see a small Gear window that gives you three options: Number of teeth; Circular pitch, px; and Pressure angle. Leave the Pressure angle setting at 20.0, since 20° is standard for off-the-shelf gears and a good place to start. Set the other options as desired for your gear. In Figure 7-29, you can see that I chose 28 teeth with a circular pitch of 24. Click Apply, and then click Close.

5. Since gears are no fun by themselves, repeat steps 3 and 4 to make at least one more gear. I created a second gear with 14 teeth.

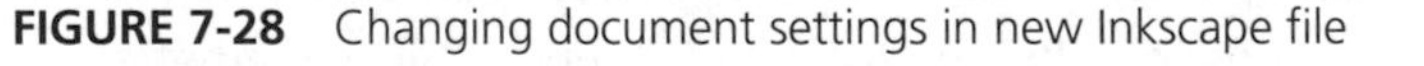

FIGURE 7-28 Changing document settings in new Inkscape file

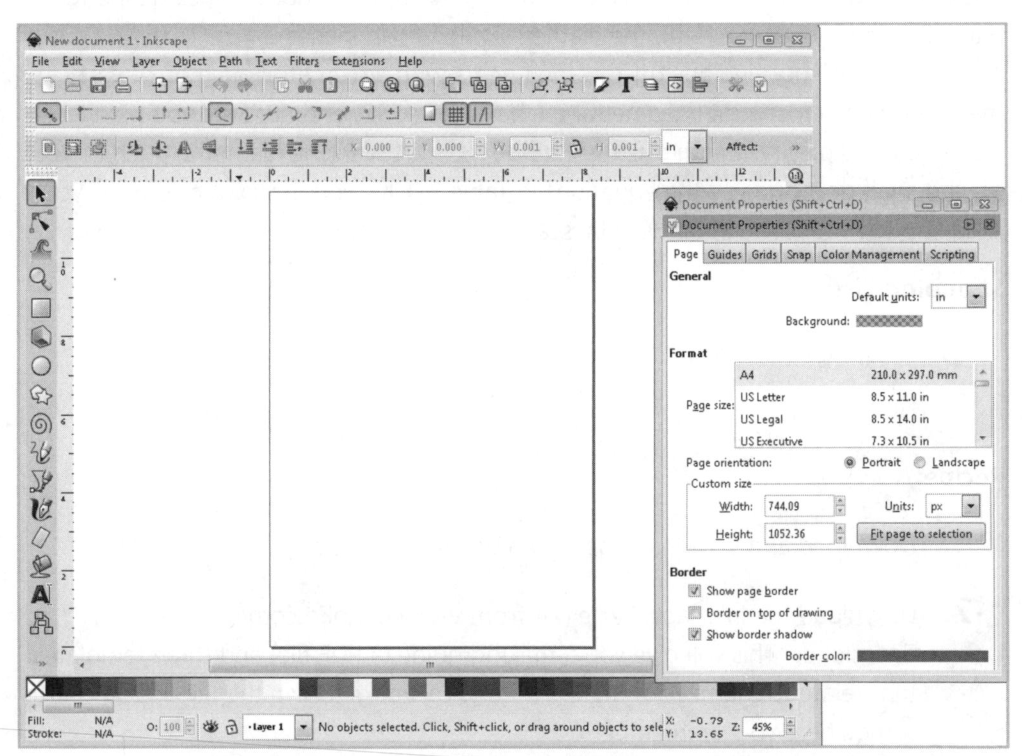

FIGURE 7-29 Using the Gear tool in Inkscape

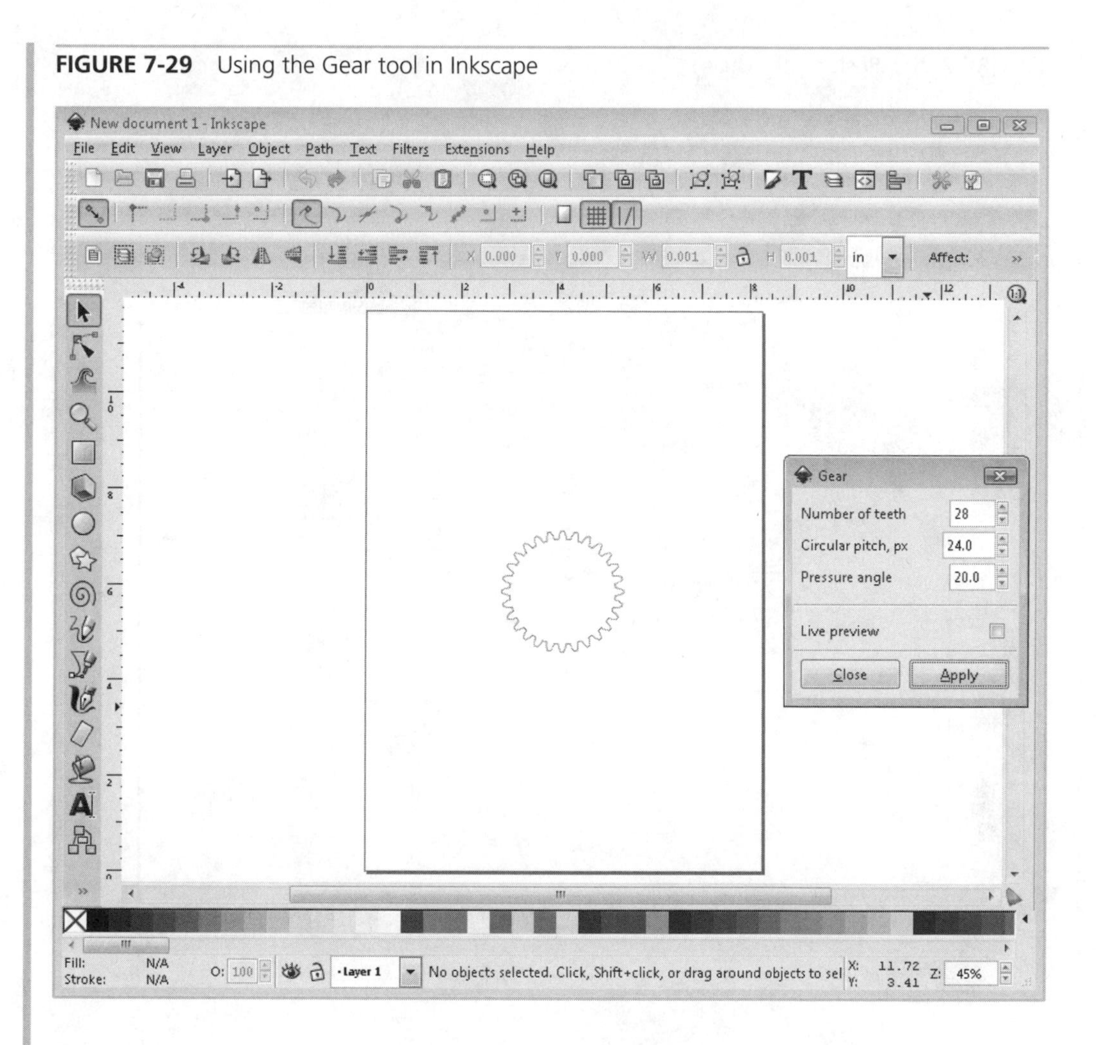

NOTE ***Remember that the pressure angle and circular pitch must be the same for the gears to mesh; change only the number of teeth!***

6. Use the Circle tool and hold down the CTRL key (on a PC) to draw a circle inside the big gear. The default circle is filled with black. Zoom in if you need to. Make sure the arrow selector is active and click the circle. Make sure inches is selected in the toolbar and the Lock button on the toolbar looks locked. Type **0.250** in the W box in the toolbar, press ENTER, and watch the H box change automatically. Your circle will resize to a diameter of 0.250 in, and your screen should look like Figure 7-30.

FIGURE 7-30 Resizing the circle

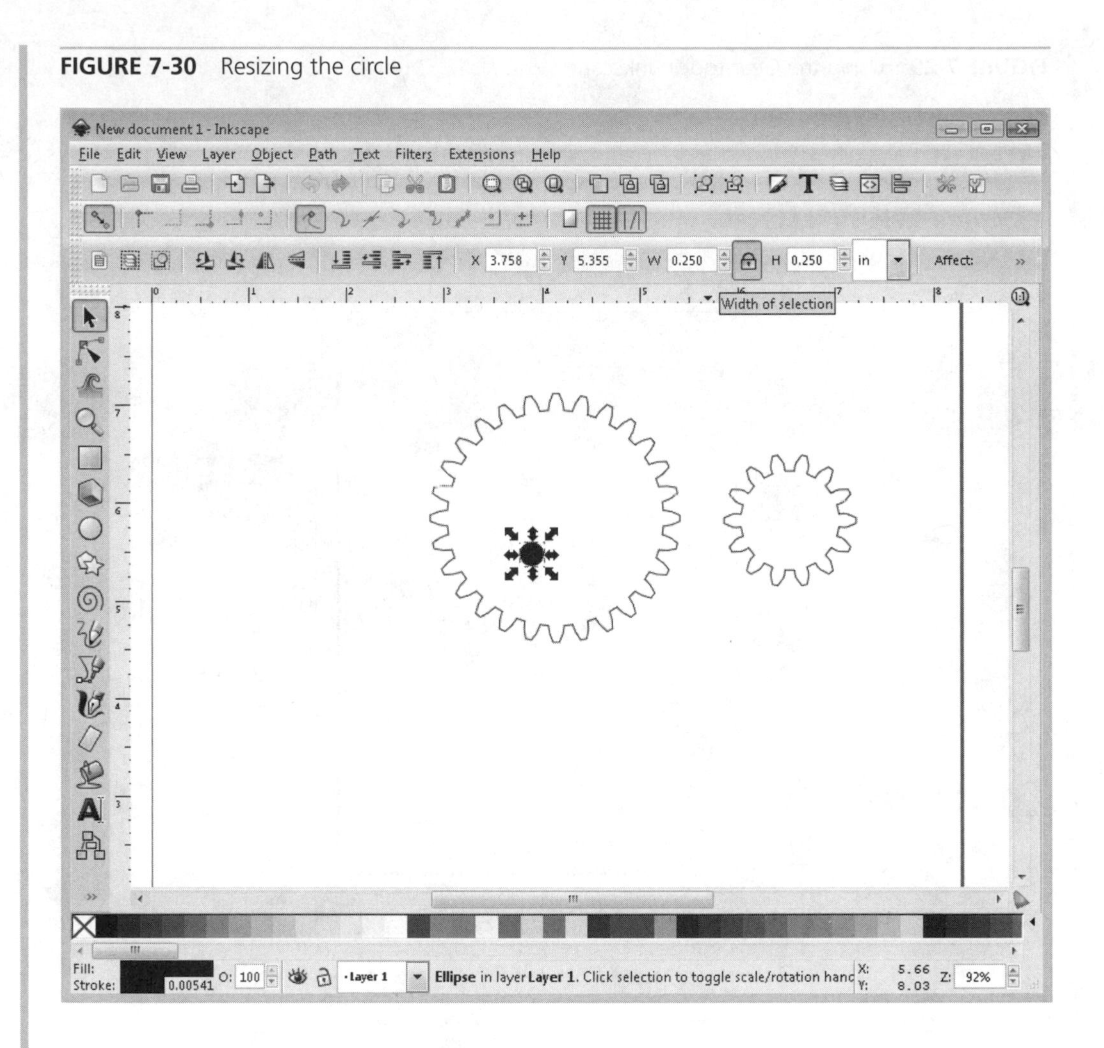

7. Click and drag a box around the big gear, small gear, and circle shape to select them all. From the menu bar, choose Object | Fill and Stroke. You will see the Fill and Stroke window, as shown in Figure 7-31.

a. In the Fill tab, click the X button for no paint.
b. In the Stroke paint tab, click the button next to the X for flat color. Leave the default color (black) for now.
c. In the Stroke style tab, change the width to 0.030mm and hit ENTER. This is what Ponoko wants the line thickness to be for laser cuts. Adjust as necessary if you're using a different laser cutter. Close the window.

FIGURE 7-31 Fill and Stroke window

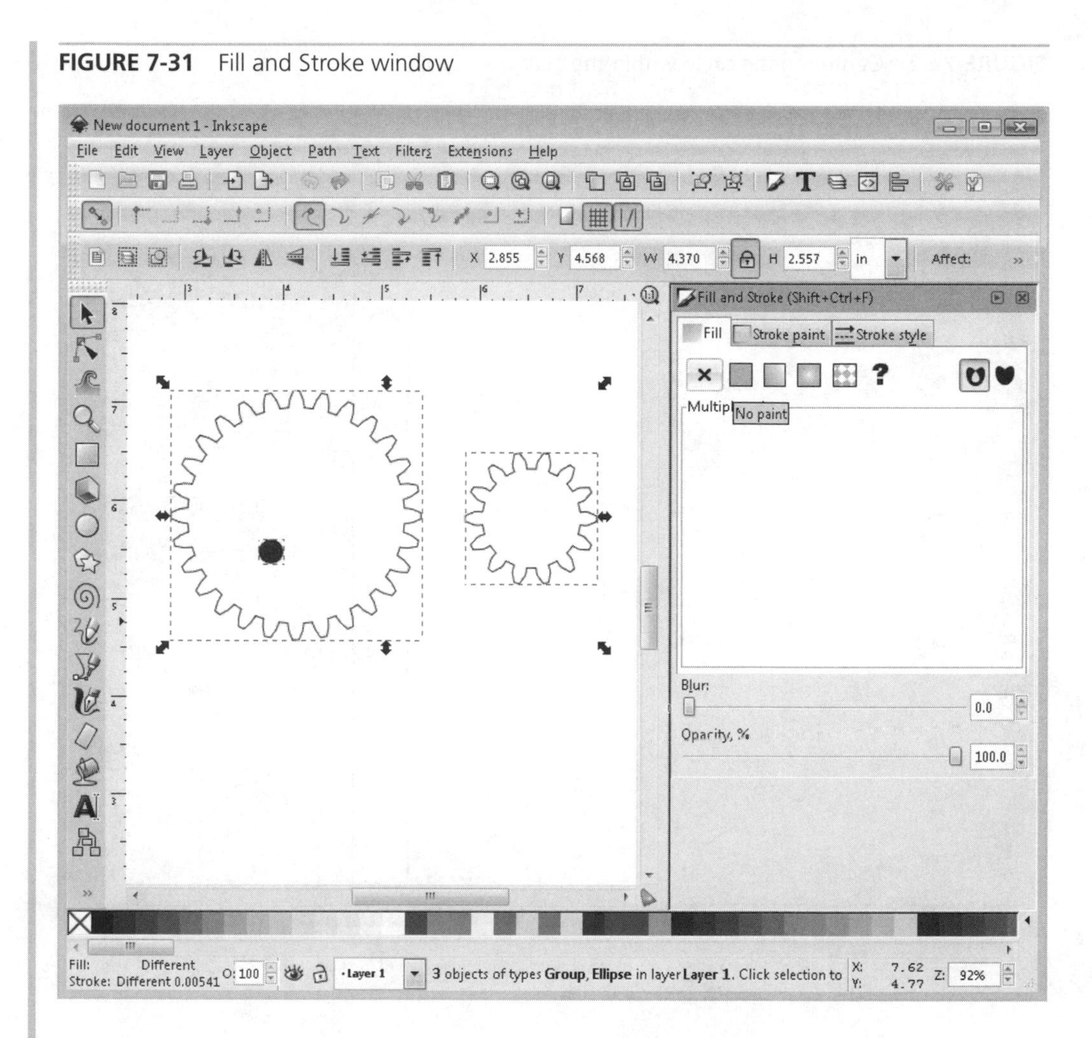

8. You need to get this circle in the exact center of the gear. Make sure the arrow selector is active. Click and drag a box around the big gear and the circle to select them. From the menu bar, select Object | Align and Distribute. Click the Center objects horizontally button (highlighted in Figure 7-32). Then click the button directly below it, which is Center objects vertically. Now you have a gear with a hole perfectly centered! Copy and paste this circle, and repeat this step to center a circle in the other gear.

FIGURE 7-32 Centering the circle within the gear

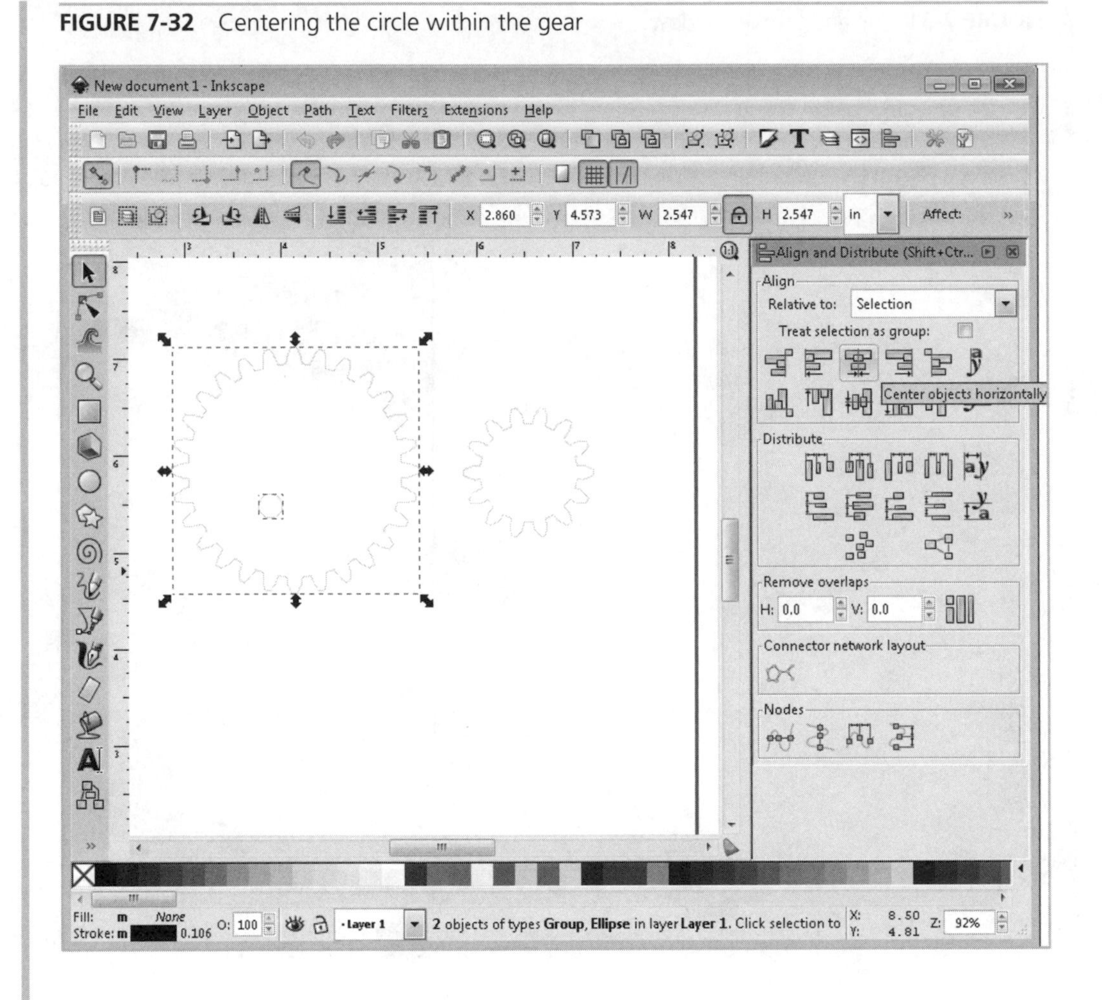

9. Now that you have your gears, you'll create a base with holes spaced the correct distance apart so you can mount the gears with 1/4 in wooden dowels and make them spin.

a. Calculate the center distance (CD) of your gears using the equations from Table 7-1. Both gears have a circular pitch of 24 pixels and a pressure angle of 20°. The big gear has 28 teeth, and the small one has 14. As explained in the project's introduction, you convert the circular pitch in pixels to a diametral pitch in inches of 11.781. If you look at Table 7-1, all you need is that number and the numbers of teeth on the two meshing gears to find the center distance (CD). Use the equation $CD = (N_1 + N_2)/2P$, and you'll find that $CD = 1.783$.

FIGURE 7-33 Placing circles for the gear base

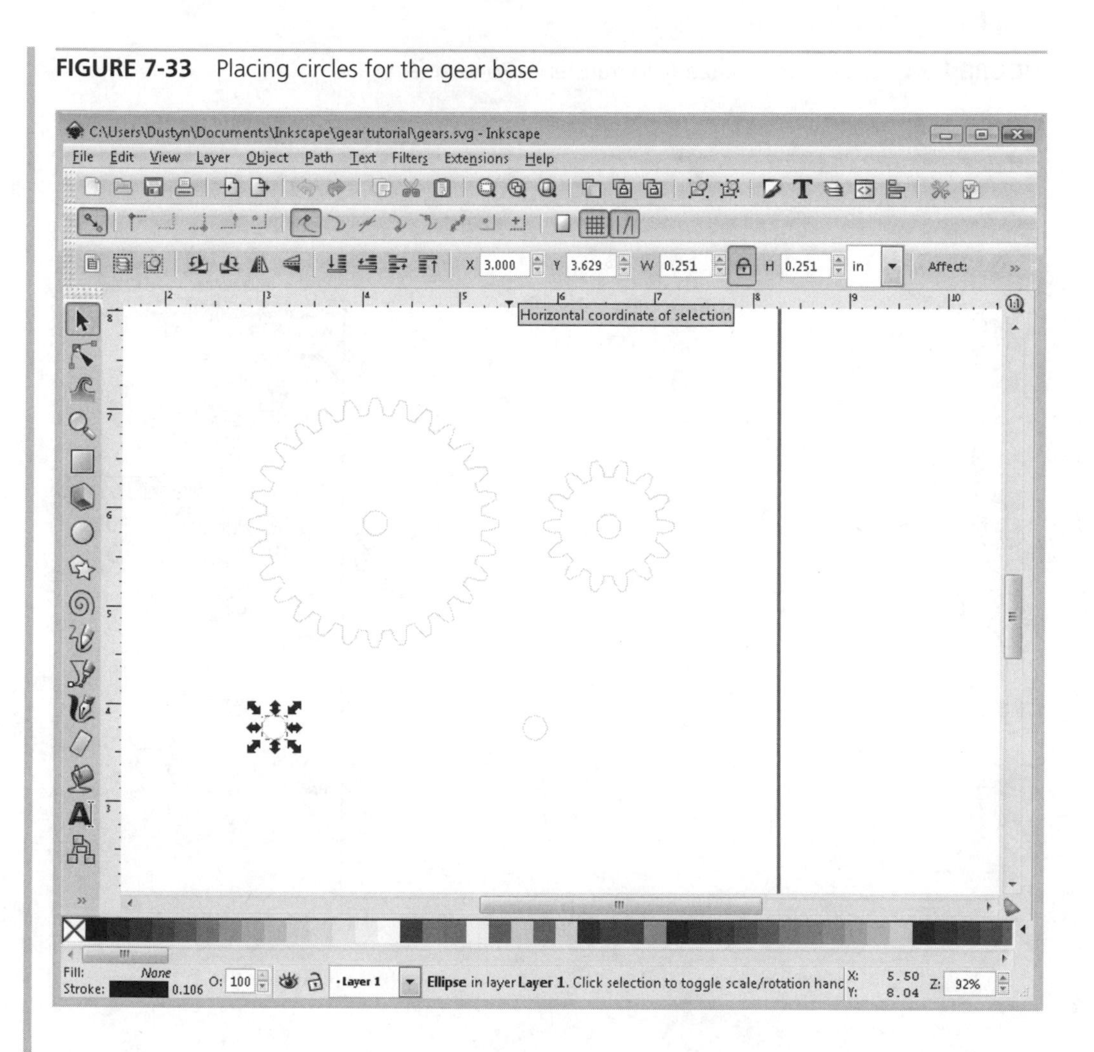

b. Copy one of the circles inside the gears, and paste two of them about 2 in apart on the lower part of the template. Select the one farthest to the left, change the x coordinate in the toolbar to 3 in, and then press ENTER. Your screen should look like Figure 7-33.

c. Use the same procedure to place the second circle to the right of the first with an x coordinate of 4.783. This is the center distance you calculated (1.783) added to the x coordinate of the first circle (3.000).

d. Draw a rectangle around the two circles to complete the base. Align the rectangle with the two circles, as shown in Figure 7-34.

FIGURE 7-34 Gears and base ready to transfer to Ponoko template

10. Now you need to prepare the file to be uploaded and ordered on the Ponoko site.

a. Ponoko uses colors to indicate how to treat the files. A blue 0.030mm line means cut it all the way through. Select everything you've drawn so far, go to the color swatches at the bottom of the screen, and hold down the SHIFT key while you click blue.

b. Open the P1.svg template you downloaded earlier. Select everything you have drawn so far, and copy and paste it into this template, as shown in Figure 7-35. Don't worry about the orange border and words; Ponoko knows to cut only the blue outlines. Save the file.

FIGURE 7-35 Transferring gears and base to the Ponoko template

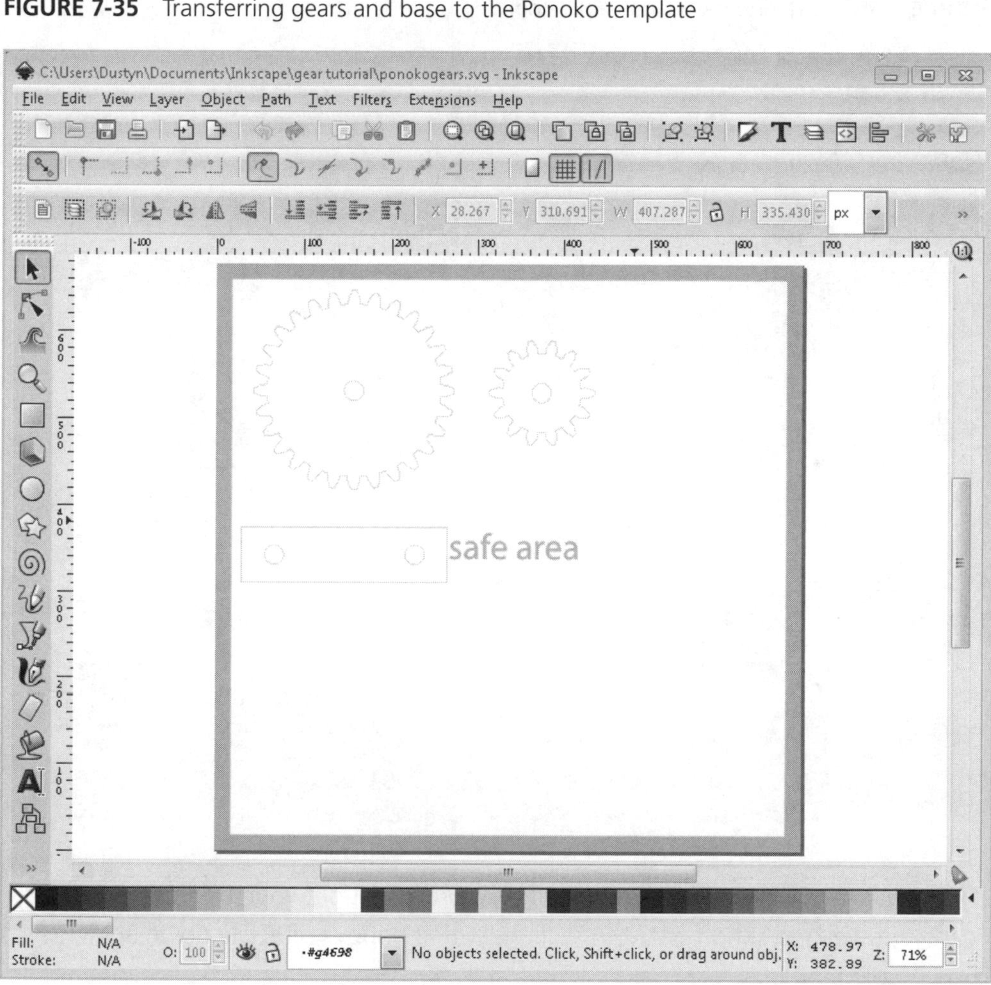

c. Go to www.ponoko.com/ and set up a free account. Then upload your file, pick a material, and arrange to have it shipped. I chose blonde bamboo, as shown in Figure 7-36, and the total cost was just $4.13 (plus shipping).

NOTE ***Once you open your free account, go to My Accounts | Preferences to set your shipping hub to Ponoko – United States (or the location closest to you). Mine was set to New Zealand by default, so my shipping charges were curiously high until I figured this out.***

FIGURE 7-36 Ordering gears on Ponoko.com

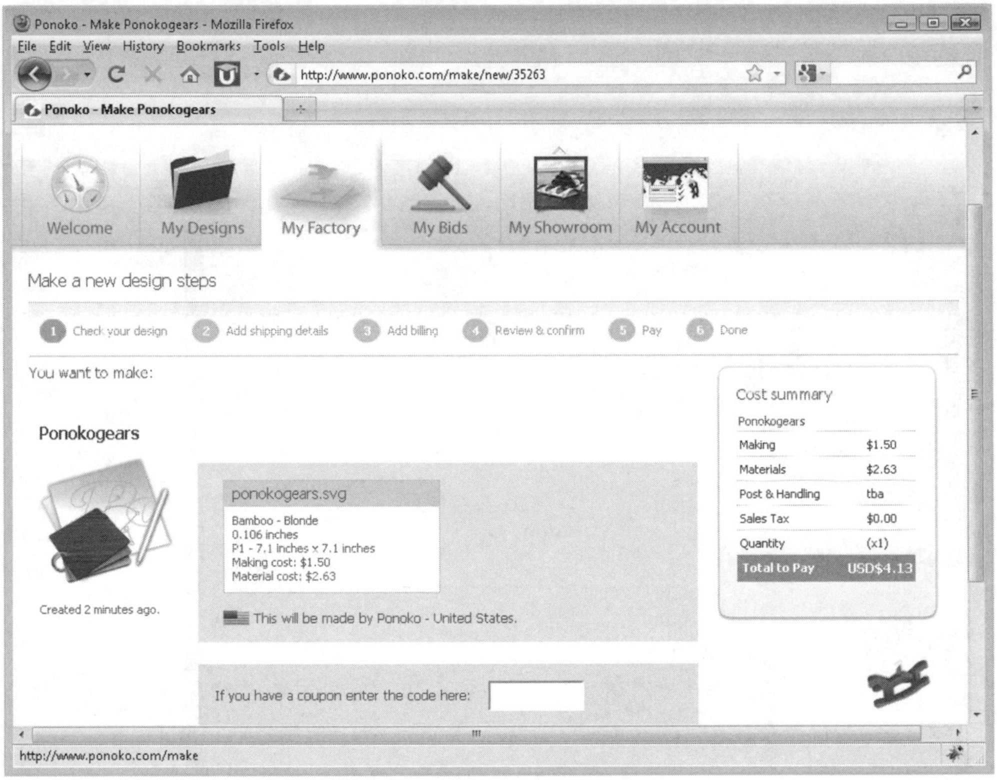

11. While you're waiting for your Ponoko order, get out your 1/4 in wooden dowel and cut off two 2 in sections with a hobby knife. File down any splintery ends.

12. The gears will come in the square template with a sticky paper protector on each side. Peel off the paper, pop out the gears, and position the two gears over the holes in the base. Insert your wooden dowels, and voila! Figure 7-37 shows my gears.

FIGURE 7-37 Final laser-cut gear assembly

Idler Gears

When you have two gears that mesh, they both turn in opposite directions when they spin. If you want to make two gears spin in the same direction, you can space them out with another gear between them. This is called an *idler gear*. It doesn't change the gear ratio of the system. It just allows you to get the input and output gears moving in the same direction (see Figure 7-38).

Idler gears are also handy when your input and output gear shafts are far apart. They don't need to form a straight line between your input and output gears, but can be offset, which allows you to vary your input and output shaft distance almost infinitely.

Compound Gears

Compound gears are formed when you have more than one gear on the same axle (see Figure 7-39). A compound gear system has multiple gear pairs. Each pair has its own gear ratio, but since a shared axle connects the pairs to each other, you multiply the gear ratios together to get the gear ratio of the system.

Compound gears are a very efficient way to gear up a weak motor to increase torque and decrease speed.

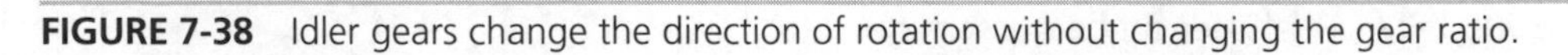

FIGURE 7-38 Idler gears change the direction of rotation without changing the gear ratio.

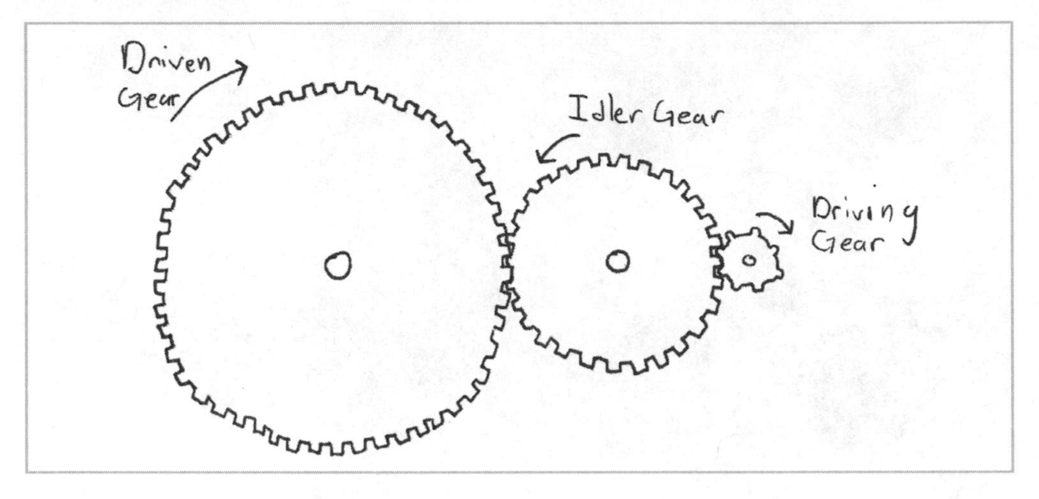

FIGURE 7-39 Compound gear system

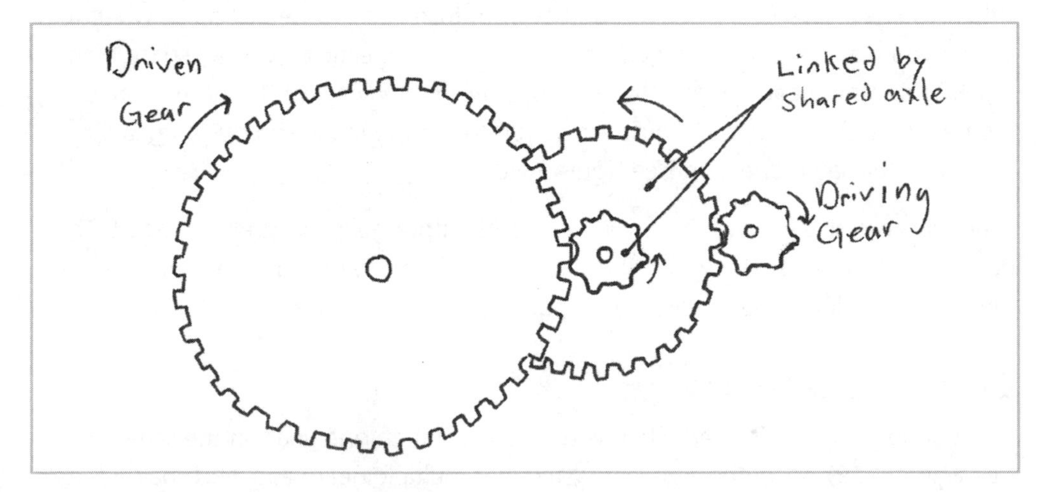

Pulleys and Sprockets, Belts and Chains

Belt or chain drives are often preferred over gears when torque needs to be transferred over long distances. Imagine how funny a bicycle would look with a bunch of gears between the pedals and the back wheel. They are also more forgiving about misalignment than gear systems are.

Sprockets, like the ones on your bicycle, are used with chains. Pulleys are used with belts, and can be flat or V-shaped with matching belts or grooved pulleys with matching toothed belts. We covered the latter type, called a timing belt pulley system, in Chapter 1. The pulleys and sprockets that come with hubs and set screws are mounted on shafts and motors to do the work. Remember that you have a mechanical advantage only if the input pulley is smaller than the output pulley, and the advantage is just the ratio of their sizes. For example, if your input pulley is half the diameter of the output, your mechanical advantage is 2:1.

It's common to include one or more tensioners in a pulley system (see Figure 7-40). *Tensioner* is the common name for a pulley that's spring-loaded and/or adjustably mounted in a slot to keep the belt tight while the mechanism runs. Tensioners are often tightened after the belt is installed, which makes installation much easier than needing to stretch the belt over pulleys that are already in position. Tensioners are similar to idler gears in that they don't change the mechanical advantage of the system; they just alter the behavior. In fact, they're often called *idler pulleys*, and commonly have bearings or bushings as hubs to allow for smooth rotation.

FIGURE 7-40 MakerBot timing belt pulley system with tensioners (image used with permission from MakerBot Industries)

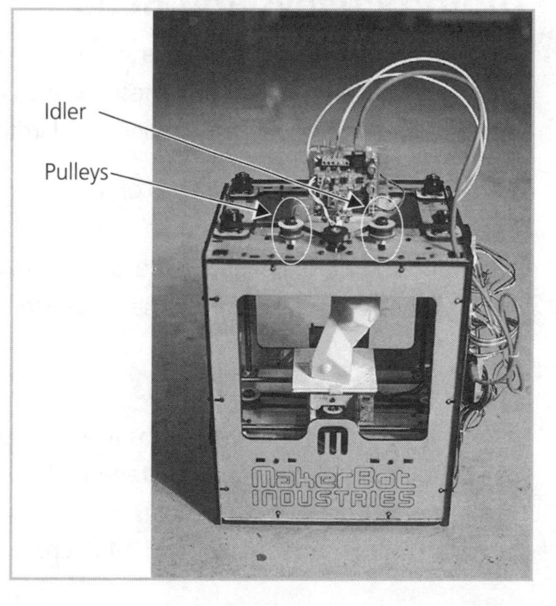

Two good sources for all these kinds of pulleys and belts are McMaster and Stock Drive Products. ServoCity is a good source for smaller sprockets and chains, especially if you're working with servo motors or the ServoCity DC motors.

Standard Pulleys and Belts

Standard pulleys provide a friction drive, so they are very sensitive to getting the belt stretched just enough to transfer motion between pulleys, but not so much that the tension causes friction or structural problems. Two pulleys connected by a belt will rotate in the same direction. To get them to rotate in opposite directions, put a half twist in the belt to create a figure 8.

Pulleys can be totally flat on the perimeter or have grooves that accommodate round or V-shaped belts. Some belts are very stiff and need a lot of tension to make them work properly, which will not bode well if you have a cardboard-and-popsicle-stick construction. So before committing to a belt, make sure you have the rest of the structure in place. There's no really good way to estimate the stiffness of a belt before you buy it, but in general, the thinner and skinnier it is, the more flexible it will be.

Timing Pulleys and Belts

Timing belts provide positive drive since the belt teeth mesh with the grooves in the timing belt pulley. You can find these in cars (see Figure 1-10 in Chapter 1), and also on a smaller scale in printers, copiers, and in the CupCake CNC (see Figure 7-40).

There are a dozen different series of sizes with names like MXL and HTD, but the series name is less important than just making sure your pulley and belt are the same series, and that your pulley is wide enough to accommodate your belt. The timing belt pulley and belt should be the same pitch, similar to meshing gears.

Sprockets and Chains

Sprockets and chains provide a positive drive similar to gears because the sprocket teeth and chain mesh together. Standard bicycle chain is 3/8 in, and you can find smaller metal chains and even plastic chains with snap-together links. Figure 7-41 shows an aluminum sprocket and 14 in chain mounted to a servo motor.

FIGURE 7-41 An aluminum sprocket and 1/4 in chain mounted to a servo motor (credit: ServoCity)

Power Screws

We talked about using screws as simple machines in Chapter 1, and screws as fasteners in Chapter 3. Power screws get their name from their intended use. Their geometry allows them to lift heavy loads, as well as precisely position anything riding on them.

There are a couple kinds of power screws: threaded rods and ball screws. You may have encountered common threaded rods, sometimes referred to as *all-thread*. These are designed for fastening things that are thick or far apart, and look just like longer versions of fastening screws. Although not designed to be used as power screws, they do the job well in MakerBot's CupCake CNC, where high precision and heavy lifting are not the main concerns. Acme threaded rods use a special geometry thread designed to lift heavy loads more efficiently.

A ball screw has a semicircular groove that spirals up the screw and allows little steel balls (housed in a ball nut) to ride up and down it. Ball screw and nut assemblies are much more expensive than other types of power screws because of their efficiency. Because the friction is so low, more of the input energy is transferred to useful work.

Regardless of the type of screw chosen, all power screws do one thing well: give tremendous mechanical advantage. As you saw from the 600:1 ratio in the car jack example in Chapter 1, this is pretty crucial in applications when you need to lift heavy loads with a low input force. Power screws have been used in this capacity for many years, and sometimes in reverse. The wooden ones in Figure 7-42 were actually

FIGURE 7-42 A diorama in a winery museum shows wooden power screws that were used to press the grapes.

hand-driven and used to squish grapes in wineries before mechanical presses were invented.

McMaster and Nook Industries (www.nookindustries.com) are two good sources for power screws.

Springs

Springs can be very useful components in your mechanisms. They can keep lids closed, return solenoids to their original position, create latches and ratchets, and more. Springs can store energy, as mentioned in Chapter 5, and are often components in the mechanical toys we'll talk about in Chapter 8. Here, we'll cover the different kinds of springs and how they can be used.

Compression Springs

When most people hear "spring," the compression spring is the type that comes to mind. You can find tiny ones inside mechanical pens and pencils, and larger ones in the shocks on mountain bikes. First, let's go over some vocabulary so you'll know what all the words mean when you shop for springs. Figure 7-43 shows how these terms apply to a compression spring.

- **Inner diameter** The diameter of the biggest rod that will fit inside the spring.
- **Outer diameter** The diameter of the outer edge of the spring.
- **Wire diameter** The diameter of the wire that is wound to make the spring.
- **Free length** The length of the spring before you do anything to it.
- **Solid height** The height of the spring when completely squished.
- **Spring rate or stiffness** The k in Hooke's law (in units of force/length), which tells you how much the spring will squish under a given weight:

$$\textit{Force } (F) = \textit{Stiffness } (k) \times \textit{Distance } (x)$$

FIGURE 7-43 Anatomy of a compression spring

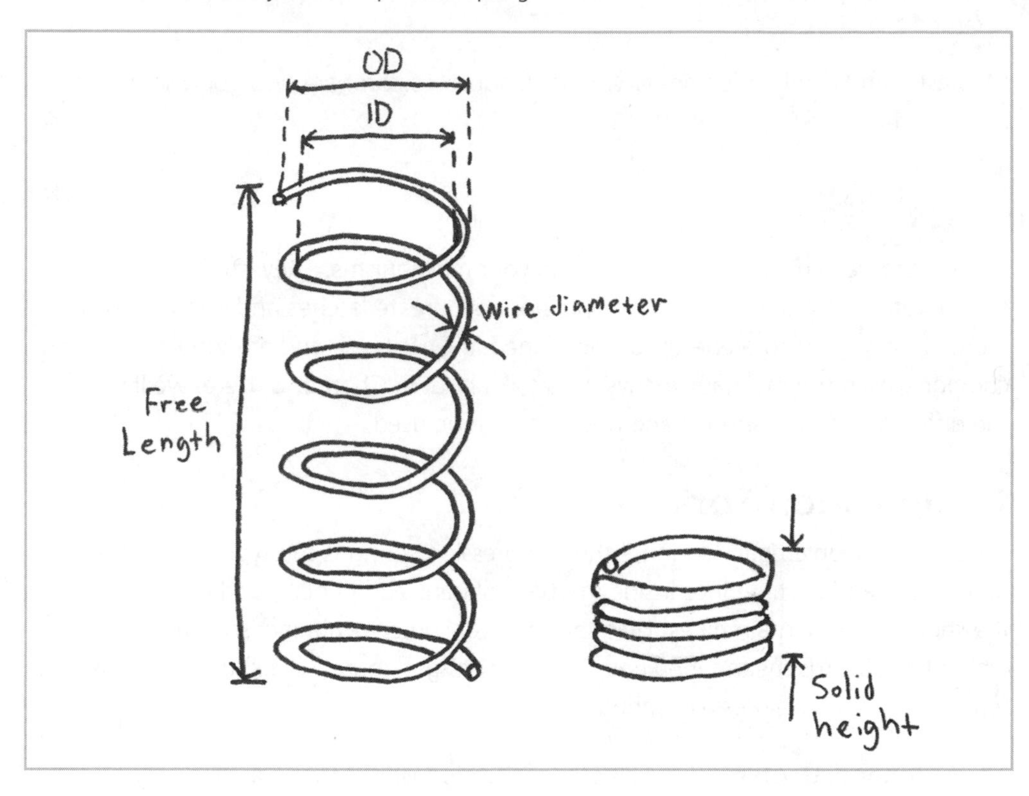

Compression springs are used as shock absorbers, return springs for solenoids, projectile launchers, belt tensioners, and return springs for jack-in-the-box latches (see Figure 7-44). It's easiest to work with compression springs that have ground ends or that are designed to sit flat. It's also a good idea to either surround the spring with a housing or mount it on a shaft to prevent it from buckling out to the side.

Tension/Extension Springs

Tension springs (also called *extension springs*) are the opposite of compression springs, but we can use most of the same vocabulary to describe them. These springs start out

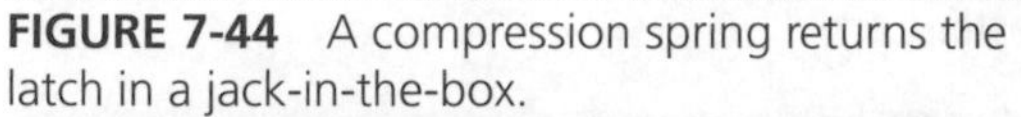

FIGURE 7-44 A compression spring returns the latch in a jack-in-the-box.

completely squished, and then resist as you pull them longer and longer (see Figure 7-45). You can stretch them only so far before they stay like that forever, so the maximum safe stretch distance is often specified as maximum extended length.

Most of us have a tension spring on our desk at all times—check inside your stapler. The tension spring inside keeps consistent force on the little staples so the next one is always ready and waiting to go.

You can also use tension springs for many of the same functions as compression springs, just mounted differently. They are generally easier to design for, since you don't need to worry about a hole or shaft to act as a guide. Instead of resting on a

FIGURE 7-45 Extension springs (credit: McMaster-Carr)

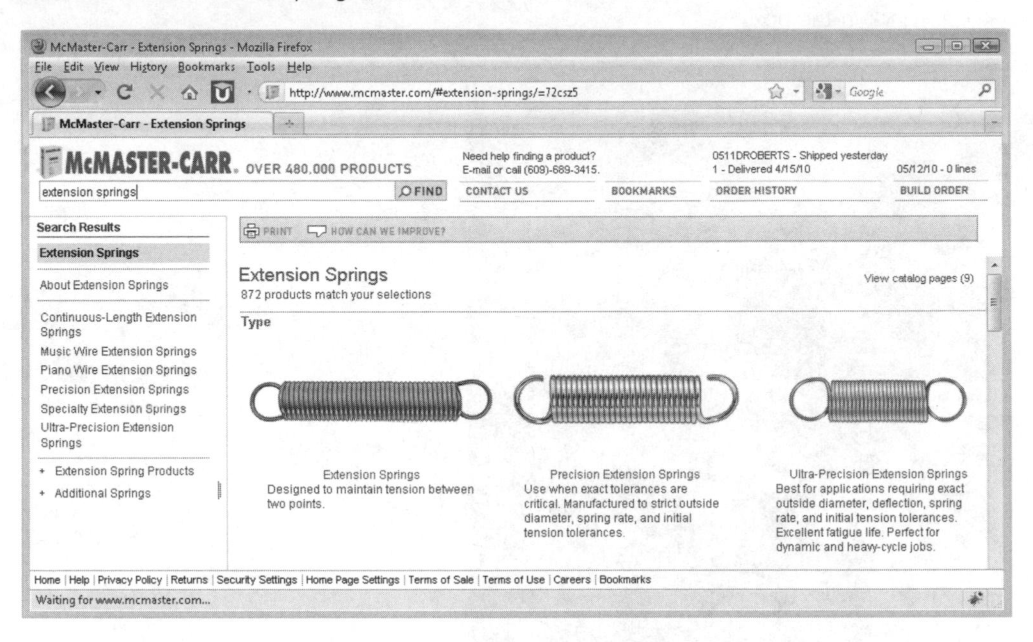

surface, these springs are often hung from something, as in fish scales and grocery store scales. You'll also see them in garage door mechanisms and around the edges of trampolines.

Torsion Springs

Torsion springs exert a torque or rotary force that's usually used to keep something shut. You've probably seen them in hair clips, mousetraps, clothespins, and clipboards. They also live inside doorknobs, allowing them to return to their original resting position after you open the door.

Torsion springs are a bit trickier to understand and buy, and there are a few different kinds. Torsion springs are categorized by the angle the legs stick out from the center spiral and the range of motion you can expect from those legs (see Figure 7-46).

Spring listings will usually give you torque only as a means of determining the strength of the spring. This is the torque at maximum deflection (closed). However, this torque

FIGURE 7-46 Shapes of torsion springs (credit: McMaster-Carr)

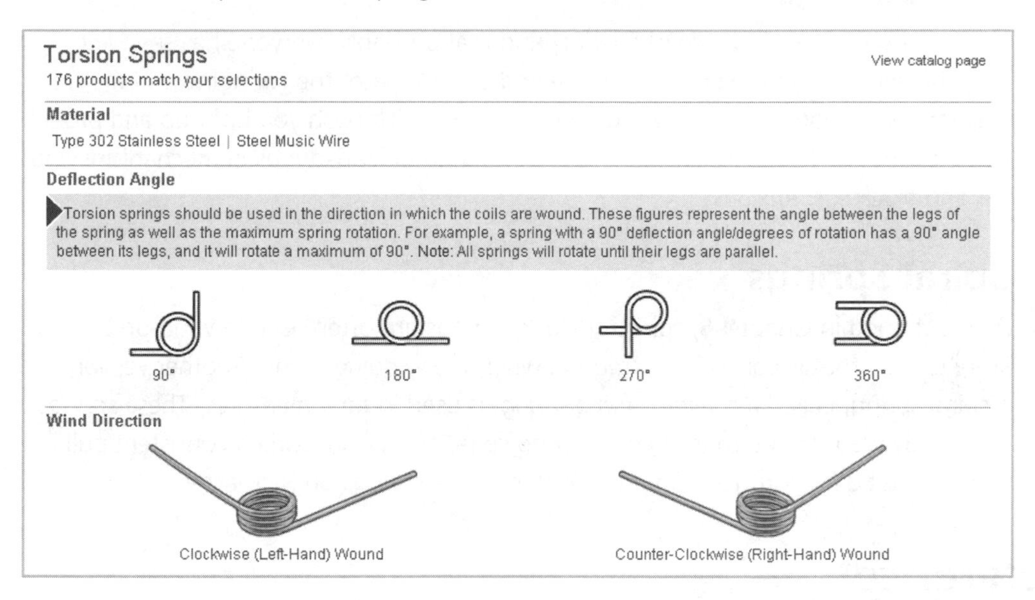

changes as you go from fully closed to fully open. Here is the equation that relates torque to how far apart the legs are:

Torque (*T*) = *Stiffness* (*k*) × *Angle* (in radians)

NOTE ***Remember that degrees × (π/180) = radians.***

To find the torque at an intermediate location, first figure out the stiffness (*k*) by using the equation and maximum angle deflection of your spring. Then you can use the stiffness multiplied by any angle and find the torque. You can also use a direct proportion. For example, if the listing says 1 in-lb at 90°, then it will have 0.5 in-lb of torque at 45°. If you want to experiment with torsion springs, revisit Project 5-1 in Chapter 5, and you'll have a new appreciation for the simplicity of a mousetrap.

Spring-lock Washers

Spring-lock washers, sometimes called disc washers, were mentioned back in Chapter 3 when we talked about putting them in bolted joints to help keep the joints from coming loose. This is the most common use of spring-lock washers. They act like little compression springs with just one revolution.

Leaf Springs

A diving board is an example of a leaf spring that probably everyone has seen and most have used. When you jump on the end of the board, the springiness of it cushions your landing and moves down, and then helps push you back up and propel you into the air. This same cushioning effect is used in leaf springs in mechanisms and car and truck suspensions.

Spiral Springs

As mentioned in Chapter 5, spiral, or clock, springs are often used in wind-up toys to store energy that is converted to motion when the winding stops. Another version of a spiral spring, called a *constant-force spring*, is used in tape measures. These springs constantly want to return to their rolled-up state, and will provide a consistent pull force in that direction. You can find constant-force springs on McMaster.

References

1. Richard G. Budynas, J. Keith Nisbett, and Joseph Edward Shigley, *Shigley's Mechanical Engineering Design* (Boston: McGraw-Hill, 2008).
2. Dennis Clark, *Building Robot Drive Trains*, ed. Michael Owings (New York: McGraw-Hill, 2003).
3. U.S. Bureau of Naval Personnel, *Basic Machines and How They Work* (New York: Dover Publications, 1971).
4. Lesley Flanigan, "Making Basic Gears: Tutorial," ITP Mechanisms and Things That Move Archives (http://itp.nyu.edu/~laf333/itp_blog/ mechanisms_and_things_that_move/).

Combining Simple Machines for Work and Fun

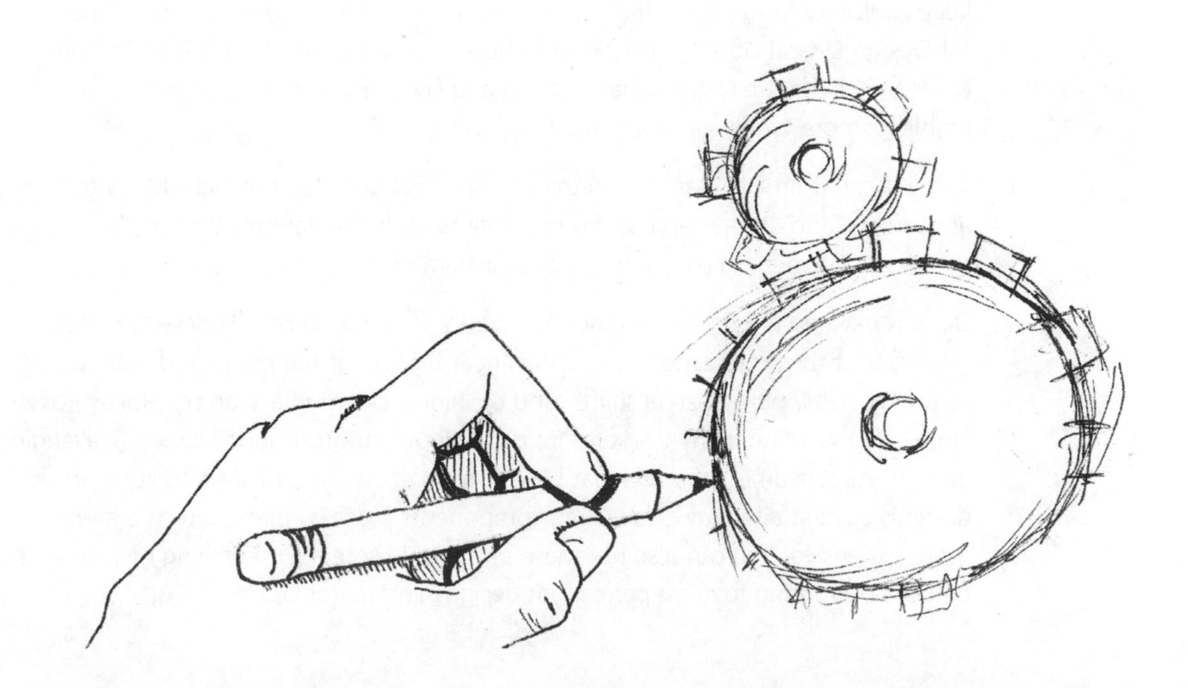

For hundreds of years, engineers and designers have been coming up with ways to convert rotary motion into useful work. You already know how to choose a motor and attach something to it, but maybe the output you're looking for isn't a simple rotary motion.

In this chapter, we'll explore how to convert rotary motion into more complicated motions by combining simple machines. First, we'll go over some general mechanisms you can use, and then we'll cover the basic types of motion and ways to convert between them. Finally, we'll look at some examples of combining simple machines to create kinetic sculpture and mechanical toys. Whether you combine simple machines for work or for fun, the applications are limited only by your imagination and, I suppose, some pesky laws of physics.

Mechanisms for Converting Motion

Rotary motion—the most common input motion—can be converted into more complicated motions through systems of cams and followers, cranks, linkages, and ratchets. All complicated machines, mechanisms, and robots are made up of combinations of simple machines like these. They help us convert rotary motion into linear, up and down, oscillating, or intermittent output motion. Cornell University keeps a library of examples that can be accessed online as the Kinematic Models for Design Digital Library (http://kmoddl.library.cornell.edu). The work of Cabaret Mechanical Theatre (www.cabaret.co.uk) and Flying Pig (www.flying-pig.co.uk) highlight more whimsical examples.

This is a great time to start taking things apart, if you haven't already. Many toys, appliances, and other everyday devices have built-in mechanisms that you can use—either directly in your projects or as inspiration for their design.

Do you have an old printer that doesn't work? Then you probably have two stepper motors that turn rotary motion into the linear motion of the print head with a system of timing belts, pulleys, steel shafts, and bushings. Do you have an old Hokey Pokey Elmo? You will find many treasures inside, including motors and linkages. Scavenging parts from consumer products that benefit from economies of scale to keep prices down is a great alternative to buying components piece by piece, which is always more expensive. You can also use these parts to do some 3D sketching of your ideas before committing to more permanent designs and materials.

NOTE **3D sketching** ***is the practice of using cheap materials and components you have around to build simple, temporary, and not necessarily functional models to wrap your brain around an idea. It's a good idea to keep some cardboard, duct tape, LEGOs, and spare parts around for such sketches. If you need more flexibility, you can try your hand at the 3D modeling programs we'll cover in Chapter 9.***

Cranks

A crank is basically a lever attached to a rotating shaft. You can use a crank as a handle to turn a shaft, as in the old bucket-raising cranks on top of a well (see the crank on the party sheep project shown later in Figure 8-17). In this case, you're using the crank as a simple wheel-and-axle machine, similar to the steering wheel in a car, as discussed in Chapter 1. If you flip this configuration, the rotating shaft can drive the crank itself. In this case, the crank can be used to convert rotary motion to *reciprocating*, or back-and-forth, motion. The *throw* of the crank is the diameter of the path it travels.

You can make cranks easily by sticking a wooden dowel into a hole in a piece of wood, or with a pair of pliers and a coat hanger or other thick wire. The crank on the left in Figure 8-1 is friction drive, since it just relies on friction and gravity to return to the original state. The crank on the right in Figure 8-1 is connected to the output shaft or lever so is not just relying on gravity.

FIGURE 8-1 Friction-drive crank that relies on gravity (left) and positive-drive oscillating crank (right)

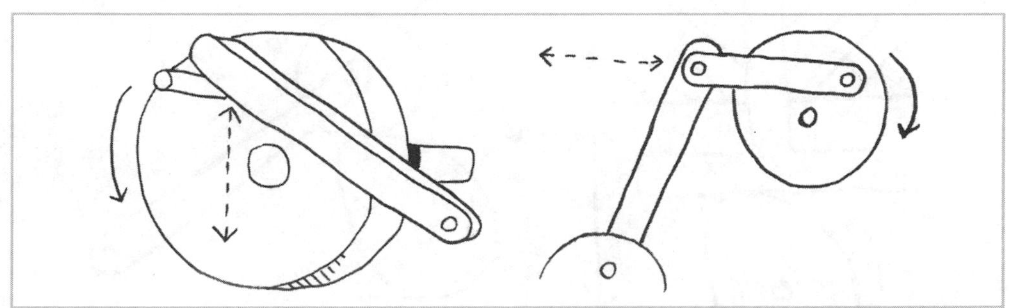

Cams and Followers

Cams are useful any time you have one or more objects you want to move in a periodic or irregular motion. In the most basic terms, a cam is any eccentric or noncircular shape that can convert rotary motion into linear motion. Cams open and close the valves on modern internal combustion engines. They are used extensively in the mechanical toys we'll talk about later in the chapter.

You can make cams yourself from wood if you have some basic tools, or find them off the shelf from WM Berg (www.wmberg.com) and other suppliers. The party sheep shown later in Figure 8-17 uses a cam attached to the rotating shaft inside to make its head nod.

Edge Cams

The most basic type of cam is called an *edge* (also *disk* or *peripheral*) cam (see Figure 8-2). The part that the cam moves is called a *follower*. The edge cam transfers motion to a follower moving against its edge. This is a friction-drive system because there is nothing locking the cam and follower together. Edge cams can be used to create linear (translating) motion or oscillating rotary motion.

The difference between the highest point on the lobe and the minimum radius on the cam is called the *throw*, and is the maximum amount of linear motion the cam can

FIGURE 8-2 Edge cam with translating follower (left) and with oscillating follower (right)

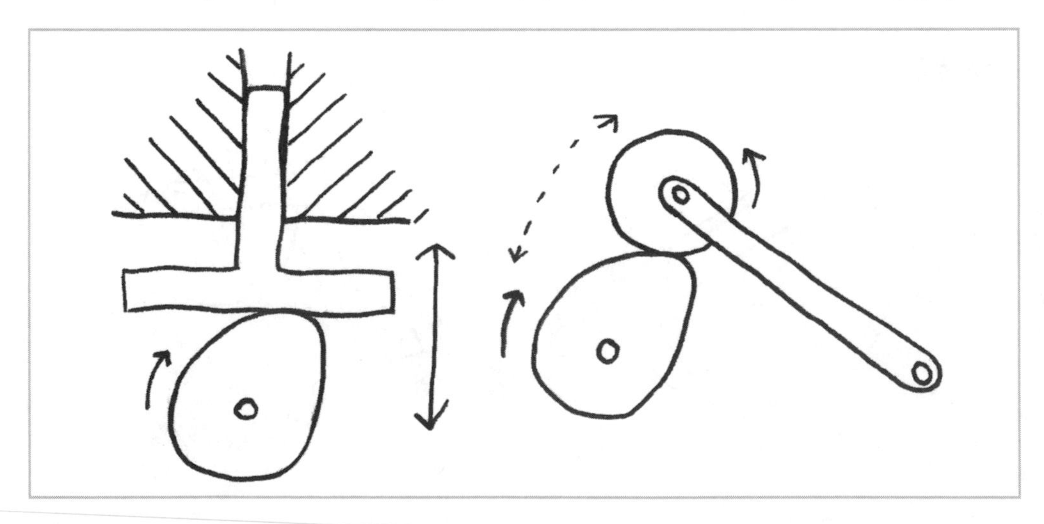

transfer to the follower. The cam can also be a disk mounted off center (eccentric cam), in which case the throw is just the difference between the maximum and minimum distances to the axis of rotation. The follower can be flat, like the translating cam in Figure 8-2 (left); it can end in a roller, like the oscillating cam in Figure 8-2 (right); or it can take the shape of a curved surface or hemisphere.

The *cam shaft* is the rotating input shaft that makes the cam spin. When the follower creates linear motion, there is usually a shaft or *stem guide* to channel this motion. The followers in Figure 8-2 rely on gravity to hold them against the cam, but could also be spring-loaded.

The cams in Figure 8-2 can rotate in both directions, but sometimes you need to make a cam that can rotate in only one direction and lock in the other. The snail cam in Figure 8-3 (left) will produce a steady rise then sudden fall when rotated counterclockwise, but will eventually lock against a follower if rotated clockwise. The ratchet-shaped cam in Figure 8-3 (right) has four of these lobes, so it will produce four such motions with just one rotation of the cam shaft. We call these motions *events*, and one complete revolution of the cam is a *cycle*. The number of events per cycle will be limited by the size of your cam. The timing of these events will also depend on the speed of rotation of the cam shaft.

Cams can also produce complex and irregular motion. The edge cam in Figure 8-4 has a *dip*, or *recess*, in addition to a lobe. The profile between the lobe and the dip with a

FIGURE 8-3 Snail cam with one lobe (left) and with four lobes (right)

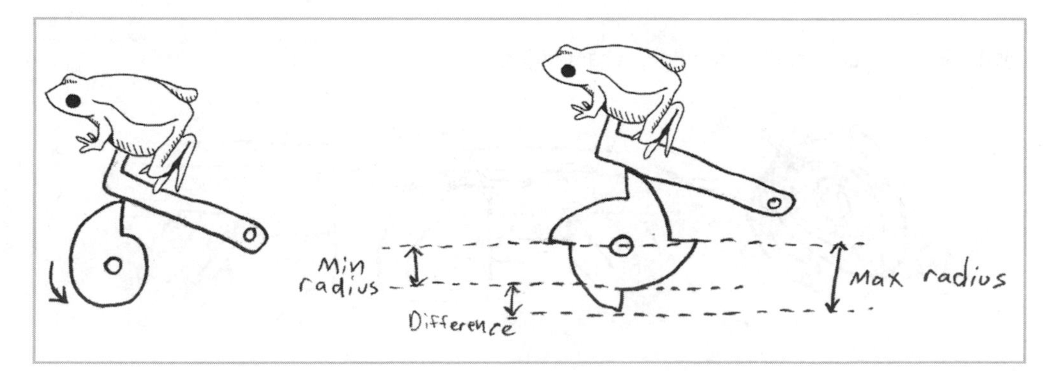

constant radius is called the *dwell*, because the follower just dwells there and doesn't create motion. The shape of the follower is important. A finer point on the follower will be more able to track intricate variations on the cam, but will need to be strong enough to survive the stress of riding in and out of all those bumps (see www.flying-pig.co.uk/mechanisms/pages/cam.html). Using a bearing or roller on the end of the follower is a good way to decrease this stress by decreasing friction. A bigger cam will give you more leverage to push on a follower and be easier to turn than a small one trying to do the same job, so don't try to do too much with a small cam.

FIGURE 8-4 Edge cam with lobe and dip

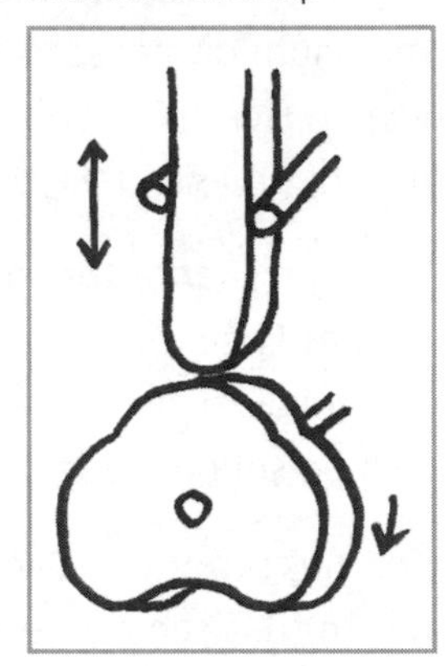

Face and Drum Cams

A *face cam* (also called a *radial* or *plate cam*) transfers motion to a pin or roller free to move in a groove on its face. These are usually designed to operate in both directions, and create a positive-drive situation in both directions because the follower is contained in the groove of the face cam.

A *drum cam* (also called a *barrel* or *cylindrical cam*) has a path cut around its outside edge in which the roller or follower sits.[1] It creates a back-and-forth motion on the follower in a plane parallel to the axis of the cam.

Figure 8-5 shows both of these cam profiles. Face and drum cams are less popular and harder to make at a hobbyist level, but are included here to show you the possibilities.

FIGURE 8-5 Face cam (left) and drum cam (right)

Linkages

A *linkage* is a connection that transfers motion from one mechanical component to another. Linkages are often the simplest, least expensive, and most efficient mechanism to perform complicated motions. They come in many shapes and configurations, and can do many jobs, such as the following:

- Change direction or otherwise alter path of motion
- Alter speed and/or acceleration
- Change timing of moving parts
- Apply a mechanical advantage
- Create elegant and efficient motion

Most linkages are *planar*, which means all the motion takes place in the same plane. Linkages usually have only one job, regardless of the number of links: create a specific output motion from a specific input motion.

Four-bar linkages are the simplest closed-loop linkage. They can create many different output motions by varying the lengths and relationships of the four segments (see Figure 8-6):

- **Ground/frame** The link that is kept stationary.
- **Coupler/lever** The link opposite the ground.

FIGURE 8-6 Types of four-bar linkages

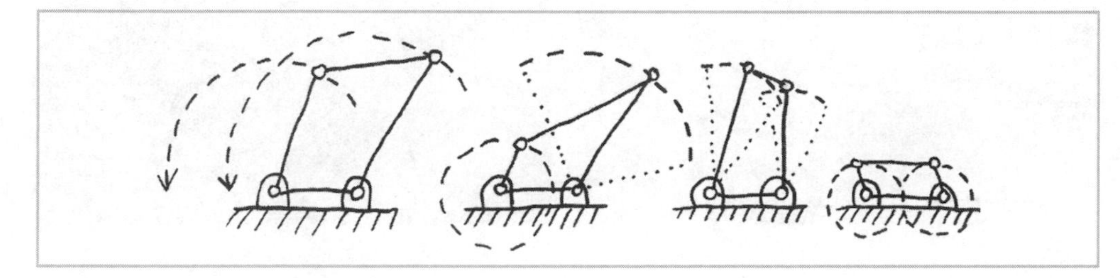

- **Crank/rocker** The driven link that joins the ground and the coupler. If it rotates through 360°, it's a crank. If it's limited to back-and-forth motion, it's a rocker.
- **Follower** The link opposite the crank/rocker. Confusingly, this is also sometimes called the *crank* or *rocker*, depending on its motion.

Within the four-bar linkage family, there are more than a dozen well-known variations. The linkage that drives windshield wipers is probably the most popular example. A close second is the four-bar linkage that opens up inside your umbrella.

In order for a linkage to have continuous motion (at least one link can rotate a full 360°), the sum of the shortest and longest links must be less than the sum of the remaining two links. This is called Grashof's law, so four-bar linkages that follow the law are called *Grashof linkages*.

Pantographs are linkages designed so that if a drawing is traced at one point, an enlarged (or miniaturized) copy will be drawn by a pen fixed to another. These were used to copy and scale line drawings hundreds of years ago. *Scissor linkages* are used in large scissor lifts and small lab jacks to raise and lower platforms (see Figure 8-7).

FIGURE 8-7 Adi Marom's short++ shoes use a scissor linkage and linear actuator to raise and lower a platform you can stand on (credit: Adi Marom).

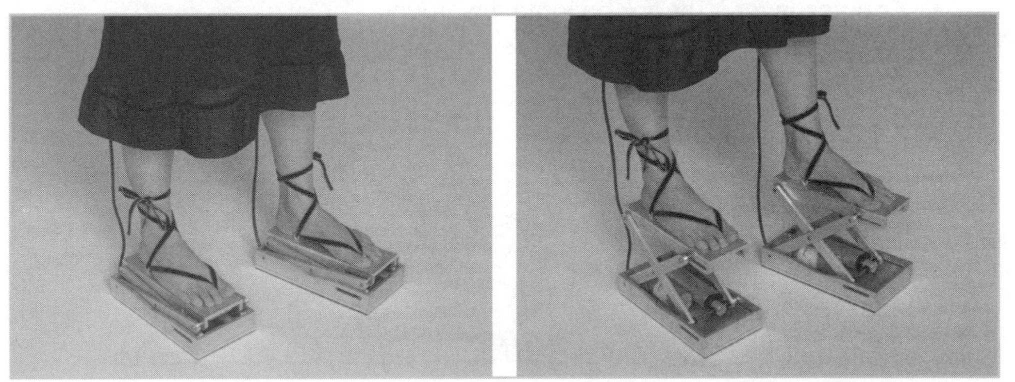

Project 8-1: I Heart Pantographs

In this project, we'll make the cardboard pantograph shown in Figure 8-8 and experiment with drawing patterns with it.

Shopping List:

- Paper fasteners
- Cardboard
- Pens or markers

Recipe:

1. Cut out six strips of cardboard in equal lengths. The ones shown in Figure 8-8 are about 6 in long by 1 in wide.
2. Attach the cardboard strips to each other with the paper fasteners (see Figure 8-8).

FIGURE 8-8 A cardboard pantograph (drawing machine)

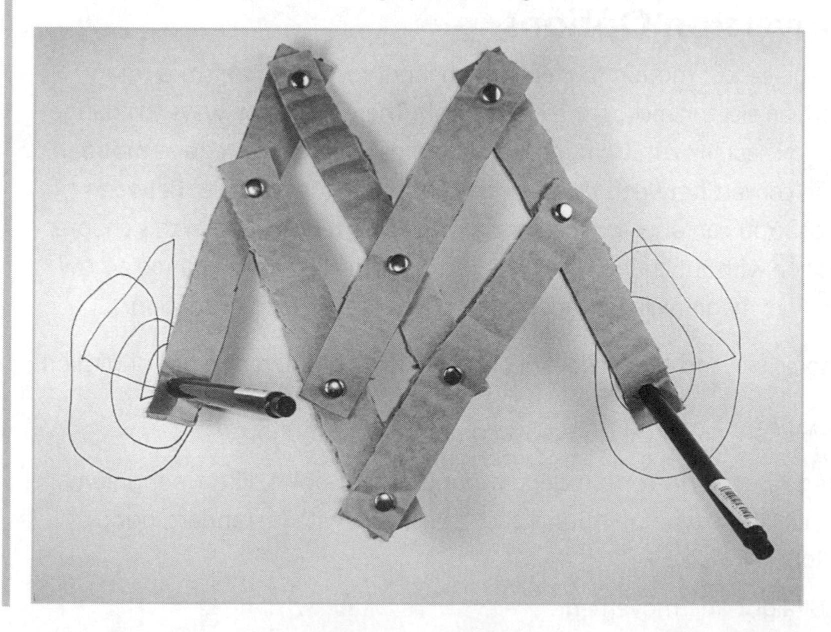

3. Make a hole in the bottom of each of the cardboard pieces on the sides and put a pen or marker in each hole.

4. Hold the center paper fastener and see how you can draw symmetrical (but mirrored!) patterns by moving the cardboard pantograph around.

5. Try replacing some of the paper fasteners with pens to draw more lines, or changing the length of some of the strips to make the patterns asymmetrical.

Ratchet and Pawl

A ratchet-and-pawl system creates a stepped motion and can be used as a locking mechanism (see Figure 8-9). A *ratchet* is a wheel with notches cut into it, similar in shape to a gear.[2] A *pawl* pushes against the notches and allows the ratchet to be driven in steps. A second pawl (*detent*) can stop the wheel from slipping backward. A ratchet-and-pawl system can also be used as a clutch to allow a shaft to rotate in only one direction.

FIGURE 8-9 Ratchet and pawl

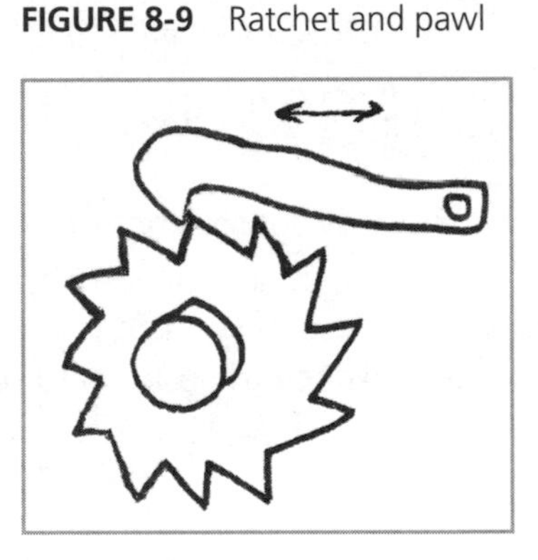

Motion Conversion Options

Most of the time, the easiest motion to create as an input for a mechanism is rotary motion, either from an electric motor or a hand crank. There are many ways to change this rotary motion to linear, intermittent, reciprocating, oscillating, or irregular motion. Sometimes you can convert between these motions as well—for example, between oscillating and linear. You can also use the simple mechanisms described in this chapter to transform a motion without changing its type. For example, you can change a slow to a fast rotary motion, magnify linear movement, or change the axis of motion.

Table 8-1 shows some ways to convert between the following different types of motion:

- **Rotary** Motion in a circle (the most common input motion).
- **Oscillating** Back-and-forth motion around a pivot point, like a pendulum in an old clock (this type of input is easy to achieve with a standard hobby servo motor).
- **Linear** Straight-line movement.

- **Reciprocating** Back-and-forth motion in a straight line.
- **Intermittent** Motion that starts and stops in a regular, predictable pattern.
- **Irregular** Motion with no obvious pattern or that doesn't fit into the other categories.

To use the table, locate the input motion you want on the top and the output motion you want on the left side. The box where they intersect shows your options for doing the conversion. (Much of the material in this table is from www.flying-pig.co.uk/mechanisms.)

We've already talked about the more general methods: gears, pulleys and belts, sprockets and chains, levers, cranks, linkages, and so on. The following are some of the trickier conversions shown in Table 8-1:

- **Scotch yoke** Used to create linear and reciprocating motion from an oscillating input (see Figure 8-10). Using a scotch yoke is an excellent way to convert the oscillating motion of a servo arm to linear motion.

FIGURE 8-10 Scotch yoke in running of the bulls (CC-BY-NC-SA image used with permission from Greg Borenstein and Scott Wayne Indiana)

TABLE 8-1 Converting Between Types of Motion

		INPUT				
		ROTARY	**OSCILLATING**	**LINEAR**	**RECIPROCATING**	**INTERMITTENT**
OUTPUT – CONVERSIONS	**ROTARY**	Gears, pulleys and belt, sprockets and chain, crank slider	Crank	Rack and pinion, linkage	Piston, bell crank	
	OSCILLATING	Crank, quick return			Linkage	
	LINEAR	Wheels, rack and pinion, scotch yoke	Scotch yoke	Scissor linkage		
	RECIPROCATING	Cam, crank, piston	Crank, cam, bell crank			
	INTERMITTENT	Geneva stop	Ratchet		Ratchet	
	IRREGULAR	Cam	Cam			
OUTPUT – TRANSFORMATIONS	**INCREASE/ DECREASE**	Gears, pulleys and belt, sprockets and chain	Gears		Lever	Lever, gears
	REFLECT	Gears	Gears	Pulley, lever	Pulley, lever	Pulley, lever
	ROTATE	Bevel gear, worm gear	Bell crank	Bell crank	Bell crank	Bell crank

- **Bell crank** Can be used to create different motion conversions and transformations depending on which part is driven and which part is doing the driving (see Figure 8-11). This is why you see it in many of the boxes in Table 8-1.
- **Geneva stop** Used to create intermittent output from a constant rotary input. In the version shown in Figure 8-12, it takes four turns of the lower circular shape to turn the star-shaped piece once. It's difficult to get the geometry just right, but you can get off-the-shelf versions from WMBerg or make your own with downloaded plans from Thingiverse (a 2D version for laser cutters from www.thingiverse.com/thing:1616 or a version for 3D printers at www.thingiverse.com/thing:1642).

FIGURE 8-11 A bell crank

- **Quick return** Used to create oscillating motion from a continuous rotary input. In the example in Figure 8-13, since the peg that drives the arm back and forth is closer to the bottom pivot point during part of the waving motion, the wave in one direction will be faster than in the other.

FIGURE 8-12 A Geneva stop mechanism

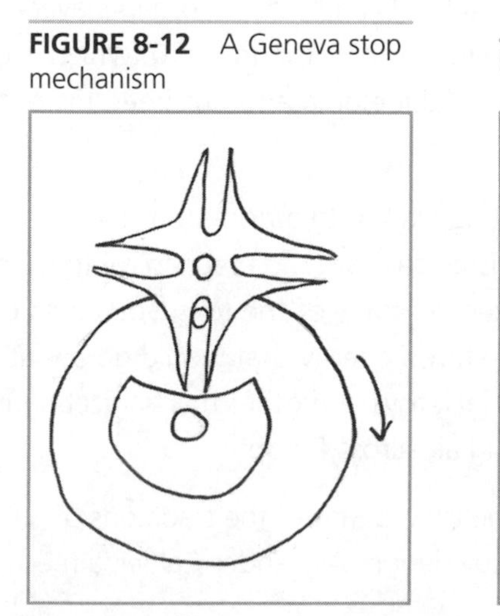

FIGURE 8-13 A quick return

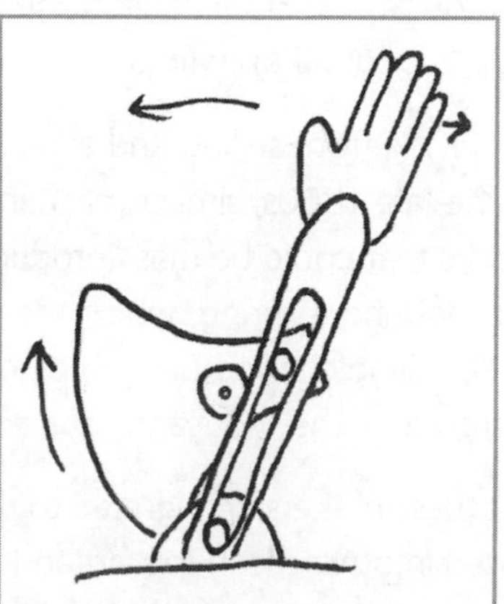

Automatons and Mechanical Toys

Automatons and mechanical toys provide some of the earliest examples of kinetic design. Early automatons were sometimes incredibly complex combinations of cams, linkages, springs, and components, often in the form of dolls or mannequins. These dolls could write poems or play the flute, based only on the interactions of mechanical parts, without any electronics, sensors, or feedback. They were powered by hand, steam, or water.

The earliest recorded automatons appeared in Egypt around the second or third century BC and were used as teaching tools to explore physical laws through movement.[3] The Greeks and Arabs were next to pick up the craft, followed by a dead period during the Middle Ages when most mechanical devices were condemned as pagan magic. Fortunately, this period ended, and by the fourteenth century, automatons began to appear in the huge cathedral clocks all across Europe.

One early example of a humanoid automaton is a robot that Leonardo DaVinci designed around 1495 (see Figure 8-14). We don't know if he actually built it while he was alive, but a few people have used his plans to create models that confirm that it works as intended. The French would-be priest Jacques de Vaucanson dropped out of his training when the Jesuit priests destroyed the angel automatons he designed for their apparent heresy. He went on to create the Digesting Duck in 1739, which looked, quacked, and crapped like a real duck through an elaborate system of cams and followers. Pierre Jacquet-Droz, a Swiss clockmaker, was another master of elaborate automatons. He created a little mechanical family with a writer, a draftsman, and an organ player around 1772. Henri Maillardet created a similar drawing automaton in 1810, which is on permanent exhibit at The Franklin Institute in Philadelphia and has been restored to create some of its original drawings.

Once making a living from selling and exhibiting elaborate automatons became impractical in the late 1800s, similar mechanisms gave way to mechanical toys, music boxes, and clocks that could be mass-produced. In many of the toys, you could crank a handle, which wound a spring, which then stored energy that would go about powering the toy. Simple string-pull jumping jack toys and other mechanical figures were manufactured by the thousands and sold all across Europe.

When some of these makers immigrated to the United States, the traditions of different countries merged into American folk art and toy design. Alexander Calder, a mechanical

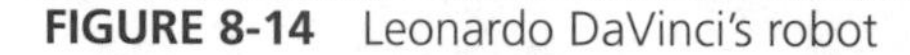

FIGURE 8-14 Leonardo DaVinci's robot

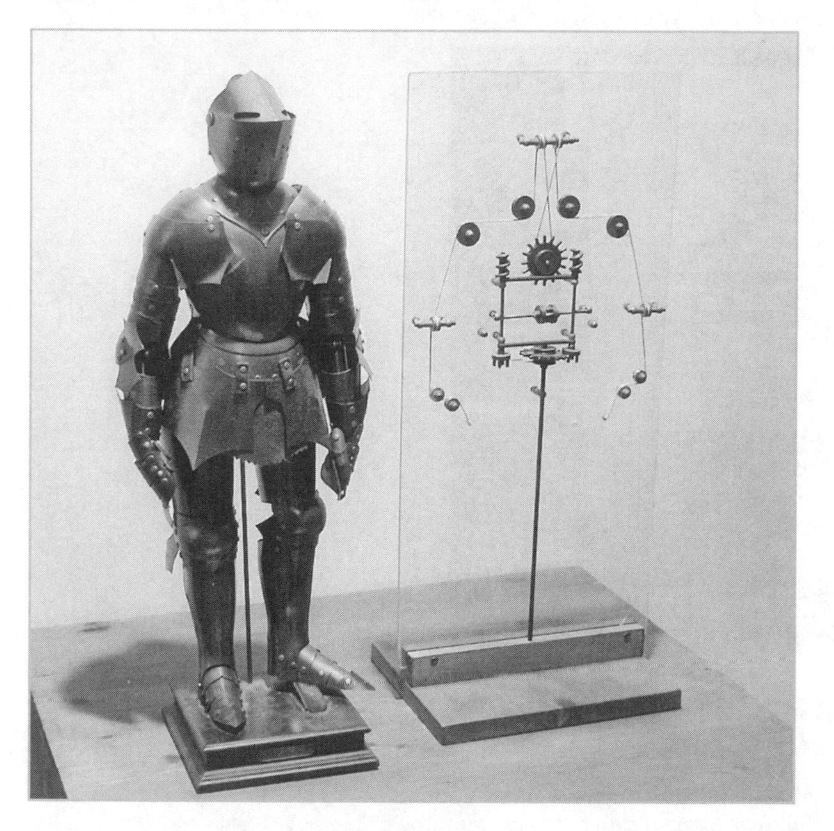

engineer turned artist, was probably the first American to popularize this kinetic art form through his mobiles and his circus, which he created in the 1920s. Calder's circus of miniature moving wire figures grew from a few characters in 1920 to eventually fill five suitcases, which he carried around to give performances. A video of his circus was created in 1961 and was most recently on exhibit at the Whitney Museum of American Art in New York City, along with a collection of his circus characters. Sam Smith, a British artist, created kinetic art and toys throughout the 1960s and 1970s that heavily influenced many modern mechanical toy makers.

More modern examples of kinetic sculpture and mechanical toys include Arthur Ganson's machines and the work of the artists of Cabaret Mechanical Theatre. Figure 8-15 shows Eun Jung (EJ) Park's Mechanical Storytelling: The Story of Grouchy the Clown automaton (see www.ejpark.com). Figure 8-16 shows an example from Andrew Jordan's Sound Creatures project (see www.andyjordan.us/).

FIGURE 8-15 Mechanical Storytelling (image used with permission from Eun Jung Park)

FIGURE 8-16 In this example from Andrew Jordan's Sound Creatures project, the spiral painted cam in the center throws its housing back and forth, creating irregular reciprocating motion from continuous rotary input.

Project 8-2: DIY Automaton—The Agreeable Sheep

Flying Pig provides a few sets of free plans to make animated creations in paper. My favorite is the Agreeable Sheep by Rob Ives. We called our version, shown in Figure 8-17, the Party Sheep due to the fluorescent card stock we used.

Shopping List:

- Size A4 or 8.5 × 11 in card stock
- Hobby knife and scissors
- Metal ruler
- Cutting mat or old magazine
- Glue stick or white glue
- Penny

FIGURE 8-17 The Agreeable (Party) Sheep (credit: Jade Highleyman, Fyebeam student resident, for putting this together and photographing it)

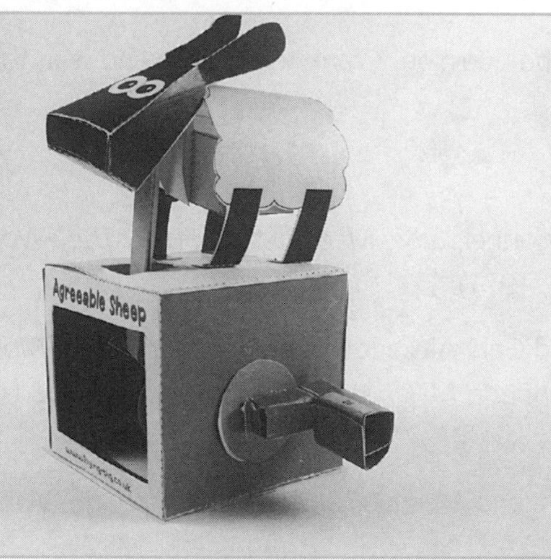

Recipe:

1. Download the model and the making instructions from www.flying-pig.co.uk/pdf/sheep.pdf.
2. Print the templates. The first three pages of the nine-page PDF are instructions. The last six are front and back versions of three templates for the sheep. If you can't print double-sided, you'll need to manually feed these into your printer in the correct orientation. Don't worry about perfect alignment.
3. Use scissors or a hobby knife to cut out all the drawings.
4. Score all the lines you will fold on by gently running the knife over the dashed lines. A metal ruler helps here to keep your scores straight.
5. Fold along all the dashed lines.
6. Follow the instructions on where to glue what. Have some patience between the gluing steps to make sure your sheep stays together.
7. Tape the penny inside where indicated.
8. Ask the sheep a yes or no question. Crank the handle and wait for an answer.

References

1. U.S. Bureau of Naval Personnel, *Basic Machines and How They Work* (New York: Dover Publications, 1971).
2. Aidan Lawrence Onn and Gary Alexander, *Cabaret Mechanical Movement: Understanding Movement and Making Automata* (London: G&B Litho Limited, 1998).
3. Rodney Peppé, *Automata and Mechanical Toys* (Marlborough, Wiltshire: Crowood Press, 2002).

9 Making Things and Getting Things Made

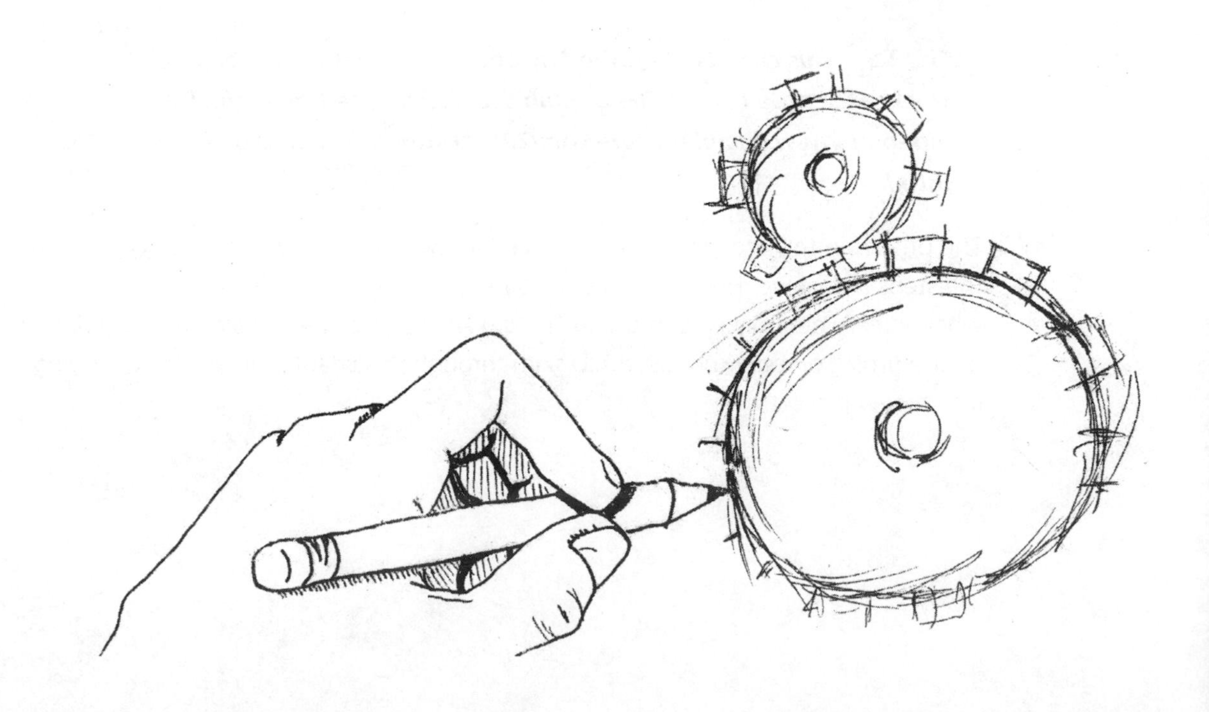

Materials, components, and general fastening techniques were covered in earlier chapters. Here, we cover how to actually make something or get someone else (or a machine) to make it for you.

The process starts with a design—whether it's a napkin sketch or a full 3D computer-generated assembly model. From there, some projects can be made by hand by sawing wood or putting together off-the-shelf components. Other projects lend themselves to modern rapid prototyping techniques that use digital files directly, including 3D printing and laser cutting. And some projects are suited to machining or other manufacturing techniques. Then you need to put everything together. Finally, since you don't really make a sound unless someone is around to hear it, the last step in any creation is sharing it. This might inspire someone else to attempt a version of your idea, therefore restarting the cycle.

The Making Things Move Ecosystem

Each phase in the "making things" cycle has methods you can do by hand (analog) and useful ways to use computerized machines and software to help (digital). We'll call this the making things move ecosystem. As shown in Figure 9-1, the phases of this ecosystem are creation, translation, fabrication, integration, and proliferation.

***NOTE** This chapter is inspired by and heavily based on Dominic Muren's Dorkbot Seattle talk on The Digifab Ecosystem. See the original slides and video at www.humblefacture.com/2010/02/dorkbot-talk-digifab-ecosystem.html.*

The phases in this ecosystem are organized in the same way you would go through the steps in real life. However, it helps to have the method of fabrication in mind when initially designing a part or mechanism—a practice called *design for manufacture* by the pros. So, it may be helpful to skim through this chapter once before digging into a new project.

FIGURE 9-1 The phases of the making things move ecosystem

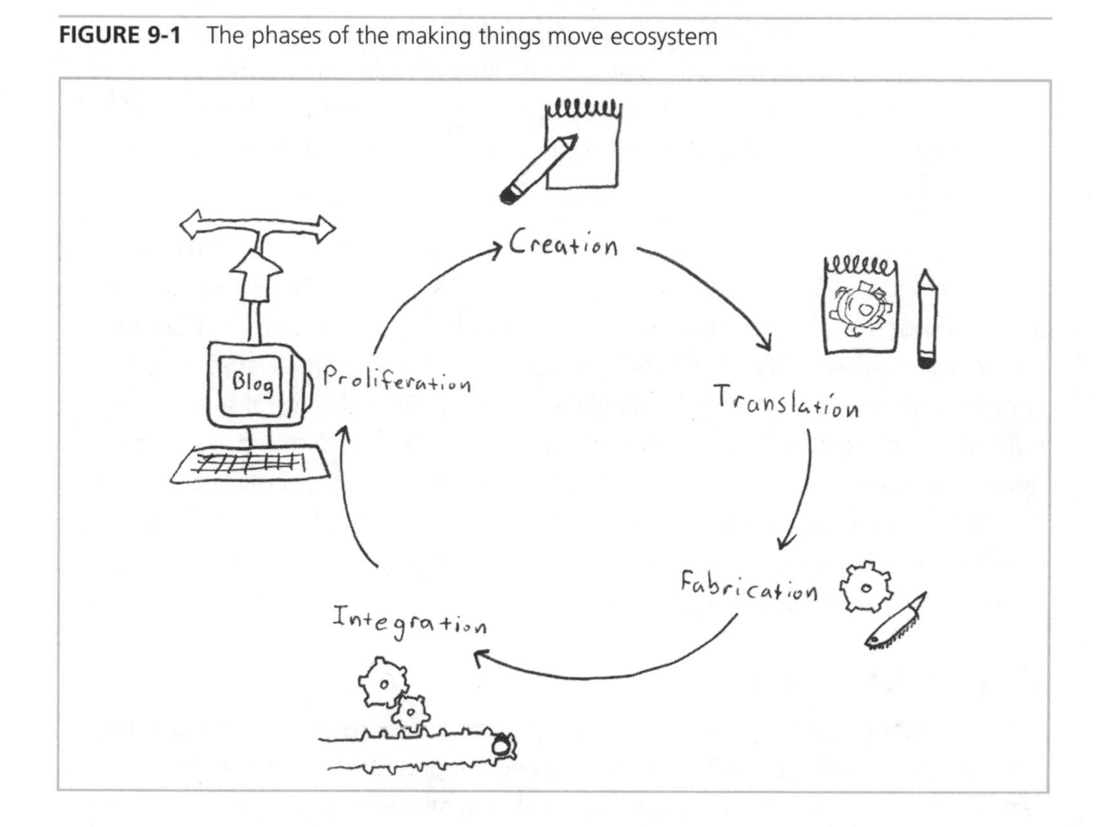

Creation

In order for a part or component to be created physically, it first needs to be described. If you plan to make something with a machine like a laser cutter or 3D printer, you need to describe your part digitally. If you plan on cutting and shaping materials by hand, you can use more analog techniques to describe your part or mechanism.

Analog Creation

The simplest way to describe something is to think about it in detail. If you need to cut off an inch of a closet rod to fit between two brackets, you probably won't use a computer program to model it. You'll just think about it and do it.

The next step up in describing your part or mechanism is sketching your ideas on paper. This helps you to plan things and get an idea of scale, and maybe of how different parts will fit together. Most mechanisms, regardless of whether they're fabricated by hand or computerized machines, begin life as 2D sketches on napkins and scrap paper.

You might also sketch in 3D. No, this doesn't mean wearing 3D glasses while you draw or standing your paper up against the wall. This means using reconfigurable and/or disposable materials to visualize your ideas in three dimensions. LEGOs are perfect for this, especially a LEGO set that has gears, motors, and a variety of other components you can use to make simple machines. LEGO sells several simple machines and motorized mechanisms kits that are perfect for this purpose. Some universities have entire rooms filled with LEGO parts to aid in the prototyping stage of creation. You can also use paper, popsicle sticks, straws, string, balsa wood, clay, hot glue, or any other material that is quick to work with, so you spend time thinking in 3D and not refining your project—yet.

Digital Creation

The computer programs you can use to visualize your designs are collectively called CAD programs. CAD stands for computer-aided design. But most of these programs can do more than aid you. They can also create digital files that you can use to make parts directly. We'll talk more about fabrication from digital files later in the chapter.

NOTE ***Whether you are going to make parts by hand or have them made by a machine, CAD software is still a handy tool to have in your prototyping toolbox. A student of mine conveyed this idea well. After using Alibre Design software to model a part in 3D that was later cut out of flat sheets of plastic, he said, "I went and 3D modeled it anyway, because it's really helpful for visualization."***

The type of object you want to make will dictate the software you use and the type of file you will create. Read on for tools for both 2D design, 3D design, and software that lets you create entire assemblies of parts.

2D Design

2D design is the digital version of sketching on paper. You're already familiar with one 2D program if you made your own gears in Chapter 7. In Project 7-1, we used Inkscape, an open source vector-drawing program similar to Adobe Illustrator. These programs can create lines that computers are able to read, which is how a laser cutter knew how to cut the gear shapes we made.

A more sophisticated tool that's still simple and affordable is QCAD (www.qcad.org). QCAD is designed to create parts and 2D plans, while Inkscape and Illustrator are primarily drawing programs.

The next step up is to use full-blown CAD packages like AutoCAD for part design, but this is overkill (and over budget!) for a lot of beginners. See Table 9-1 for a comparison of 2D and 3D modeling programs.

3D Design

Most 3D parts start life as 2D sketches that are pushed, pulled, or otherwise formed into 3D models on your computer screen. Some programs use a kind of wire mesh frame to create objects, others use solid shapes, and a few use more of a direct mathematical language.[1] Solid modeling programs talk to fabrication machines the best, but designers with any computer science or programming experience might prefer the math-based ones.

Table 9-1 lists the computer programs available for 2D and 3D modeling. In order to navigate the large number of options, look for the asterisks (*), which indicate favorites, the notes that include ease of use, and the x's that indicate on which platforms the software will run.

> **NOTE** ***You can use Boot Camp, Parallels, or VMware Fusion to run Windows-only programs on a Mac.***

An exciting feature of some of the 3D modeling programs listed in Table 9-1 is that you can create assembly files that include multiple parts, and relate the parts to each other just as they do in real life. This allows you to move parts around on the screen to mimic their real functions, and make sure the pieces don't jam into each other when they move, so there's plenty of space for the range of motion you want.

TABLE 9-1 Programs for 2D and 3D Modeling

			COMPATIBILITY			
		COST	**WINDOWS**	**MAC**	**LINUX**	**NOTES**
2D						
	*Inkscape	Free	x	x	x	Open source.
	QCAD	24 euro	x	x	x	Open source. Offers a free demo version for trial, but the software terminates after 10 minutes of use and can be used only up to 100 hours.
	Adobe Illustrator	$599	x	x		Considered the standard vector-drawing tool. Dimensioning plug-ins can help make clear 2D plans.
	AutoCAD	$3,975	x			Good for 2D parts and plans; not the best for 3D modeling. AutoCAD LT is a stripped-down version of AutoCAD, available for $1,200.
3D						
MESH	Blender	Free	x	x	x	Steep learning curve and more suited to animation and character modeling than machine design. Open source software is promising but not very useful for machine design.
MESH	Art of Illusion	Free	x	x	x	Free, open source 3D modeling and rendering studio. Decent as open source software goes, but clumsy for machine design. Very basic program; not easy to model anything but the simplest shapes.
MESH	Google SketchUp	Free	x	x		Good for 2D and 3D part and assembly design. Very easy to learn and useful to visualize concepts.

TABLE 9-1 Programs for 2D and 3D Modeling *(continued)*

			COMPATIBILITY			
		COST	WINDOWS	MAC	LINUX	NOTES
	Rhino 3D OSX	Free		x		The Mac version is free during development, but plug-ins designed for the Windows version will not work on the Mac version. It's generally intuitive to learn, easy to model basic shapes, and very capable for single part design.
	Rhino 3D	$995	x			$195 for students and teachers. Free evaluation version that will save 25 times.
	Vectorworks (Machine Design)	$1,795	x	x		
	3ds Max	$3,495	x			Free 30-day trial.
	Maya	$3,495	x			Free 30-day trial.
SOLID	BRL-CAD	Free	x	x	x	Open source.
SOLID	HeeksCAD	Free	x			Open source.
SOLID	*Alibre Design Personal	$99	x			When you download it, you get a 30-day free trial of the Pro version. When that expires, you can pay $99 to continue with the Personal version, which is still very capable.
SOLID	SolidWorks	$3,995	x			A great, professional solid modeling program.
SOLID	Autodesk Inventor	$5,295	x			A free student version is good for 12 months, but it will watermark all files as "not for commercial use."
SOLID	Pro/ENGINEER	$4,995	x			Considered a top-of-the-line, solid modeling program.
MATH	OpenSCAD	Free	x	x	x	Open source.
MATH	TopMod	Free	x	x	x	Open source.

Using assemblies also allows you to download CAD files of all kinds of off-the-shelf components directly from McMaster and other vendors. You can then assemble them on your screen before buying anything, so you make sure everything fits together perfectly. With 63,000 component CAD files currently available on McMaster alone, this can save you a ton of time.

Assemblies allow you to visualize what your final mechanism will look like, while keeping the part files separate from each other. This way, one part can represent an off-the-shelf motor, another part can be exported for 3D printing, and another can be made into a drawing to send to a laser cutter.

Project 9-1: Download and Open a 3D Model of a Part

Of the programs listed in Table 9-1, only a few include the built-in ability to make assemblies of parts: SolidWorks, Autodesk Inventor, Pro/ENGINEER, and Alibre Design Personal. Let's step through an example using the most affordable option: Alibre Design. Alibre's mission is "... providing full parametric CAD technology to anyone that needs it, versus only to those in the relatively unique financial position to afford traditional CAD systems."

Recipe:

1. Download Alibre Design from www.alibre.com. It starts with a free 30-day trial of the Pro version, and then you can choose to purchase the Personal edition (currently $99) to maintain functionality.

2. Go to the McMaster site (www.mcmaster.com) and find the part you want to download. As shown in Figure 9-2, I chose a standard 1/4-20 by 1 in long stainless steel socket head cap screw (92196A542). Check in the sidebar to see if 3-D Model is an option (if not, find another part that does have this option).

3. Click 3-D Model, and you'll get a drop-down menu that gives you options of 3D models or 2D technical drawings to download. Choose 3D STEP, IGES, or SAT—all of these work with Alibre Design. Download the part and save it somewhere you'll remember.

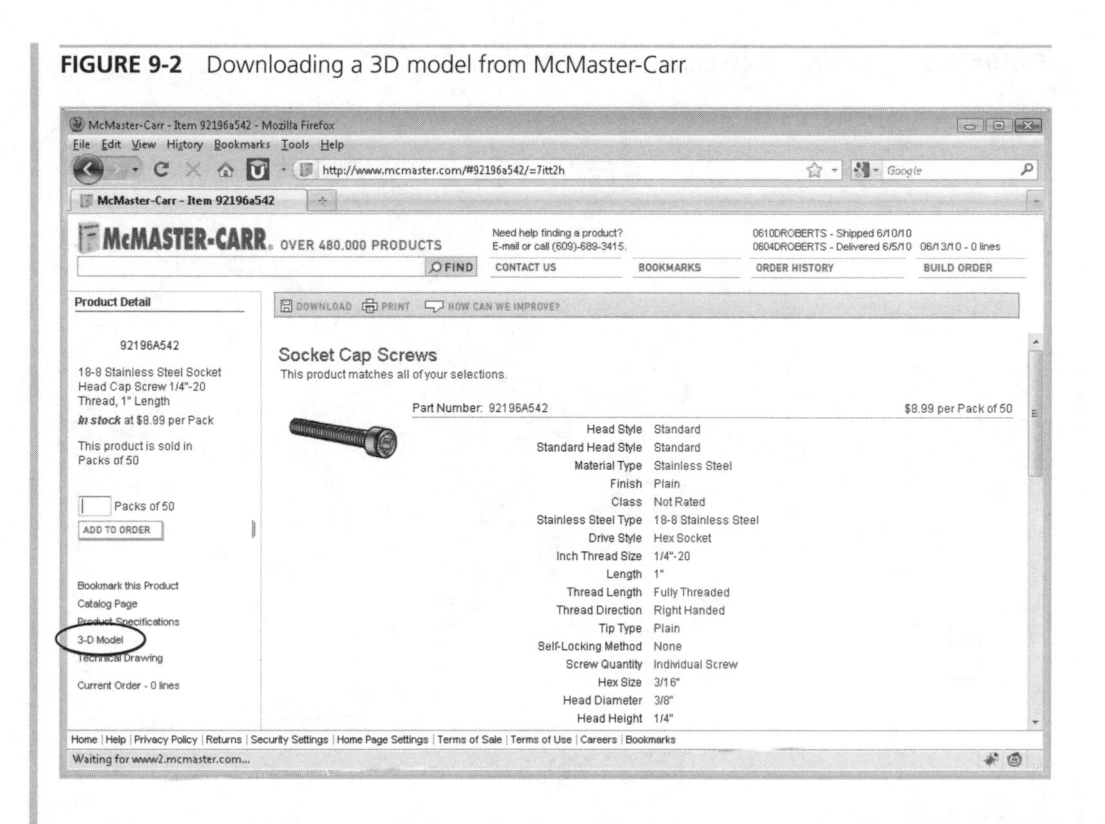

FIGURE 9-2 Downloading a 3D model from McMaster-Carr

4. Open Alibre Design. From the main screen, choose File | Import and look for the file you just downloaded. Open the file, and then click OK when asked about Import File Options.

5. Voila! The CAD file of the part should pop up on your screen, as shown in Figure 9-3. Use the Rotate button to spin the model around, and the Measuring tool (select Tools | Measurement Tool) to confirm the file was imported correctly.

FIGURE 9-3 Importing a CAD model into Alibre Design

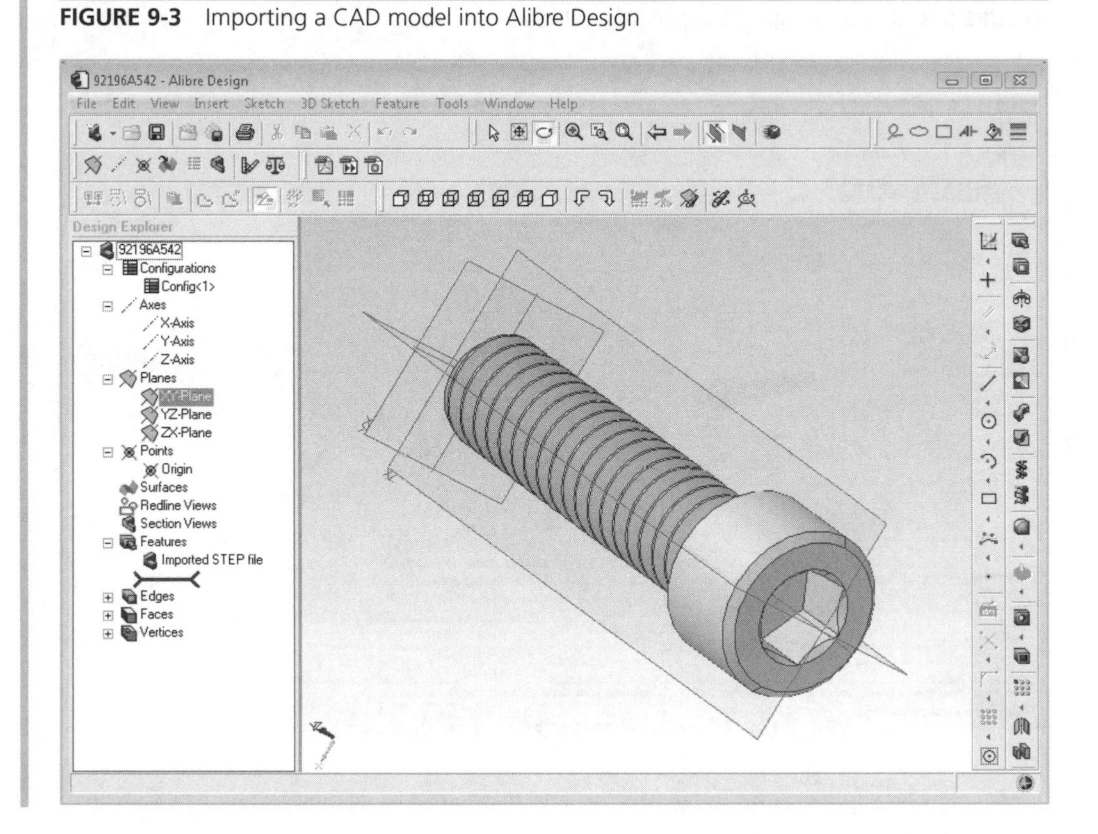

CAUTION ***Just because you can create a virtual assembly, doesn't mean you can create an actual assembly. As a frustrated intern at Make magazine put it (http://blog.makezine.com/archive/2010/01/interns_corner_makey_robots_sonar.html), "I'm trying to get the Arduino into the robot body. Suddenly I learn a profound lesson regarding computer-aided design. In real life, circuit boards cannot morph through walls into their desired resting place. In the computer, it happens all the time. With a simple motion of the mouse, the Arduino circuit board has glided into place, right through the aluminum robot body ... but in real life, it won't fit. There is no possible angle or tilt that will get the Arduino into the robot. Out come the vise-grips and hacksaw. I saw, bend, and twist off the offending aluminum tabs. This is reality-aided design."[1]***

Translation

In order to make anything, you need to translate your idea, sketch, or computer model it into something makeable. If you are doing this by hand, translation may be as easy as making a pencil marking on some wood before cutting. If you are using a digital fabrication technique, you might need to save the file in a different format than the default, or use software to translate a model or drawing into code that a fabrication machine can understand.

No matter which method you are using to make your part, you need to choose the material for it. This is an important step in translating the design from your paper or computer into something real. For example, you can't make a 3D printed part out of wood (yet), but you could laser-cut layers of wood and glue them together to create a 3D model. Refer to Chapter 2 for an extensive list of materials and their uses. In the fabrication discussion later in this chapter, we'll cover more ways to cut and work with different materials.

Analog Translation

If you have a design drawn on a piece of paper and you want to cut it out of a flat piece of material, you have at least three ways to do this:

- Trace the design on tracing paper with a pencil. Turn over the traced drawing and tape it to the material. Use the back of a spoon to press on the drawing lines and transfer the pencil to the material.
- Use the previous technique to transfer the drawing onto card stock, so you can cut it out and use it as a template.
- Take the original sketch (or a photocopy), spray the back lightly with spray mount adhesive, and then stick it to the wood, aluminum, cardboard, or other material you're working with. Now you have a template you can use to make your cuts.

CAUTION ***If you use spray mount adhesive, make sure your workspace has plenty of ventilation, and you may want to wear a mask. Spray mount doesn't taste good, and it has a tendency to get everywhere.***

Digital Translation

If you created a part using a solid modeling program (like Alibre Design), you can skip right to tool-path generation. If you used a mesh modeler (like Rhino), you may need to check or clean up your design first before sending a file to a fabrication machine. You can also create 3D objects to make with 2D methods by slicing or unfolding the model. Here are the digital translation possibilities:

- **Cleanup** MeshLab (an open source program for processing 3D meshes) and Blender allow you to clean up 3D files generated in mesh modeling programs. Sometimes models generated by these programs can be nonmanifold. This means that a fabrication machine might not know which surface is the inside and which is the outside, or be otherwise confused.
- **Unfolding** If you've designed a 3D part that you want to make out of 2D material or fabric, a few programs can figure out the unfolding or slicing for you. If you used Google SketchUp, you can download an Unfoldtool plug-in for free from http://sketchuptips.blogspot.com/2007/08/plugin-unfoldrb.html. Pepakura Designer is a low-cost program that breaks down 3D models into 2D panels that can be folded from paper to create the 3D object. Lamina Design and Rhino offer more unfolding options and flexibility (for a slightly higher price).
- **Tool-path generation** Tool-path generation programs can take 3D models and break them down into tool paths and layers that machines like laser cutters and 3D printers understand. Some options are ReplicatorG, Skeinforge, Pleasant3D, and SketchUp SliceModeler.

Fabrication

There are two ways to make something: do it yourself or get someone (or something) to do it for you. The problem with doing it yourself is that it can take a long time. To determine the actual amount of time it will take to make something, consider the *rule of pi*: multiply how long you think it will take by pi (3.14). This rule of pi is surprisingly accurate. So in the spirit of getting things done, whenever possible, get someone or something to make it for you.

Remember that DIY doesn't have to mean do it *all* yourself. Hack together off-the-shelf parts instead of making things from scratch. You can still breathe life into your mechanism during the integration phase, but if you spend too much time in fabrication, you may never get there. That said, there will be times when you need to make some simple parts yourself or modify store-bought parts to fit your needs. We'll go through a handful of useful tools for manual fabrication, and then cover a lot of ways to get custom parts made from digital files. This section covers subtractive methods (cutting away material) and additive methods (creating objects by adding sequential layers of material).

Analog Fabrication

Measure twice, cut once. Actually, make that measure once, go back and check your measurement calculations, measure again, and then cut. We'll cover a variety of ways to drill and shape materials, most of which you can do without expensive tools. However, if you need a tool you don't own and don't want to buy (a lathe, for example), look into local shared workspaces and shops, especially if you live near a big city—there's TechShop in San Francisco, 3rd Ward in New York City, and The Hacktory in Philadelphia, to name a few. Check the list of hackerspaces at www.hackerspaces.org for more. You can also find local machine shops to make custom parts for you.

Drilling

A portable drill and/or Dremel are handy tools to keep around. A Dremel tool is good for small holes in thin material, but a portable handheld drill is better for drilling bigger or deeper holes quickly, since it has much more torque.

The first step in drilling anything is to put on your safety glasses. Then secure the part you're working on by clamping it down to your working surface (try a C-clamp or two, such as McMaster 5133A15). If you're drilling into wood, you can just place the tip of the drill bit where you want the hole and start drilling. If you're drilling into metal or even plastic, it's a good idea to use a center punch (like McMaster 3498A11) or other sharp, hard object to make a little dimple for your drill bit to start. This will prevent the always frustrating outcome of the drill bit skipping or walking away from the intended starting point.

The next step up from using a hand drill or Dremel is to use a drill press (see Figure 9-4). The Dremel company makes a setup called the 220-01 WorkStation, which you can use to mount your Dremel tool and create a benchtop drill press for around $40. This allows you to drill holes perpendicular to your work surface as well as at set angles. More heavy-duty drill presses, such as those that sit on your table or stand on the floor, can be found at McMaster for a higher price tag.

FIGURE 9-4 A drill press

A deburring tool (such as McMaster 4289A35) is a handy tool to have available when you are drilling holes in metal. It has a sharp, pivoting head that cleans off the burrs, or little chips of metal, that your drill will leave at the opening and exit of a hole. If you don't have one, a small circular file or countersink tool will do.

Project 9-2: Drill a Centered Hole Without a Lathe

The first time you try to drill a hole in the center of a rod or shaft, you'll realize that a hand drill or even a drill press is not the best tool for the job. It's almost impossible to mark the true center of a rod, let alone drill exactly into that mark. A lathe is designed to work with circular parts, and it is the best tool to use if you need a hole exactly in the center of something. If you don't have access to a lathe, here is an example of how to best use a drill press to drill a hole in the center of something. (This project is based on Vik Olliver's blog post at http://vik-olliver.blogspot.com/2010/02/drilling-down-middle.html.)

Shopping List:

- Drill press with a vise (preferably with a small notch in the middle of the jaws) mounted to the base
- Drill bit
- Safety glasses
- Cylindrical part to drill a hole through
- WD-40 or other lubricant/coolant (if your item is metal)

Recipe:

1. Put on your safety glasses and clear your workspace.
2. Put the drill bit in the chuck the wrong way around and tighten the chuck around the smooth base section.
3. Lower the drill press so the bit can be clamped in the vise.
4. Adjust the vise as necessary to accommodate the bit, tighten the vise, and bolt it to the drill press base (see Figure 9-5).

FIGURE 9-5 Drill bit clamped in vise (credit: Vik Olliver)

5. Loosen the chuck so the bit is free. Slowly raise the drill press.
6. Place the rod, shaft, bolt, or whatever you want to drill a hole in the center of into the drill chuck. Tighten it and make sure it is still aligned with the bit by lowering it down for inspection (see Figure 9-6).
7. If your part is metal, dab some WD-40 or other lubricant/coolant onto the drill bit.
8. Turn on the drill press. Using high speed and very little pressure, lower the drill press with your part in the chuck onto the bit. It might vibrate and skip a little initially, but keep pushing until the bit starts drilling and finds the center.
9. Slow down the drill and apply more pressure.
10. Turn off the drill press and inspect your work.

FIGURE 9-6 Part in drill chuck ready to go (credit: Vik Olliver)

CAUTION ***If your part is long and you need to back the drill bit out to clean it off, do so with something other than your finger to avoid getting cut or burned from touching a hot drill bit.***

11. If your part is longer than the drill bit, remove it from the chuck and reverse it. Double-check to make sure you didn't knock the drill bit off center, and then mount the piece in the chuck in the opposite direction. Repeat the preceding steps to complete the through hole.

Working with Round Parts

A tube cutter is the most economical way to cut tubes without deforming them. If a C-clamp and a pizza cutter had a kid, it would be a tube cutter. The clamp keeps the tube or rod positioned against a sharp, rotating blade, so you can cut into the tube evenly without crushing it and get a clean edge.

If you will be doing a lot of work with circular parts and shafts, a lathe can make your life much easier. Professional lathes can cost tens of thousands of dollars, but small hobbyist-style lathes start around $700 (from Micro-Mark at www.micromark.com/MICROLUX-7X14-MINI-LATHF,8176.html, for example) and will handle most small jobs easily. A lathe is ideal for drilling holes in the center of rods and shafts or removing a tiny bit of material from the outside of a shaft so it fits perfectly into a wheel or bearing. Metal lathes are meant for precision work, but wood lathes are designed for more artistic-type work, such as fence posts and baseball bats.

Cutting

You can use many tools to cut things. After scissors, a close second is an X-Acto knife, followed by the knives on multitools (such as Leatherman products). A Dremel tool with a cutting wheel can handle small jobs in wood, plastic, and softer metals like brass and aluminum. Tin snips or sheet metal snips (like McMaster 3902A1) are good for thin metal jobs, and a hacksaw can be used for larger or thicker pieces. For even larger pieces, a band saw is a common piece of equipment to have in a shop. It comes in vertical and horizontal configurations, and can be used to cut just about any material if you can adjust the speed and the blade.

CAUTION ***When cutting paper or cardboard (or anything), keep all body parts out of the path of the knife. This may seem obvious, but if you've ever held some paper with your thumb flared out and then cut right into it, you know that safe cutting practices don't always come naturally. Save the dissection for biology class and make safety a priority before you cut anything.***

Casting and Molding

To cast a part, you first need to create a mold around it that will have the imprint of the part in it. The original part is then removed, sometimes by cutting the mold or with the help of some kind of mold release. Then the cavity in the mold can be filled to create a *positive* cast that matches the original part. This is generally accomplished by pouring a liquid plastic compound into the mold and letting it cure (harden) before removing it from the mold. Casting is an excellent way to clone an off-the-shelf or 3D printed part. It can be very economical if you need to make several copies of the same part.

One good method is to use silicone rubber for the mold (like Mold Max from www.smooth-on.com), and then a liquid plastic casting compound like Smooth-Cast 300 to make the positive cast. Smooth-Cast has a pretty fast setting time, a 1-to-1 mixing ratio, and it's easy to work with. You can use dyes to create any color you want, and it can be painted and machined when cured. The Compleat Sculptor (www.sculpt.com) is a great source for all of these materials, and the shop frequently holds classes in its New York City store.

For the preparation of the negative mold, the main objective is to make sure there are no air bubbles and that the mold can be separated in a logical way. Before the silicone rubber is poured around the original part, you can glue small wooden dowels to it to create channels that will allow air bubbles to escape. A larger, primary dowel also needs to be glued to the part somewhere in the center to create a pour hole.

Working with Wood

Wood can but cut and manipulated with many of the tools mentioned earlier, but there are some additional tools that are specific to woodworking. Planers come in hand-operated and powered versions. They take ragged pieces of wood and shave off the top surface until it is parallel with the bottom surface.

CAUTION ***Don't saw or plane used or scavenged wood. Old leftover nails and screws might be embedded, and that could be dangerous.***

Although hacksaws and coping saws are good for small jobs by hand, power tools like table saws, jigsaws, routers, miter saws, and circular saws can make big or repetitive jobs much easier. Low-end versions of these will neither leave you broke nor take up much room—even table saws come in extra small.

Make sure you have files and sandpaper around to finish your cuts and avoid splinters.

Working with Metal

A few key tools and machines are used to work with metal. We've already talked about drill presses, which you can use with a variety of materials. A more advanced tool that's similar to a drill press is a mill.

As mentioned in Chapter 7, a milling machine is a fancier version of a drill press where the base moves in the x, y, and z axes, so you can do more than just drill straight down. Although you may never use this machine yourself (it's large and expensive), it's helpful to keep it in mind when you're designing parts you'll need to have custom made. You can use mills with normal drill bits and also with endmills, which are like drill bits with the tip cut off so they can create holes with flat bottoms and cut along the side of the bit as well. If you do want to try your hand at milling, you can get a very capable mini-mill from LittleMachineShop.com for around $650 plus the cost of some basic tools.

Just as you can wrap a paperclip around a pencil with a little work, you can bend many shapes, sizes, and thicknesses of metal in various ways if you have the right tools. All metal forming works like this. You just need the piece of metal and something to wrap around or form it to. Clamps help to hold a metal piece to the form initially, and a rubber mallet or other nonmarring hammer can help convince the metal to bend as you want. For 90° bends, you can clamp sheet metal in a vise and bend it by hand or with the help of a hammer.

Working with Plastics

The same tools and techniques for working with wood and metal also work with plastic. Many power tools (such as jigsaws and circular saws) have blades specifically made for plastics.

If you're drilling into plastic, especially a type that cracks easily, you can get drill bits made for plastics that will decrease that probability. For cutting thin plastics, scoring it with a knife or other sharp blade can give you a clean edge to break the part along.

Digital Fabrication

For digital fabrication, there are a growing number of ways you can use digital files to create parts directly, both in 2D and 3D.

2D Methods

One method of 2D digital fabrication that everyone is familiar with is your regular desktop inkjet or laser printer. You can print designs on paper to cut out of other material, or print designs onto thicker card stock that you can actually use to make things (see Project 8-2).

The next step up from using a machine to print on something is to use a machine to cut out something. You can do this on vinyl with a vinyl cutter. These machines start at about twice the size and a few times the cost of your average home inkjet printer.

The next step up in 2D digital fabrication is a computer numerically controlled (CNC) router. A CNC router allows you to create a digital design using CAD software, and then upload it to the machine, and the router will cut your material by following the lines and contours in your model. ShopBot (www.shopbottools.com) makes low-cost versions for small businesses and hobbyists, which are used mainly for wood, but the entry-level price is still more than most individuals can handle. However, if you have a community shop near you, you might find one there.

If you made your own gears in Project 7-1, you're already familiar with a very popular 2D digital fabrication technique: laser cutting. In that example, we used Ponoko to cut gears for us, but you can find plenty of other shops online that do similar custom work. The type and thickness of material you can cut depend on the strength of the laser. In Figure 9-7, you can see how an Eyebeam resident, Ted Southern, used a laser cutter to cut fabric patterns that were later assembled into a prototype spacesuit glove (see www.finalfrontierdesign.com).

3D Methods

As with 2D digital fabrication work, CNC routers and CNC mills can be used for 3D projects. Routers are designed to do mostly 2D work, but they do have a third axis for

FIGURE 9-7 Spacesuit glove patterns cut out with Eyebeam's V-660 Universal Laser Systems cutter (left) and assembled spacesuit glove prototype (right) (credit: Nikolay Moiseev and Ted Southern).

small 3D thicknesses. CNC mills can make all kinds of shapes in all kinds of sizes out of a variety of materials.

3D printing is the new trend when it comes to manufacturing things quickly. The various systems and machines accomplish it in different ways, but the end result is a real 3D thing that started life as just a CAD model. Engineering and product design companies use these machines to visualize parts and assemblies and troubleshoot designs before final production runs, as well as to impress potential investors. You can use these machines to print actual usable parts in Projects 10-1 and 10-3 in Chapter 10.

In the open source, low-cost arena, MakerBot Industries (www.makerbot.com) is leading the pack. Their machines make parts by depositing super-thin strands of plastic in layers that stack on top of each other until the part is complete. Industrial-scale machines made by companies like Stratasys do the same thing, but not for around $1,000.

If you don't have access to a real 3D printer, you certainly have access to virtual ones. Plenty of companies will give you instant online quotes when you upload a model (check www.solidconcepts.com and www.shapeways.com) and can be good solutions if you just need one or two parts made.

Another method for making functional parts is called stereolithography (SLA). It uses light to cure a special plastic resin in layers, so a solid part rises up out of a pool of goop. All these parts end up a whitish or yellowish tint, since the base material needs to be light-curable. Other 3D printing machines use different kinds of powders along with some kind of binder or heat to melt the powder together in layers. A commercial example is Z Corp, and an awesome example is CandyFab from Evil Mad Scientist Laboratories, which prints 3D objects out of layers of melted sugar.

Integration

Integration is where all the off-the-shelf motors, nuts, and bolts come in to create a moving thing out of your pile of parts. This is usually the most fun and frustrating step in making things move. The rule of pi applies here as well.

NOTE ***Here are some words of wisdom to keep in mind: If it moves and it shouldn't, use duct tape. If it doesn't move and it should, use WD-40.***

Analog

You're familiar with tools used to assemble things by hand. Screwdrivers, hammers, clamps, wrenches, and the like need no introduction. As discussed in Chapter 7, shims of various materials and sizes are always good to have around as well, as they fill in gaps.

Digital

While you can simulate an assembly digitally through 3D modeling software, the only digital way to assemble real-world parts is with a robot. Since we don't have thousands of dollars for assembly-line robots and pick-and-place machines, this means we're usually stuck with analog assembly. You can automate this process a bit by making exploded views of assemblies if you used CAD software in your design phase.

Proliferation

Proliferation is the phase where you share what you've done. Show it, teach it, get feedback, sell it, make it better, and then start on the next iteration—or inspire someone else to—and close the loop on the making things move ecosystem.

FIGURE 9-8 Closed and open sharing models

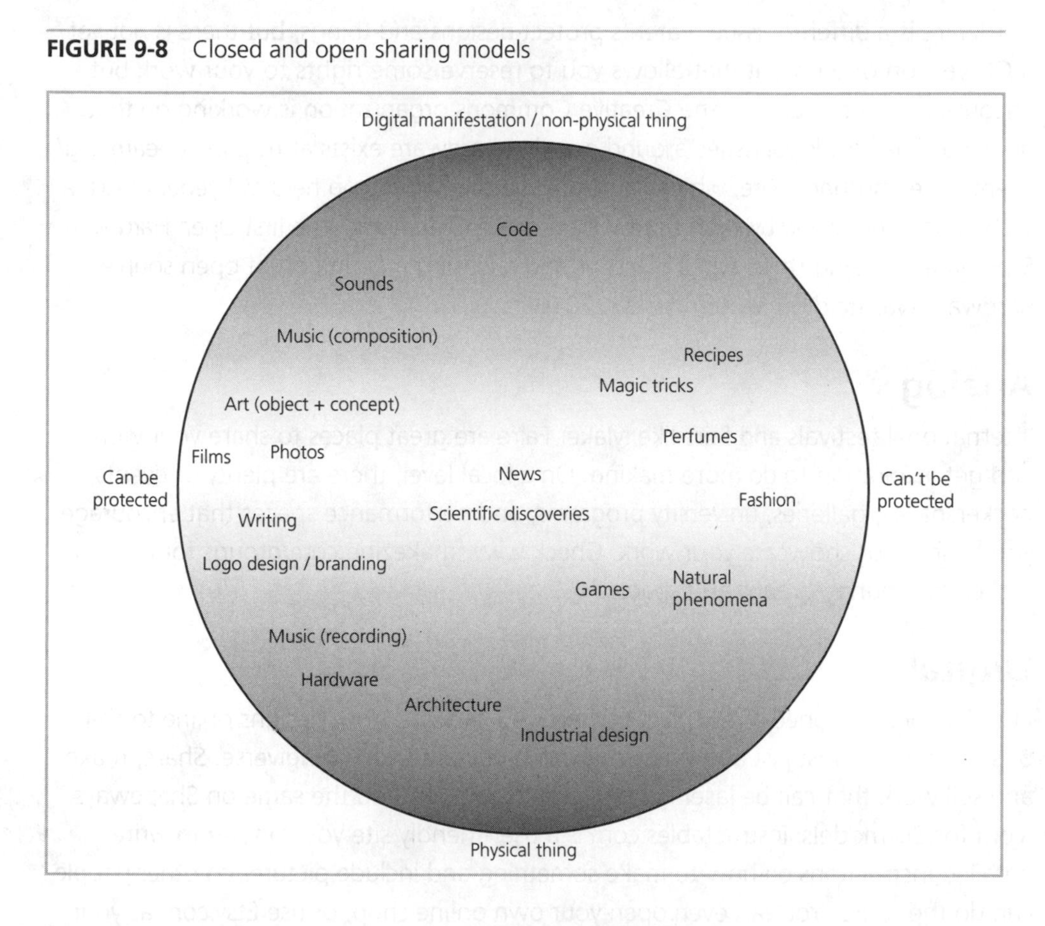

Some things can be protected by copyright or patents, and some can't. Figure 9-8 (inspired by a graphic in a talk by Johanna Blakely at TEDx USC; see www.boingboing.net/2010/05/26/why-the-absence-of-c.html) provides an overview of the situation. However, there is a growing open-culture movement of people and organizations who *choose* to share their work in less restrictive ways.

If you've ever used Mozilla's Firefox browser, you're familiar with the product of an open source software project. For digital files like photographs and online content, Creative Commons is a nonprofit organization that increases sharing and improves collaboration. Using Creative Commons (CC) licenses on your work (instead of © for copyright) can allow you to share, remix, and reuse other people's work legally. If you try to upload a 3D model to Thingiverse (www.thingiverse.com), you'll get to a section where you can choose a CC license for your work to encourage sharing.

Hardware is a different issue. Patents protect designs and things, but there is not yet a CC version of a patent that allows you to reserve some rights to your work but encourage sharing. Luckily, the Creative Commons organization is working on that. A good summary of legal issues around opening hardware exists at http://eyebeam.org/events/opening-hardware, which is the archive of a workshop held at Eyebeam Art + Technology Center, led by Ayah Bdeir with Creative Commons. The first Open Hardware Summit addressing these types of issues and releasing a definition of open source hardware was held on September 23, 2010.

Analog

International festivals and fairs like Maker Faire are great places to share your work and get inspiration to do more making. On a local level, there are plenty of events, hackerspaces, galleries, university programs, and performance spaces that encourage you to share or showcase your work. Check www.makezine.com/groups for a list of maker community groups and spaces.

Digital

I've already mentioned a few places where you can post your designs online to share or sell. You can post just about any makeable digital file on Thingiverse. Share, make, and sell work that can be laser cut on Ponoko.com, and do the same on Shapeways .com for 3D models. Instructables.com is a user-friendly site you can use to write detailed instructions on how to make something and include pictures so other people can do the same. You can even open your own online shop, or use Etsy.com as your storefront.

Reference

1. Saul Griffith, "Simply Cad," *Make Magazine* (Sebastopol, CA: O'Reilly, Volume 6).

Projects

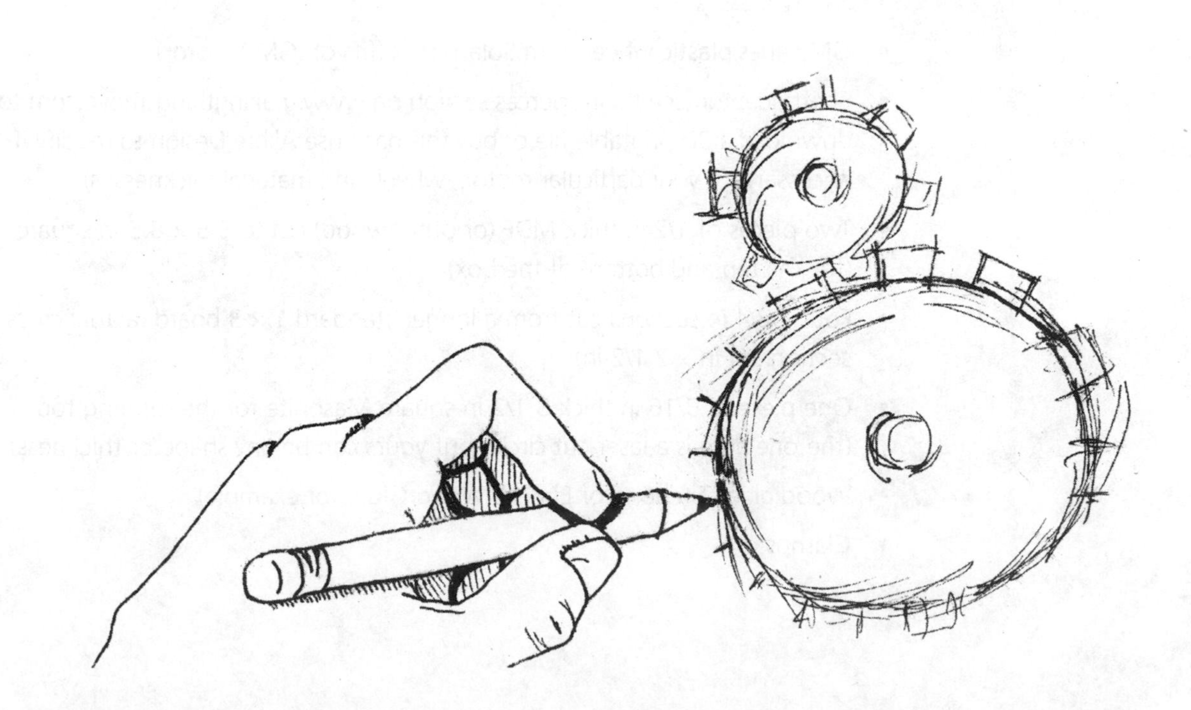

The projects below draw on the skills you've learned in several of the chapters in this book.

Project 10-1: Not Lazy Susan

In this project, we'll use a lazy Susan (also called a turntable or thrust bearing) to create a rotating platform. To make things interesting, we'll use an infrared (IR) LED and phototransistor to make the table rotate with just a wave of your hand.[1,2] You can use this as a table to magically serve food to dinner guests, or to make a fun interactive centerpiece for the next wedding you plan.

If you did Project 6-7 with a transistor and Project 6-8 with a photocell, this will be easy. If not, don't worry—we'll still go step by step. We'll build the box first, then the circuit, and finally integrate the two. As always, it's a good idea to skim through the steps and take a look at the pictures before getting started.

Shopping List:

- Hardware
    - Safety glasses
    - GM series plastic wheel from Solarbotics (fits on GM9 motor)
    - Shaft adaptor (see the resources section on www.makingthingsmove.com to download a 3D printable file or buy this part; use Alibre Design to modify if necessary for your particular motor, wheel, and material thicknesses)
    - Two pieces of 1/2 in thick MDF (or other wood) cut to 8.5 × 8.5 in square (for the top and bottom of the box)
    - Four 7 3/4 in sections cut from a longer standard 1 × 3 board (actual cross section 3/4 in × 2 1/2 in)
    - One piece of 3/16 in thick 8 1/2 in square Masonite for the rotating top (the one here is a laser-cut circle, but yours can be any shape or thickness)
    - Wood glue (Titebond or Elmer's Wood Glue, for example)
    - Clamps

- Files, sandpaper, or (preferably) a Dremel tool
- Lazy Susan (McMaster 6031K17)
- Six #4 flat head wood screws 1 1/4 in long (McMaster 90031A117)
- Eight #4 flat head wood screws 1/2 in long (McMaster 90031A110)
- Phillips head screwdriver
- Drill (portable or drill press) and drill bits: 1/2 in, 1/4 in, 1/8 in, and 1/16 in
- Epoxy putty

- Electronics
  - DC gearhead motor with about 12 in hook-up wire leads soldered to terminals (an old Solarbotics GM9 used here)
  - Corresponding battery (9V used here) and snap or holder with wire leads (like RadioShack 270-324)
  - Four AA batteries and holder (like SparkFun PRT-00552)
  - Breadboard (like All Electronics PB-400)
  - Jumper wires (like SparkFun PRT-00124) or hook-up wire to make your own
  - Black electrical tape (McMaster or any hardware store)
  - TIP120 Darlington transistor (Digi-Key TIP120-ND or Jameco 32993)
  - Phototransistor (All Electronics PTR-1)
  - IR LED (All Electronics ILED-8)
  - On-off toggle switch (like SparkFun COM-09276) or other SPST switch with about 4 in hook-up wires soldered to the legs
  - Diode (SparkFun COM-08589)
  - 22KΩ resistor (RadioShack 271-1339)
  - 100KΩ resistor (RadioShack 271-1311)

Box Recipe:

1. Use a hacksaw, band saw, or table saw, or ask your local lumber or hardware store to cut the wooden squares and side pieces. Sand, file, or Dremel any rough or splintered edges.

FIGURE 10-1 Clamp the box sides while the glue dries.

2. Arrange the four wooden sides of the box as shown in Figure 10-1. Apply wood glue to the mating surfaces, and clamp two at a time together until dry. You should have two L-shaped pieces now.

3. Apply wood glue to open ends and clamp together these two L-shaped pieces to complete a box. If your boards aren't perfectly square, you might benefit from additional clamps, as shown in Figure 10-2. Let the glue dry while the pieces are clamped.

4. When the glue is dry, coat the bottom of this box with wood glue, and then stick it to one of the wooden squares as the base. Let it dry.

FIGURE 10-2 Using additional clamps to square up the box sides while the glue dries

5. Put on your safety glasses. In the other wooden square, drill 1/8 in clearance holes in each corner for the #4 wood screws you'll eventually use to close the box. You can just lay the square on top of your box and eyeball locations with a pencil. Refer to Figure 10-3 to see these holes. Use a scrap piece of wood behind the square so you don't drill into your table, or use a drill press. Either way, clamping down the material before you drill is a good idea.

6. Find and mark the center of the board. Drill a hole with the 1/8 in drill bit. Chase this with the 1/2 in drill bit (the pilot hole makes it easier to drill a hole with the larger bit).

FIGURE 10-3 Mounting the lazy Susan bearing

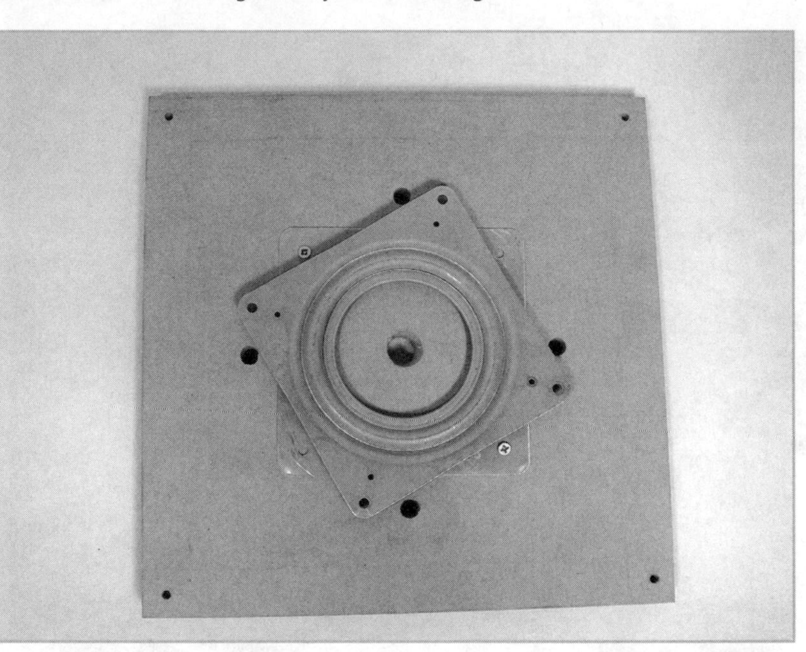

7. Center the lazy Susan bearing on the top wooden square. Mount the bearing with the shiny ring side down with four 1/2 in long #4 wood screws. If you have trouble driving them in by hand with the screwdriver, mark and drill 1/16 in pilot holes, and then try again. The mounted bearing with two screws is shown in Figure 10-3.

8. In order to eventually mount the rotating tabletop, you'll need to get access to the four mounting holes in the corners of the turntable that you haven't used yet. To drill access holes, rotate the bearing 45° or so, until you can look through the holes and see only wood underneath. Mark these locations, and then use the 1/4 in drill bit to drill these holes. The access holes are the four unused holes in Figure 10-3 around the edge of the lazy Susan.

9. Prepare the rotating tabletop by cutting it or ordering it online. See links in the resources section of www.makingthingsmove.com if you want to download the digital file to make a template, cut it yourself, or buy it.

10. Mount the motor. First, slide the shaft adaptor on your motor and stick it through the centered hole you drilled earlier. It should look like Figure 10-4.

11. Place the plastic wheel on the shaft adaptor. The wheel should be centered on the bearing and will be resting on top of it (and the shaft adaptor) at this point (see Figure 10-5). Adjust the position until the wheel is centered and the shaft adaptor is not rubbing against the inside of the 1/2 in hole you drilled earlier.

FIGURE 10-4 Shaft adaptor centered in the box top

FIGURE 10-5 Wheel/bearing/shaft adaptor/motor sandwich, centered

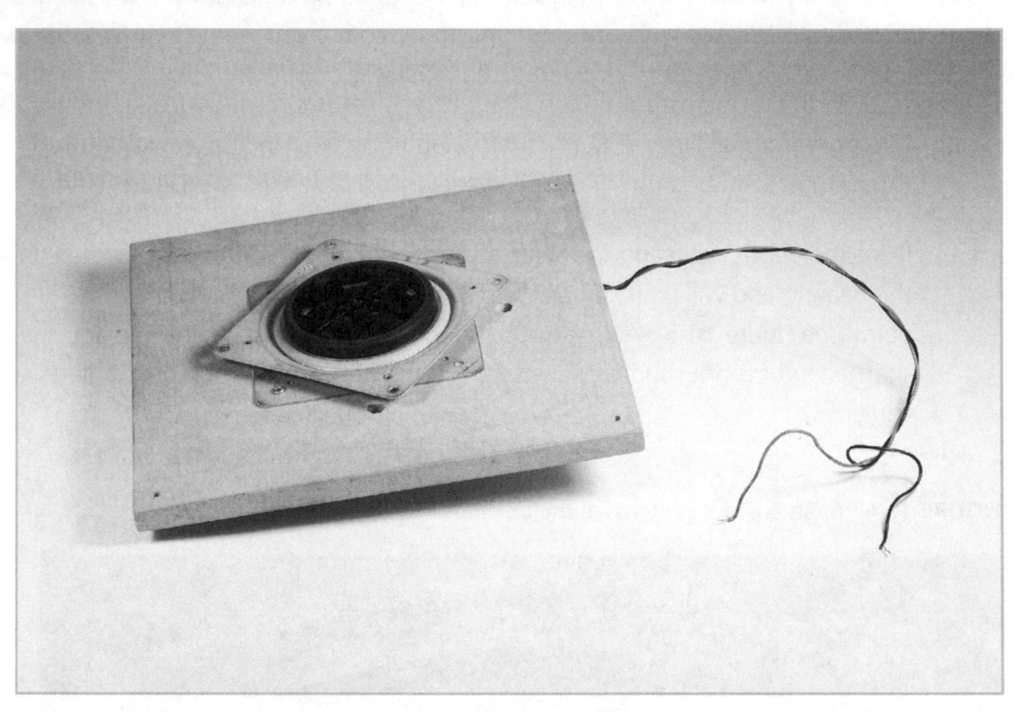

12. Turn the box top over and mount the motor using two of the 1 1/4 in long #4 wood screws. You may want to drill 1/16 in pilot holes in the wood after you've marked the locations to drill. Your motor should sit flat on the board, as shown in Figure 10-6. In order to make that happen, we removed the strap that held on the motor and used a diagonal cutter to remove the hook that the strap attached to. Also notice how the wires are insulated with hot glue and strain-relieved. This is important, especially for this project.

13. Check to make sure your motor, shaft adaptor, and wheel are still centered on the bearing. Place the rotating tabletop down on your work surface. Rotate the bearing and wheel on the box top so you can see the exposed holes in the lazy Susan you're about to use to mount the rotating tabletop through the 1/4 in access holes you drilled earlier.

FIGURE 10-6 Motor mounted on the underside of the box top

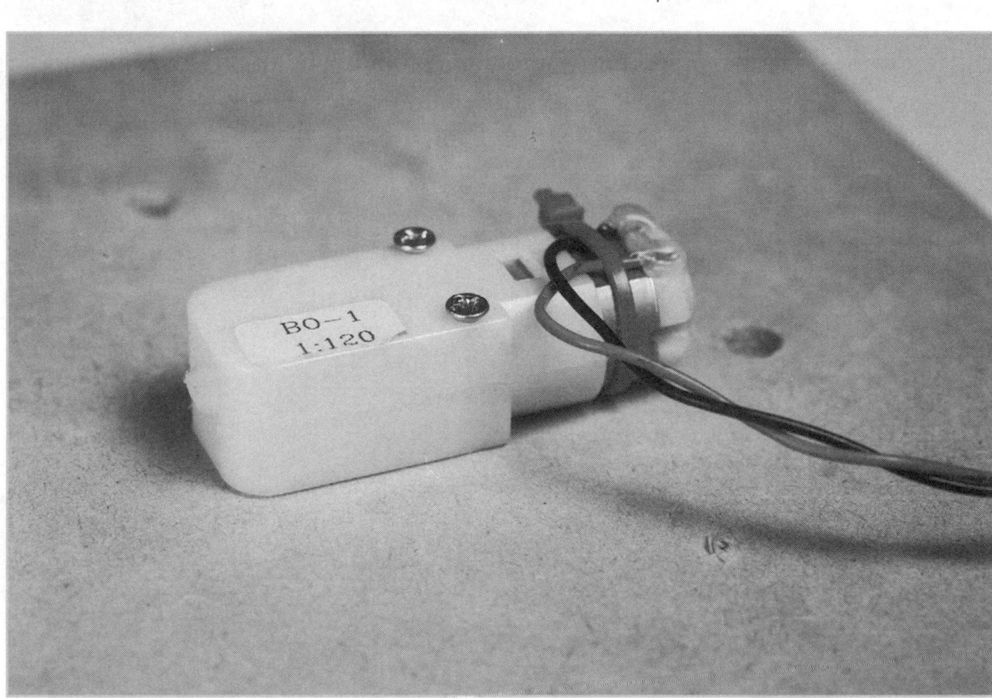

14. If you're using the laser-cut Masonite, you can just drop in the 1/2 in long #4 wood screws in each hole and mount the tabletop. The stackup should look like Figure 10-7. If you cut your own tabletop, you may want to mark and drill 1/16 in pilot holes to make this mounting step easier. Alternate between the four screws as you screw into the tabletop to ensure you're sandwiching the plastic wheel evenly and not distorting the bearing.

15. Good work! Set aside the tabletop and box assemblies for now and move on to the electronics.

FIGURE 10-7 Rotating tabletop mounted on box top

Electronics Recipe:

1. Connect the 5V power and ground on one side of the breadboard. On the other side of the breadboard, connect a 9V battery to power and ground. Make sure the ground is linked between both ground columns on the breadboard. The circuit will look like Figure 10-8 when finished, so keep this configuration in mind as you work through the project. Build the circuit on your desk before putting it into the box in the integration later.

FIGURE 10-8 The completed circuit

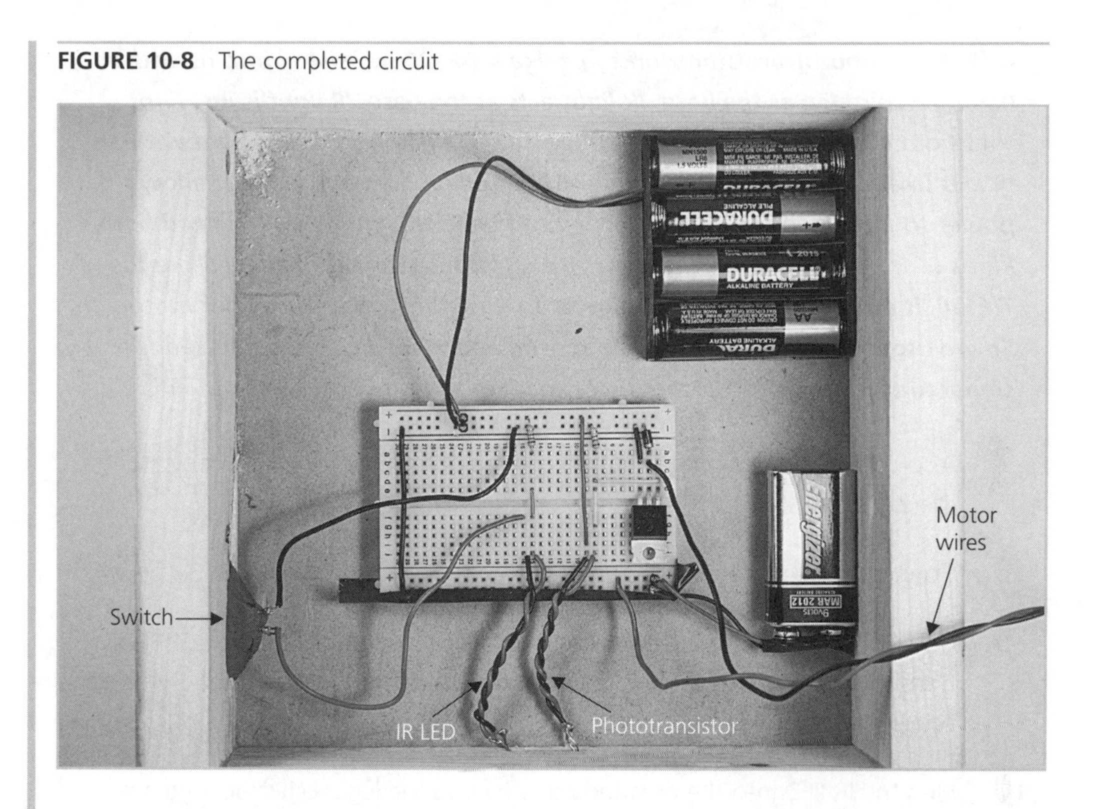

2. Plug the TIP120 transistor into the breadboard so each leg has its own row.

3. Connect the collector (middle pin) of the TIP120 transistor to ground through the diode. Make sure that the diode is pointing in the right direction, with the stripe mark closest to the middle of the board.

4. Connect the emitter pin of the TIP120 transistor to ground.

5. Plug the phototransistor into the breadboard and bend the legs so it faces off to the side. It's a good idea to solder hook-up wire to the phototransistor legs so you can adjust the position later. The wires coming from the phototransistor legs can be seen in Figure 10-8 (the pair of twisted wires closest to the TIP120 transistor).

6. Connect the base pin of the TIP120 transistor to the long leg of the phototransistor. This is the phototransistor's emitter, and the short leg is the collector.

NOTE ***A phototransistor works just like a TIP120 transistor, but instead of having a third leg as the base, IR light acts as the base. IR light is just light that's out of the range of colors we can see. We'll generate this light with an IR LED in a minute. When you shine IR light on a phototransistor, it allows power to flow through it from collector to emitter. So why do we need both kinds of transistors? Even though the phototransistor here behaves like the TIP120, it doesn't allow enough power to flow through it to run our motor. So we use the phototransistor as a sensor, and when light hits it, it sends a signal to the bigger TIP120 transistor to let motor power flow.***

7. Also connect the long leg (emitter) of the phototransistor to ground through the 22KΩ resistor.

8. Connect the short leg (collector) of the phototransistor directly to 5V power.

9. Connect one of the motor wires to the collector (middle leg) of the TIP120 transistor on the breadboard. Connect the other motor wire to the 9V battery power column.

10. Plug the IR LED into the breadboard and bend the legs so it's facing the same direction as the phototransistor. It's also a good idea to solder hook-up wires to the IR LED and use those to plug into the breadboard, so you can adjust the position of the LED later. The wires from the IR LED legs are the twisted pair in Figure 10-8 right at the center of the breadboard, on the bottom.

11. Connect the short leg of the IR LED to ground through the switch (in the off position). Connect the long leg of the IR LED to the 5V power column through a 100Ω resistor.

12. Flip the switch to on. Put your hand or a white piece of paper in front of the IR LED and phototransistor. The closer you get, the more light from the LED will bounce off your hand into the phototransistor, and the faster the motor will spin! If you get far enough away, it won't spin at all. Don't believe me that the IR LED is actually on? If you have a camera phone, pretend you're about to take a picture of your setup. Although our eyes can't see IR light directly, the

camera phone can, and you should see a faint glow on your screen. (In the pictures in this section, I cheated and used a white LED for emphasis; your IR LED won't look that bright.)

Integration Recipe:

1. Pick up your whole circuit and plop it into the box. Make pencil marks where the LED and phototransistor touch the front of the box. Do the same thing with the switch. Remove the circuit from the box.

2. Use the 1/4 in drill bit to drill holes for the LED and phototransistor to look through opposite the marks you made in step 1 (see Figure 10-9). This is easiest to do with a portable hand drill while supporting the box on the corner of a table. Clean up any splinters with sandpaper or a Dremel tool.

FIGURE 10-9 IR LED (left) and phototransistor (right) peeking through drilled holes

3. Drill a hole for the switch opposite the mark you made in step 1 using the 1/2 in bit. It's best to start with a 1/8 in pilot hole here first. Use the Dremel tool or a file to remove some material from the inside of the hole—just enough so the switch can be flipped from the outside (see Figure 10-10).

4. Once those holes are drilled, you can return the whole circuit to the inside of the box. Make sure your motor wires are still attached and none of your other connections came loose.

5. Stick the IR LED and phototransistor through the holes you drilled, and test to make sure your circuit still works. We found that ours worked best with the phototransistor wrapped in black electrical tape and pushed through the hole so it was flush with the outside of the box. We also sanded the surface of the IR LED a little to diffuse the light, and stuck that out farther past the outside of the box (see Figure 10-7).

FIGURE 10-10 A view of the switch from the outside

6. Use the epoxy putty to secure the switch, and put electrical tape under any other components you want to secure.

7. Once everything works, it's time to put the top on the box! Use four 1 1/4 in long #4 wood screws and mount the rotating tabletop assembly to the corners of the wooden box. Your project should now look like Figure 10-11.

8. Find a vase of flowers or a plate of food to put on the rotating tabletop. Then when someone asks you if it's a lazy Susan, just wave your hand and show them how not lazy it actually is!

FIGURE 10-11 Not lazy Susan

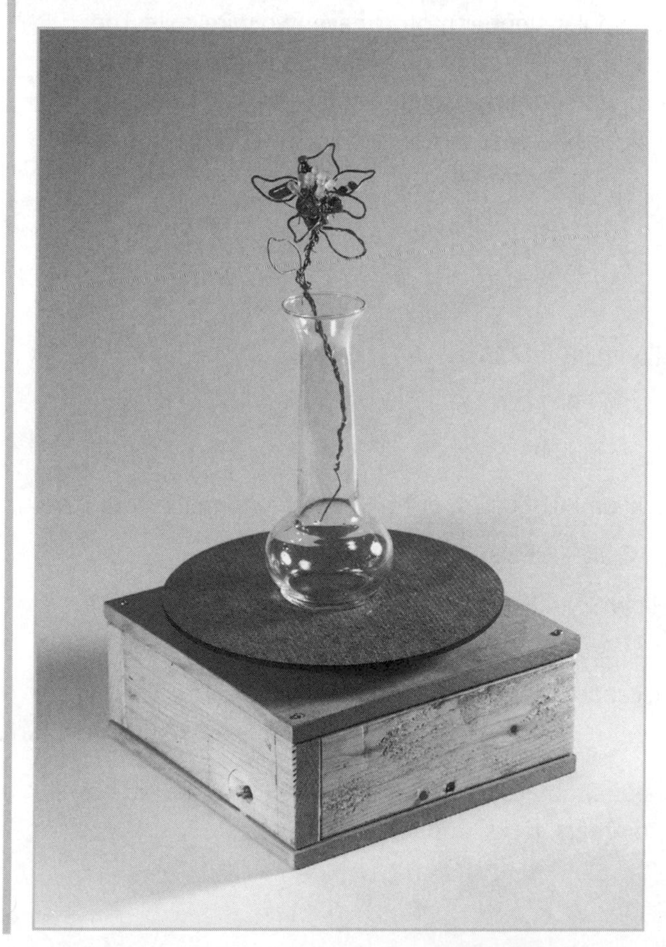

Project 10-2: Wind Lantern

In this project, we'll build a small, vertical-axis wind turbine, or VAWT for short. These are not as efficient as their horizontal-axis cousins, but they are better suited to urban environments where wind can come from all different directions.

Normally, when you give electricity to a motor, it spins. The same is true in reverse: If you give a motor a spin, it acts as a generator and creates electricity. The wind lantern will use energy from the wind to turn a motor and the resulting energy to light up some LEDs within the base. The wind lantern will use this electricity to create a flickering, glowing indicator of the wind.[3,4]

You already know that LEDs and diodes allow current to flow through them in only one direction. You also know that bipolar stepper motors have two wire coils. The challenge here is to design a circuit that directs energy generated in each coil through an LED in the correct direction, no matter which way the lantern spins. To do this, we'll build a rectifier circuit for a bipolar stepper motor (see Figure 10-18 in the recipe steps).

Shopping List:

- Electronics
  - Stepper motor (SparkFun ROB-09238)
  - Male header pins (SparkFun PRT-00116)
  - Breadboard (like All Electronics PB-400)
  - Jumper wires (like SparkFun PRT-00124) or hook-up wire to make your own
  - Eight diodes (SparkFun COM-08589)
  - One or more LEDs (yellow SparkFun COM-09594 used here, but choose any color)
  - One or more 1,000 μF capacitors (SparkFun COM-08982)
- Hardware
  - 1/4 in acrylic plastic sheet about 15 × 30 in or equivalent (size based on Ponoko P3 template) for gears, disks, and sail holder pieces

NOTE ***See the resources section of www.makingthingsmove.com for links to a file you can download and cut yourself (either with a laser cutter or to use as a template), as well as a link where you can buy the pieces.***

- 10 in wide aluminum flashing (usually sold in rolls; you need about 2 ft length for this project)
- 5mm bore shaft collar with set screw (McMaster 57485K65)
- 7 1/2 in bore shaft collars with set screws (McMaster 6166K25)
- 18 in length of 1/2 in outer diameter aluminum tube (McMaster 1658T45 is 8 ft long but a good value if you have the means to cut it down to 18 in—a hacksaw will work)
- Two flanged sleeve bearings for 1/2 in shaft diameter (McMaster 2938T12)
- Thrust bearing cage assembly for 1/2 in shaft diameter (McMaster 5909K31) with two matching washers (McMaster 5909K44)
- Three female threaded standoffs, 4 in length, 1/4 in -20 screw size (McMaster 92230A350)
- Six socket head cap screws, 1/4 in -20 thread, 3/4 in length (McMaster 92196A540)
- Six lock washers for 1/4 in screw size (McMaster 92146A029)
- Six flat washers for 1/4 in screw size (McMaster 92141A029)
- Four M3 screws 40mm long (McMaster 91292A024)
- Four M3 lock washers (McMaster 92148A150)
- Four M3 washers (McMaster 91116A120)
- Set of inch and metric hex keys (like McMaster 7324A18)
- Hacksaw (like McMaster 4077A1)
- Deburring tool (like McMaster 4289A35) and/or rounded file
- Coarse sandpaper
- Epoxy putty

Recipe:

1. Order or make the gears, disks, and sail holders. See the Resources section at www.makingthingsmove.com for templates and links to where you can order these.

2. Put on your safety glasses and cut an 18 in length of the aluminum rod with a hacksaw. Use a deburring tool or file on the inside and outside of the end of the rod to smooth it and avoid cutting yourself.

3. Make sure your aluminum rod fits through the flanged sleeve bearings, thrust bearing and washers, and the shaft collars. Remember that issue of tolerances you learned about in Chapter 2? Look at the tolerances of all the parts:

   - The aluminum rod has a ±.025 in outer diameter tolerance, which means it can range from 0.475 to 0.525 in.
   - The shaft collars don't give a tolerance for their inner diameters.
   - The flanged sleeve bearings say +.001 to +.002 in for the inner diameter. This means they will be between 0.501 to 0.502 in.
   - The thrust bearing says 1/2 in +0.002 to +0.007, which means the inner diameter can range from 0.502 to 0.507 in.
   - The thrust washers don't give any tolerance for the inner diameter.

   This means that the outer diameter of the aluminum rod needs to be smaller than the smallest possible part it needs to fit into, which is the 0.501 in sleeve bearing. As you can see here, we have a good possibility for overlap in an inconvenient direction.

4. If your aluminum rod is too big for the sleeve bearing, put on your safety glasses, dust mask, and gloves (aluminum dust is not good for you). Grab the aluminum rod with the sandpaper and rotate it while you're squeezing until you see aluminum dust coming off. Continue this until the rod fits through all the components.

NOTE ***If you're lucky enough to have access to a lathe, it could be a big time-saver when you have a lot of aluminum to shave off. A bench grinder will work faster than sanding by hand, but it will be harder to maintain the round shape of the rod.***

5. Assemble the base (refer to the full assembly in Figure 10-17 as you go through the steps). Start with the two disks, the hex standoffs, and the 1/4-20 screws, lock washers, and washers. Install the standoffs by sandwiching the acrylic disk, a washer, and a lock washer on each end with a 1/4-20 screw (see Figure 10-12).

6. Install one of the flanged sleeve bearings in the center hole of the base disk. The base is the one without the four holes to mount the motor.

FIGURE 10-12 Detail of installing the standoffs

FIGURE 10-13 The part stackup inside the base

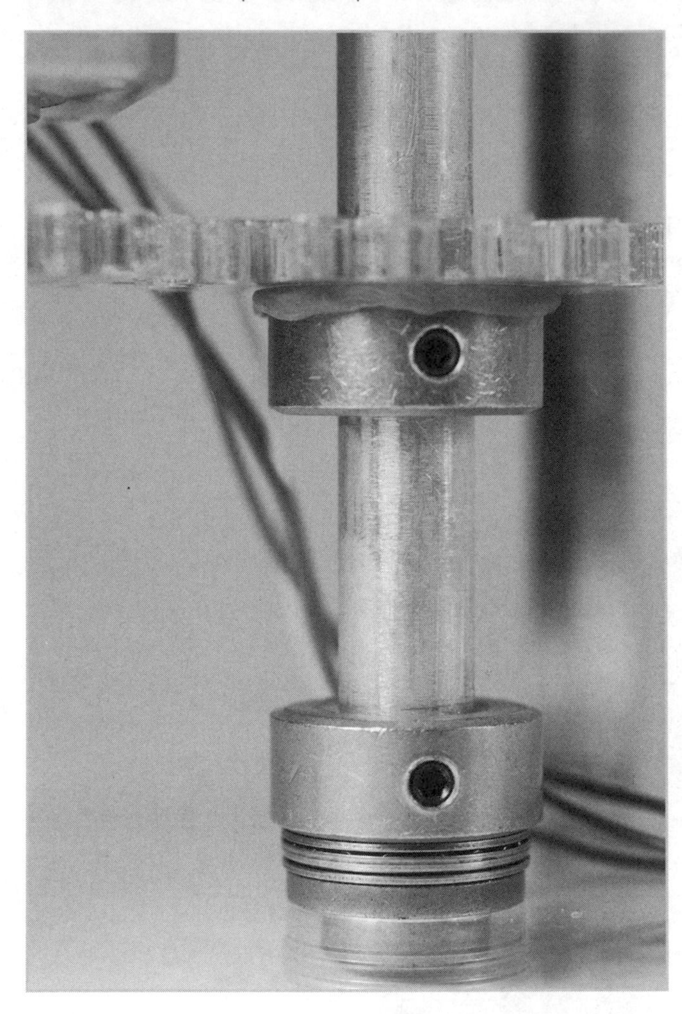

7. Rest a thrust washer, thrust bearing, and then the other thrust washer on top of the flange. The inside of the base will start to look like Figure 10-13.

8. Slide the aluminum rod in from the top. Before it hits the sleeve bearing on the bottom, it should slide through the other sleeve bearing, a 1/2 in shaft collar, a laser-cut gear, two more 1/2 in shaft collars, and finally the thrust washer, bearing, washer stack.

9. Pull up slightly on the aluminum rod so it's not hitting your work surface. Use your Allen key set to tighten the set screw in the lowest shaft collar. At this point, the shaft collar is resting on the thrust bearing and attached to the aluminum rod, so you should be able to spin the rod.

10. Lift the next shaft collar from the bottom up with the gear to about the halfway point inside the base. Tighten the set screw. This shaft collar will be attached to the gear with epoxy putty later, but *do not* do this yet.

11. Secure the top sleeve bearing with the top shaft collar, as shown in Figure 10-14.

FIGURE 10-14 Top sleeve bearing secured with shaft collar under the top disk

12. Before you continue up the rod, this is a good time to mount your motor. First, cut the wires to about 8 in long and solder a set of four male headers to the wires (just as in Project 6-9). Red and green should be next to each other on one side, and blue and yellow on the other.

13. Remove the screws that hold the motor together. Use the longer M3 screws from the shopping list to mount the motor from the back, on the underside of the top disk. Sandwich an M3 washer and lock washer with each screw (see Figure 10-15).

14. Slide the other gear onto the motor shaft and use the 5mm shaft collar to secure it temporarily. Adjust the height of both shaft collars until the gears are at the same height and mesh well. Now you can break out the epoxy putty and secure the gears to their respective shaft collars.

FIGURE 10-15 Stepper motor mounted

15. Continue up the aluminum rod. Slide on a 1/2 in shaft collar, one of the plastic sail holders, and then another 1/2 in shaft collar. Pull the lower shaft collar up so it's not resting on the top of the base and secure it to the rod with its set screw. Then pinch the plastic sail holder with the shaft collar on top of it, and secure the assembly with a set screw. When you rotate the whole assembly by the shaft, it should rotate smoothly, and the sail holders should rotate with the shaft.

16. Cut out three sails for your wind turbine to catch the wind. There's no right answer here, and you have a few different slots in the sail holders, so just use scissors to cut the aluminum flashing in a length you think will work. Then cut 1/2 in tabs into each corner to slide into the slots. Bend over the tabs to secure the sails (see Figure 10-16).

FIGURE 10-16 Bend the tabs in the aluminum flashing over to secure the sails.

FIGURE 10-17 Fully assembled wind lantern

17. Do the same shaft collar, sail holder, shaft collar assembly on the top of the sail to finish this section of the build. Your project should look like Figure 10-17 without the breadboard at this point. It should spin with very little friction when you turn it by hand with the aluminum rod.

18. Now, the electronics. We need to create a circuit like the one shown in Figure 10-18. Use the eight diodes and jumper wires to create this circuit on your breadboard as shown in Figure 10-19. It will tell any electricity generated in each coil to go to the same place: the power column on the right side of the breadboard. Make sure all your diodes are facing the right direction, and don't forget to jump the ground columns across the board.

FIGURE 10-18 Rectifier circuit for a bipolar stepper motor

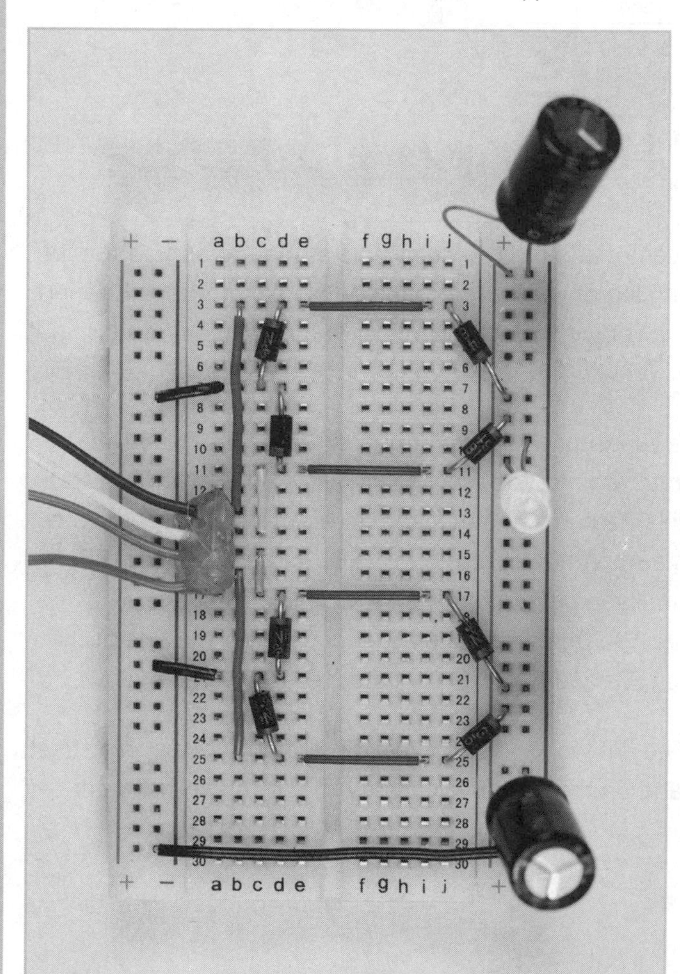

FIGURE 10-19 Rectifier circuit created on a breadboard

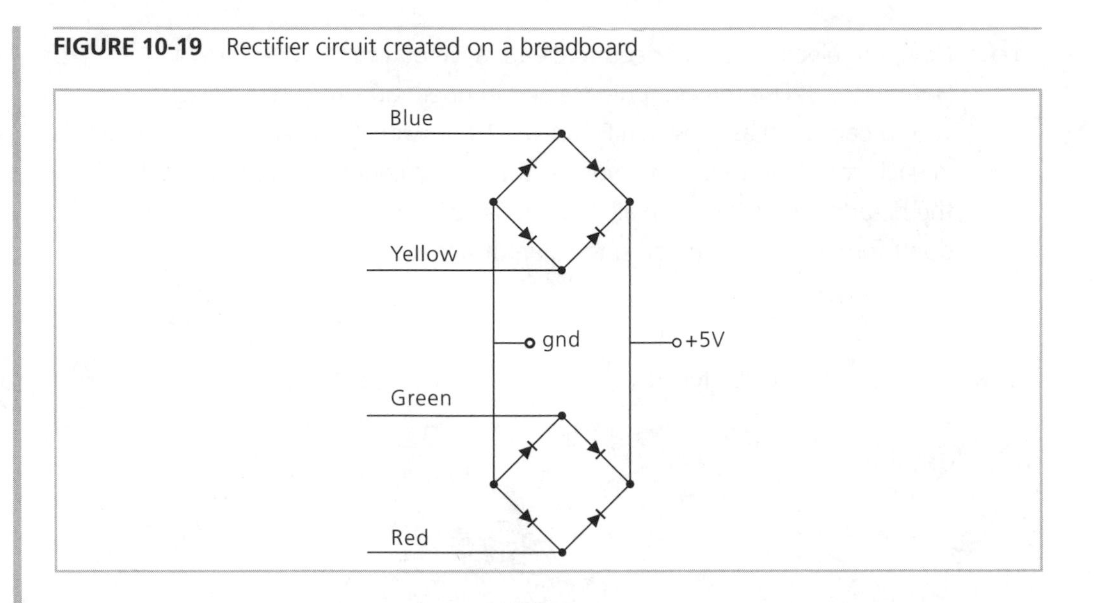

19. Notice the LED in the center and the two capacitors at the top and bottom of the board. Plug the long leg of your LED into the power column and the short one into ground. Before you add the capacitors, give the wind lantern a spin and watch the LED flicker!

20. Try adding at least one capacitor as shown in Figure 10-18. The negative marked side should go to ground, the other to power. The capacitor will store energy while the wind lantern is creating it, and release it when it is not. The resulting effect here is a smoother flicker on the LED. Try adding more LEDs and more capacitors until you get a smooth glow when you spin the aluminum rod. In Figure 10-20, we placed diffuser paper over the side of the lantern and used three LEDs and two capacitors to create a pleasing glow.

21. Now take it outside! See if it works with real wind. We had success on a street corner in Manhattan and on the roof of Eyebeam Art + Technology Center's two-story building.

FIGURE 10-20 Wind lantern at work

Project 10-3: SADbot: The Seasonally Affected Drawing Robot

SADbot was created in collaboration with Ben Leduc-Mills for the window gallery at the Eyebeam Art + Technology Center. The main idea was to use solar energy to power a drawing machine that could interact with people outside the window through light sensors. You can re-create this project to install in your own window at home.

SADbot gets its name from the source of its power: the sun. Since the motors are solar-powered, they will run only if it has been sunny enough to store solar energy in the battery. When the batteries run low and the SADbot motors stop running, SADbot appears sad because it has to wait for the sun to come out before it can keep drawing.

Shopping List:

- Electronics
    - Multimeter
    - Arduino with USB cable and AC adaptor
    - Soldering iron, stand, and solder
    - Three small breadboards (like All Electronics PB-400)
    - Jumper wires (like SparkFun PRT-00124)
    - Hook-up wire: red, black, and white (SparkFun PRT-08023, PRT-08022, and PRT-08026)
    - Two stepper motors (SparkFun ROB-09238)
    - Two EasyDrivers (SparkFun ROB-09402)
    - Male header pins (SparkFun PRT-00116)
    - Four photocells (1KΩ – 10kΩ: SparkFun SEN-09088)
    - Four 1KΩ resistors (SparkFun COM-08980)

NOTE ***You can also use a 1KΩ – 10kΩ photocell (SparkFun SEN-09088). In that case, you should use a 1KΩ resistor (SparkFun COM-08980) to get the best response.***

- Benchtop power supply for testing
- 12V 5Ah SLA battery (PS-1250 F1 from Microbattery.com, www.microbattery.com)
- Solar charge controller (SKU 06-1024 from Silicon Solar, www.siliconsolar.com)
- 12V 7W solar battery charger panel (Silicon Solar SKU 9358)

- Hardware
    - Large plywood or other wooden board to use for canvas (around 3 ft × 2 ft will work well)
    - Eight M3 screws, 20mm length (McMaster 92095A185)
    - Eight M3 lock washers (McMaster 92148A150)
    - One pack M3 washers (McMaster 91116A120)
    - Drill (either portable or drill press) and drill bits: 3/8 in, 1/8 in
    - Diagonal cutters (like SparkFun TOL-08794)
    - Two 3D printed pulleys

***NOTE** See the resources section on www.makingthingsmove.com for a link to a file you can download, as well as a link to where you can buy the pulleys.*

- Spring clamp (like McMaster 5107A1) that will hold the marker
- Black marker
- Monofilament fishing line
- Large white paper

Recipe:

1. Prepare two stepper motor/EasyDriver/breadboard assemblies as you did in Project 6-9. Use a benchtop power supply to get 12V power and ground to the GND and M+ pins on each of the EasyDrivers. Set them up to interface with the Arduino as shown in Figure 10-21.

2. For the left motor:
    - Arduino GND to GND on left EasyDriver
    - Arduino pin 11 goes to DIR
    - Arduino pin 12 goes to STEP

FIGURE 10-21 Arduino setup to run two stepper motors

3. For the right motor:

- Arduino GND (one of the two pins) to GND on right EasyDriver
- Arduino pin 6 goes to DIR
- Arduino pin 7 goes to STEP

4. Make sure the two stepper motors work. Type in the following code, verify it, and upload it to the Arduino.

```
/*
Driving two stepper motors with an Arduino through
SparkFun's EasyDriver v4.3
By Ben Leduc-Mills and Dustyn Roberts
Created: 2010.06
*/

#include <Stepper.h>  //import stepper library
```

```
#define STEPS 200  // 360/1.8 (step angle) = 200 steps/revolution

//declare new stepper objects from stepper library (one per motor)
Stepper right_motor(STEPS, 6, 7);  //6=DIR, 7=STEP
Stepper left_motor(STEPS, 11, 12);  //11=DIR, 12=STEP

void setup()
{
  //set motor speeds (in RPM)
  right_motor.setSpeed(200);
  left_motor.setSpeed(200);
}

void loop()
{
  //step each motor every time through the loop
  right_motor.step(10);
  left_motor.step(10);
  delay(10);  //gives the motor a chance to get to new step
}
```

5. If the code works, your motors should just start spinning. Attach some tape flags to the motor shafts to help indicate what's going on.

6. Play around with the voltage setting on the power supply. You'll notice that the motors actually run at much lower than 12V. In fact, our motors would still spin as low as 3.7V and draw about 120 mA of no load current while there. At the full 12V, they ran at 240 mA no load, and jumped to about only 250mA when we attempted to stall the motor by hand. Since we're using solar energy to charge batteries in this project, it's good to know that the motors can handle some variability and aren't too hungry for current.

7. Now make the motors respond to photocells. First, get out the third breadboard and wire up the photocells as you did in Project 6-8. Each photocell should have one leg connected to the power column and one leg connected to the ground column through a resistor (see Figure 10-22).

8. The leg going to ground should also go to one of the ANALOG IN pins on the Arduino. From left to right, connect the ground legs of the photocells to pins 0, 1, 2, and 3 on the Arduino, which correspond with up, down, left, and right in the code, respectively.

FIGURE 10-22 Two steppers wired to photocells

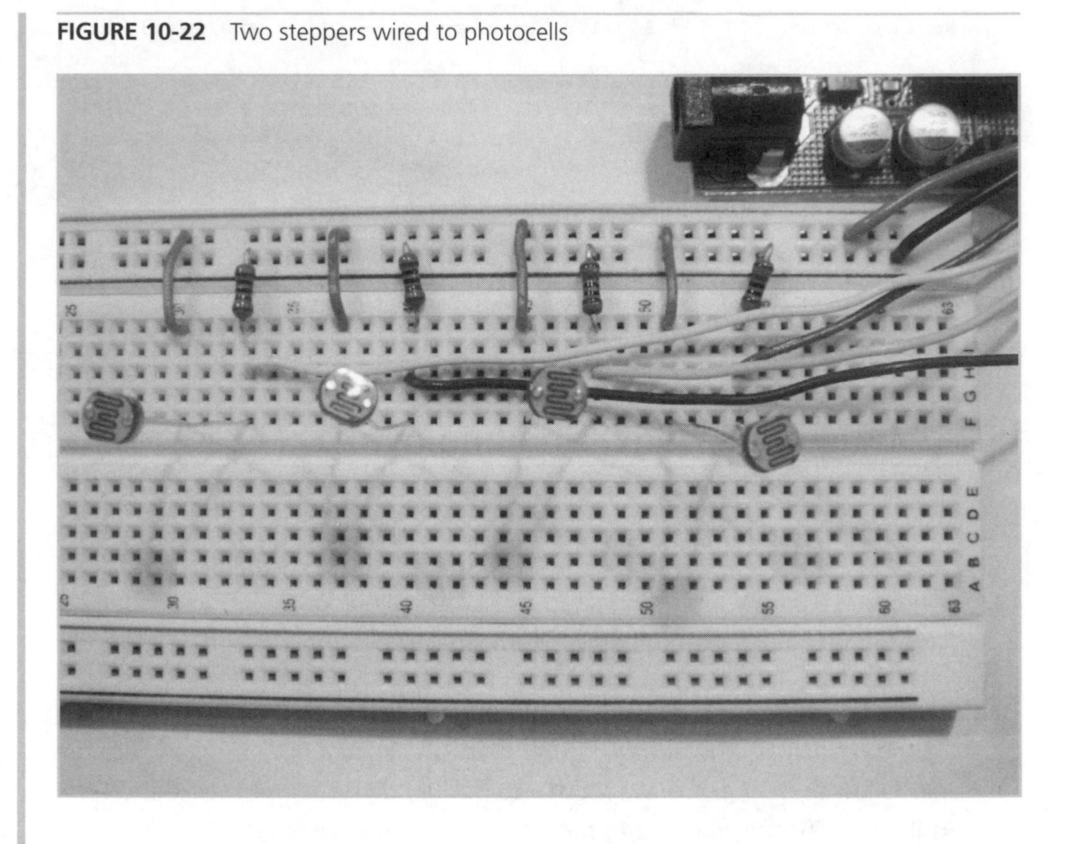

9. Jump power and ground to the breadboard from the Arduino GND and 5V pins. See Figure 10-23 for the full setup.

10. Now we will try some code that uses the photocells to move the stepper motors. Type in the following code, verify it, and upload it to the Arduino.

```
/*
Using photocells to drive two stepper motors
with an Arduino through SparkFun's EasyDriver v4.3
CC-GNU GPL by Ben Leduc-Mills and Dustyn Roberts
Created: 2010.06
*/
```

FIGURE 10-23 Full SADbot circuit-testing setup

```
#include <Stepper.h>  //import stepper library

#define STEPS 200  // 360/1.8 (step angle) = 200 steps/revolution

//declare new stepper objects from stepper library (one per motor)
Stepper right_motor(STEPS, 6, 7);  //6=DIR, 7=STEP
Stepper left_motor(STEPS, 11, 12);  //11=DIR, 12=STEP

int distance; // how far motors should go
int lowest;  // variable to store lowest photocell value
int i;  // for looping

// variables for 4 photocell values
int photo_up;
int photo_down;
```

```
int photo_left;
int photo_right;

void setup()
{
  Serial.begin(9600);  //start serial printout so we can see stuff

  // set motor speeds (in RPM)
  right_motor.setSpeed(200);
  left_motor.setSpeed(200);
}

void loop()
{
  //read and print all photocell values from analog pins 0-3
  photo_up = analogRead(0);
  Serial.print("up");
  Serial.println(photo_up);

  photo_down = analogRead(1);
  Serial.print("down");
  Serial.println(photo_down);

  photo_left = analogRead(2);
  Serial.print("left");
  Serial.println(photo_left);

  photo_right = analogRead(3);
  Serial.print("right");
  Serial.println(photo_right);

  delay(1000); //give me time to read them in the monitor

  //store photocell values in an array
  int photoValues[]= {photo_up, photo_down, photo_left, photo_right};

  lowest = 9999; //set this higher than possible photocell values

  //loop to find lowest photocell value
  for(i = 0; i < 4; i++)  //4 = number of photocells
    {
    Serial.println(photoValues[i]);  //prints out photoValue array
```

```
      //assign actual photocell value to "lowest" variable if it's lower
      //than whatever "lowest" is set to (starts at 9999)
      if (lowest >= photoValues[i] )
        {
        lowest = photoValues[i];
        }

      //print it out to confirm that the lowest value is being selected
      Serial.print("lowest:");
      Serial.println(lowest);

      delay(1000);  //wait one second before looping so we can read the values

      }//end for

    distance = lowest;  //set travel distance variable = lowest value

    //find the sensor that matched the lowest, go that direction
    //see below for what the up, down, left, right functions do
    if (lowest == photoValues[0])
      {
       up( distance );
      }
    else if (lowest == photoValues[1])
      {
       down( distance );
      }
    else if (lowest == photoValues[2])
      {
       left( distance );
      }
    else if (lowest == photoValues[3])
      {
       right( distance );
      }
}//end loop

/*
Here are the directional functions.  Loop size = distance.
Positive step numbers are clockwise, negative counterclockwise
*/

void up(int distance) {
  for( i = 0; i < distance; i++){
    right_motor.step(10);
    left_motor.step(-10);
  }
}
```

```
void down(int distance) {
  for( i = 0; i < distance; i++){
    right_motor.step(-10);
    left_motor.step(10);
  }
}

void left(int distance) {
  for( i = 0; i < distance; i++){
    right_motor.step(-10);
    left_motor.step(-10);
  }
}

void right(int distance) {
  for( i = 0; i < distance; i++){
    right_motor.step(10);
    left_motor.step(10);
  }
}
```

11. Try covering up the photocells one at a time.
    - When you cover the up photocell, the left motor should turn counterclockwise and the right motor should turn clockwise.
    - When you cover the down photocell, the left motor should turn clockwise and the right motor should turn counterclockwise.
    - When you cover the left photocell, both motors should turn counterclockwise.
    - When you cover the right photocell, both motors should turn clockwise.
12. Get out your plywood board and put on your safety glasses. The stepper motor data sheet from SparkFun indicates that the motor mounting holes are 31mm apart in a square, with the shaft at the center. Make pencil marks up on the corners of your board where you want the motor shafts to go, and then measure and make marks at each corner of a 31mm square centered on that motor shaft mark. Plan your motor mount so it looks like Figure 10-24.

FIGURE 10-24 Motors mounted at the top corners of the plywood canvas

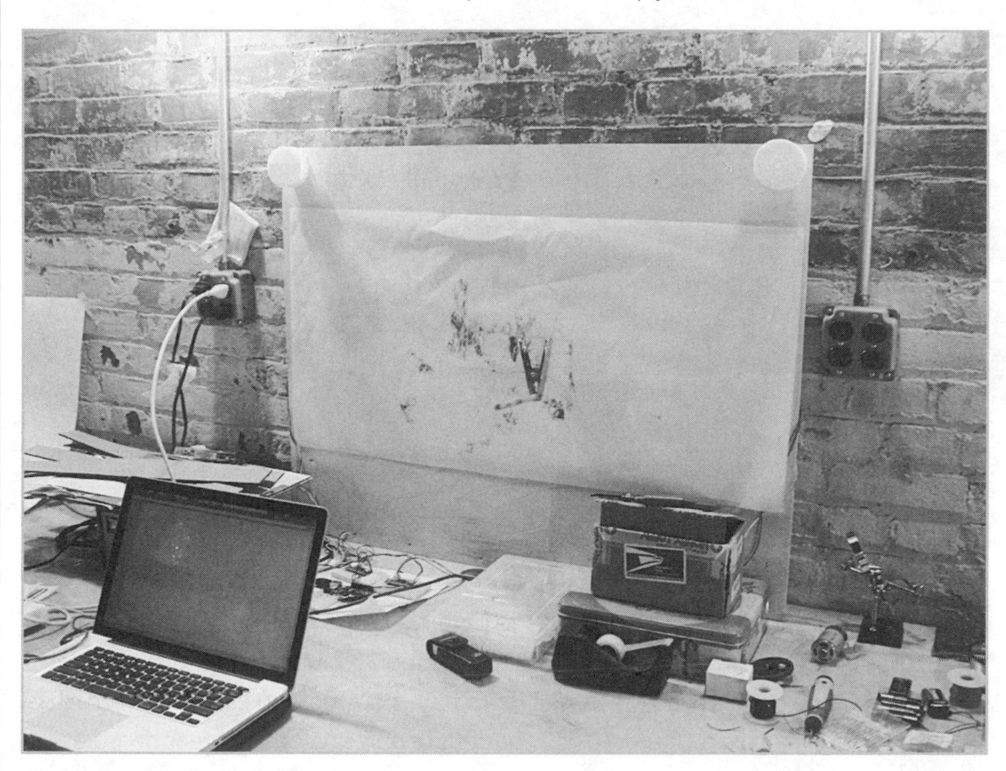

13. Use the 1/8 in drill bit to drill out the clearance holes for the M3 screws the motor will mount with. Use the 3/8 in drill bit to drill out the center hole. Now mount your motors with the M3 screws, lock washers, and washers, as shown in Figure 10-25. You may need fewer washers, depending on the board thickness you're using.

14. Once both motors are mounted, confirm the circuit still works as intended and none of the wires in your circuit have come loose. Press the pulleys onto each motor shaft.

FIGURE 10-25 Detail of motor mounting

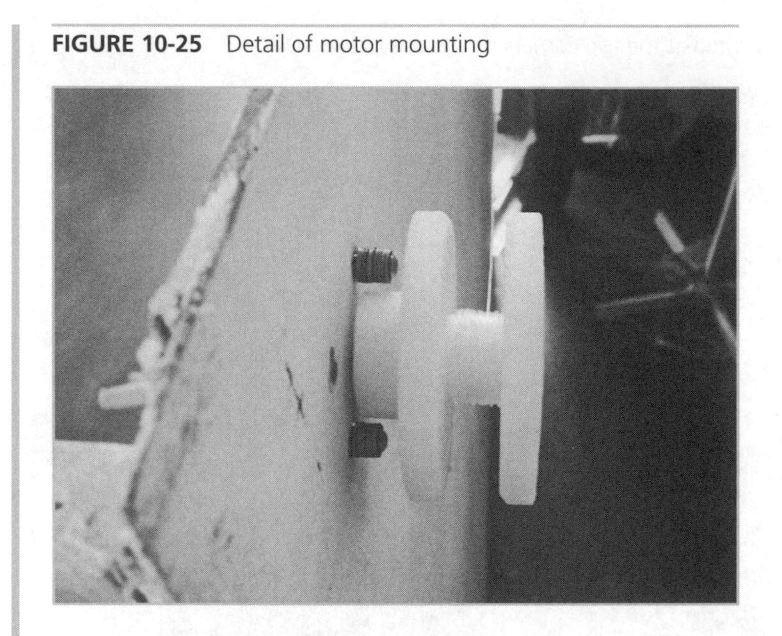

15. Now let's get SADbot running off the battery charged by the solar panels through the charge controller (the Arduino should plug into the wall through the AC adaptor). First, simplify the wiring by choosing one EasyDriver board as the power hub (see Figure 10-26). Designate one side of the breadboard as the hub, and then use small jumper wires to jump the GND and M+ pins of the EasyDriver to this hub. Connect the power and ground columns of this hub to GND and M+ on the other EasyDriver breadboard with long lengths of hook-up wire.

16. Cut some more hook-up wire and connect this power hub to the two far-left screw terminals on the charge controller (refer to Figure 10-27).

17. Solder hook-up wire to the battery terminals (red for positive and black for negative). Use the screw terminals in the center of the charge controller to hold the stripped ends of these wires.

FIGURE 10-26 Simplify wiring so the power hub is on one EasyDriver board.

FIGURE 10-27 Charge controller indicating connections

18. Cut off the RC plug that comes on the solar charger wires. Separate the two wires and strip the insulation off the ends to expose about 1/4 in of wire. To figure out which one is positive and which one is negative, use your multimeter. The black lead should be in the COM connection, and the red lead in the voltage-measuring connection, just as for testing batteries (see Figure 5-1 in Chapter 5). Now touch the red lead to one of the solar panel wires and the black lead to the other. If the reading on your multimeter is positive, you guessed right. If not, you guessed wrong.

19. Do this with both solar panels and squish both negative wires in the charge controller on the remaining negative terminal. Each positive wire gets its own screw terminal on the far right of the charge controller (see Figure 10-27).

20. If you've done everything right, your motors should be moving! Make sure the switch on the charge controller is on. You'll want to put your solar panels in a sunny spot so the battery can charge as the motors use up the initial charge.

21. Now let's make SADbot draw something interesting. Cut two approximately 5 ft lengths of fishing line and tie each on through the hole in the center of the pulleys. Tie the other ends of the line to a spring clamp and clamp onto a marker. Mount the white paper on your canvas board with tape or clamps (refer to Figure 10-24).

22. Let's tell the stepper motors to draw in random directions by default, and then to behave as they did before when someone is covering a photocell. This will create a drawing like the ones SADbot made in Eyebeam's window gallery (see Figure 10-28). You can install SADbot the same way if you just solder some hook-up wire to each photocell and position them however you want.

23. Type, verify, and then upload the following code to the Arduino.

```
/*
SADbot v.03

SADbot will draw a random distance in random direction until a
photocell is blocked. When SADbot detects a photocell has been blocked, it
will draw towards it. Stepper motors are driven through SparkFun's
EasyDriver v4.3
```

FIGURE 10-28 SADbot's first drawing in the Eyebeam window gallery

```
CC-GNU GPL by Ben Leduc-Mills and Dustyn Roberts
Created: 2010.06
*/

#include <Stepper.h> //import stepper library

#define STEPS 200  // 360/1.8 (step angle) = 200 steps/revolution

//declare new stepper objects from stepper library (one per motor)
Stepper right_motor(STEPS, 6, 7);  //6=DIR, 7=STEP
Stepper left_motor(STEPS, 11, 12);  //11=DIR, 12=STEP

int distance;  // how far motors should go
int lowest;  // to store lowest photocell value
int i;  // for looping
```

```
// variables for 4 photocells
int photo_up;
int photo_down;
int photo_left;
int photo_right;

// Set canvas size. 1000 steps is roughly .4 inch
#define CANVASWIDTH 32000
#define CANVASHEIGHT 20000

//total distance for bounds checking
//SADbot starts at center (canvaswidth/2 and canvasheight/2)
float totalWidth = CANVASWIDTH /2;
float totalHeight = CANVASHEIGHT /2;

int randomDirection;
int randomDistance;

void setup()
{
  Serial.begin(9600);  //start serial printout so we can see stuff

  // set motor speed (in RPM)
  right_motor.setSpeed(200);
  left_motor.setSpeed(200);

  //use random seed to get better random numbers
  //*set to an analog pin that you're not using
  randomSeed(analogRead(4));

}// end setup

void loop()
{
  //read and print all sensor values from analog pins 0-3
  photo_up = analogRead(0);
  Serial.print("up");
  Serial.println(photo_up);

  photo_down = analogRead(1);
  Serial.print("down");
  Serial.println(photo_down);

  photo_left = analogRead(2);
  Serial.print("left");
  Serial.println(photo_left);
```

```
  photo_right = analogRead(3);
  Serial.print("right");
  Serial.println(photo_right);

  delay(1000); //give me time to read them in the monitor

  //before drawing, check our totalHeight and totalWidth
  Serial.print("totalHeight:");
  Serial.println(totalHeight);
  Serial.print("totaWidth:");
  Serial.println(totalWidth);

  delay(1000);  //give me time to read them in the monitor

  //store photocell values in an array
  int photoValues[]= {photo_up, photo_down, photo_left, photo_right};

  lowest = 9999; //set this higher than possible photocell values

  //loop to find lowest photocell value
  for(i = 0; i < 4; i++) //4 = number of sensors
    {
    Serial.println(photoValues[i]);  //prints out photoValue array

    //assign actual photocell value to "lowest" variable if it's lower
    //than whatever "lowest" is set to (starts at 9999)
    if (lowest >= photoValues[i] )
      {
      lowest = photoValues[i];
      }

    //print it out to confirm that the lowest value is being selected
    Serial.print("lowest:");
    Serial.println(lowest);

    delay(1000);  //wait one second before looping so we can read the values

    }//end for

  distance = lowest;  //set travel distance = lowest value

  //if lowest value indicates a covered photocell, draw towards lowest
  if (lowest < 550 )
    {
    //find the sensor that matched the lowest, go that direction,
    //but only if SADbot is within the bounds of the canvas
    if ((lowest == photoValues[0]) && ((totalHeight + distance) <
CANVASHEIGHT))
```

```
      {
      up( distance );
      totalHeight += distance;  //increment totalHeight variable
      }
    else if ((lowest == photoValues[1]) && ((totalHeight - distance) > 0))
      {
      down( distance );
      totalHeight -= distance;  //decrement totalHeight variable
      }
    else if ((lowest == photoValues[2]) && ((totalWidth - distance) > 0))
      {
      left( distance );
      totalWidth -= distance;  //decrement totalWidth variable
      }
    else if ((lowest == photoValues[3]) && ((totalWidth + distance) <
CANVASWIDTH))
      {
      right( distance );
      totalWidth += distance;  //increment totalWidth variable
      }
    }//end if

    //otherwise, no one is covering any sensors, draw according to random
    else
      {
      //pick random number 1 through 9 to map to direction
      randomDirection = random(1, 9);
      Serial.print("random direction:");
      Serial.println(randomDirection);

      //pick random number 1 through 200 to map to distance
      randomDistance = random(1, 200);
      Serial.print("random distance:");
      Serial.println(randomDistance);

      //directions for any randomDirection value generated
      switch (randomDirection)
        {
        case 1:  //go up
          if((totalHeight + randomDistance) < CANVASHEIGHT)
            {
            up(randomDistance);
            totalHeight += randomDistance;
            }
          break;
```

```
        case 2: //go down
          if((totalHeight - randomDistance) > 0)
            {
            down(randomDistance);
            totalHeight -= randomDistance;
            }
          break;

        case 3: //go left
          if((totalWidth - randomDistance) > 0)
            {
            left(randomDistance);
            totalWidth -= randomDistance;
            }
          break;

        case 4: //go right
          if((totalWidth + randomDistance) < CANVASWIDTH)
            {
            right(randomDistance);
            totalWidth += randomDistance;
            }
          break;

        case 5: //go upRight
          if(((totalWidth + randomDistance) < CANVASWIDTH) && ((totalHeight
+ randomDistance) < CANVASHEIGHT))
            {
            upRight(randomDistance);
            totalWidth += randomDistance;
            totalHeight += randomDistance;
            }
          break;

        case 6: //go upLeft
          if(((totalWidth - randomDistance) > 0) && ((totalHeight +
randomDistance) < CANVASHEIGHT))
            {
            upLeft(randomDistance);
            totalWidth -= randomDistance;
            totalHeight += randomDistance;
            }
          break;
```

```
        case 7: //go downRight
          if(((totalWidth + randomDistance) < CANVASWIDTH) && ((totalHeight
- randomDistance) > 0))
            {
            downRight(randomDistance);
            totalWidth += randomDistance;
            totalHeight -= randomDistance;
            }
          break;

        case 8: //go downLeft
          if(((totalWidth - randomDistance) > 0) && ((totalHeight -
randomDistance) > 0))
            {
            downLeft(randomDistance);
            totalWidth -= randomDistance;
            totalHeight -= randomDistance;
            }
          break;

        default:  //just in case
          left(0);

      } //end switch

  } //end else

} //end loop()

/*
Here are the directional functions.  Loop size = distance.
Positive step numbers are clockwise, negative counterclockwise
*/

void up(int distance)
{
  for( i = 0; i < distance; i++) {
    right_motor.step(1);
    left_motor.step(-1);
    }
}

void down(int distance)
{
  for( i = 0; i < distance; i++) {
    right_motor.step(-1);
    left_motor.step(1);
    }
}
```

```
void left(int distance)
{
  for( i = 0; i < distance; i++) {
    right_motor.step(-1);
    left_motor.step(-1);
    }
}

void right(int distance)
{
  for( i = 0; i < distance; i++) {
    right_motor.step(1);
    left_motor.step(1);
    }
}

void upRight(int distance)
{
  for( i = 0; i < distance; i++) {
    right_motor.step(2);
    left_motor.step(-.2);
    }
}

void upLeft(int distance)
{
  for( i = 0; i < distance; i++) {
    right_motor.step(.2);
    left_motor.step(-2);
    }
}

void downRight(int distance)
{
  for( i = 0; i < distance; i++) {
    right_motor.step(-.2);
    left_motor.step(2);
    }
}

void downLeft(int distance)
{
  for( i = 0; i < distance; i++) {
    right_motor.step(-2);
    left_motor.step(.2);
    }
}
```

24. Enjoy! Play with the photocells so you can interact with SADbot and make interesting drawings.

References

1. Tom Igoe, "Phototransistors Photocells" (http://itp.nyu.edu/physcomp/sensors/Reports/PhototransistorsPhotocells).

2. Dan O'Sullivan and Tom Igoe, *Physical Computing: Sensing and Controlling the Physical World with Computers* (Boston: Thomson, 2004).

3. Jeff Feddersen, NYU/ITP Sustainable Energy class notes, spring 2010.

4. Jeff Feddersen, "Anatomy of a Stepper Motor" (http://itp.nyu.edu/sustainability/energy/texts/Rectification.pdf).

Appendix: BreadBoard Power and Arduino Primer

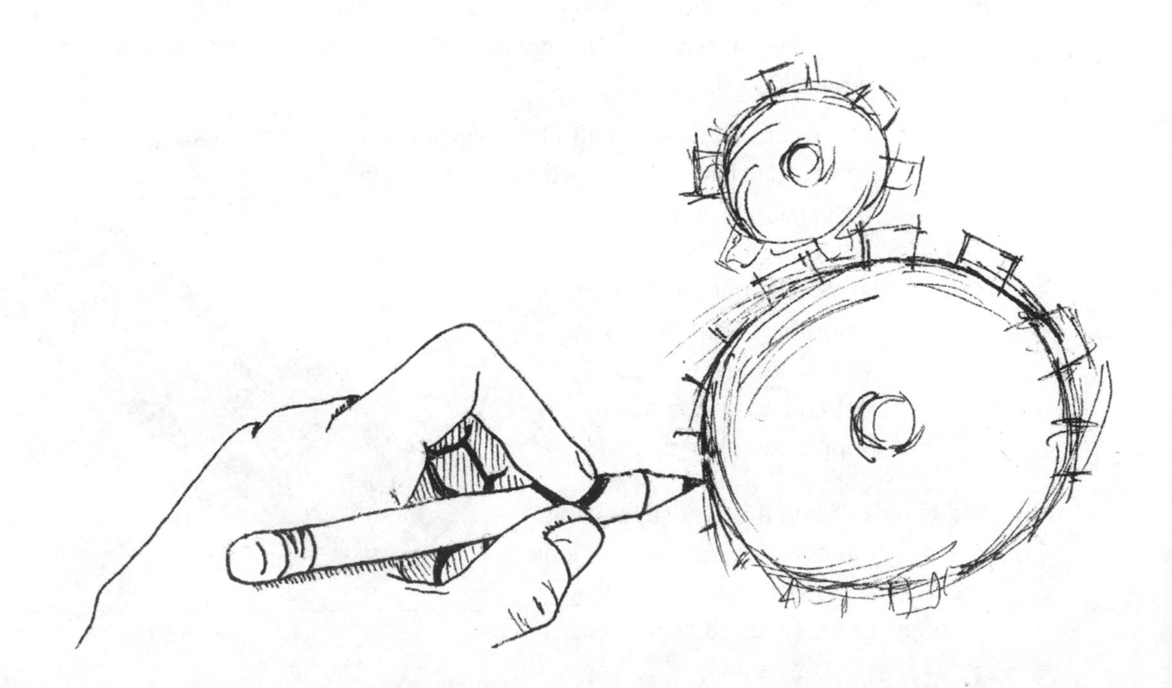

This appendix includes sections on getting power to your breadboard as well as setting up the Arduino hardware and software.

Getting Power to Your Breadboard

Directly plugging a battery pack into your breadboard is not the best way to get power to it. This method, used extensively in Chapter 6, is not technically a good practice because the four AA batteries can give you anywhere between 4V to 6.4V, depending on if the batteries are rechargeable and how new they are. However, chips like the H-bridge used in Project 6-5 are more comfortable with a steady voltage between 4.5V to 5.5V. And most motors will want a different voltage than your logic circuit. The following methods help get a more regulated power supply into your breadboard:

- Use hook-up wire to bring regulated power from a benchtop supply into your breadboard (see Project 10-3 for reference).
- See Tom Igoe's notes on soldering a power supply connector to plug directly into your breadboard and use with an AC adaptor (http://itp.nyu.edu/physcomp/Labs/Soldering).
- Use one of SparkFun's (www.sparkfun.com) breadboard power supplies (like PRT-09319, shown in Figure A-1) to give you a reliable 5V supply. You'll want to solder on some male header pins (PRT-00116) in order to plug it into a breadboard.
- Use Adafruit's (www.adafruit.com) adjustable breadboard power supply (version 1.0). It comes as a kit you need to assemble, but can give you smooth power from 1.25V to 20V and up to 1.25mA, depending on the power supply you plug in.

If you can't find a power supply that exactly matches the voltage your logic circuit or motor wants, there are a couple of tricks you can use. Linear

FIGURE A-1 SparkFun's breadboard power supply (PRT-09319)

regulators take a high-input voltage and make it lower. The LM7805 5V regulator (SparkFun COM-00107) is popular and converts an 8V to 15V input to a 5V output (see an example at http://itp.nyu.edu/physcomp/Labs/Breadboard). Step-up regulators do the opposite: They take a low voltage and make it higher. These are a bit more involved to use, so it's best to go with a ready-made module like SparkFun's PRT-08999, which can take in 1V to 4V and output 5V. (For an example of how to work with boost regulators, see www.ladyada.net/make/mintyboost/process.html.)

Arduino Primer

An Arduino is like a mini-computer that acts as the brains for your project once you get past just wanting to turn something on and off. Why Arduino? A handful of other development boards out there do basically the same thing. They have a microcontroller (mini-computer) on board, places to plug in stuff, and a way to get power onboard. These include MIT's Handyboard, Phidgets, the Make Controller, and others. We work with Arduino in this book for several reasons:

- **Price** At $30 per board and free software to run it, you can get up and running without making a huge investment.
- **Compatibility** It works on Windows, Mac OSX, and Linux systems.
- **Convenience** The board is very flexible and can be used to control motors, blink LEDs, and do so many other things that I won't even try to go through them here. An Arduino can be the backbone of any project that requires more smarts than a simple switch.
- **Ongoing support and development** The hardware and software are both open source and extensible, which means that you or anyone else can make improvements that might be folded into later generations of the Arduino. The Arduino system is very popular and has sold more than 150,000 modules, so it has extensive online documentation, sample code, help forums, sample projects, and so on to get you going and keep you going. The Arduino team members use and teach the platform themselves, so they are committed to constant improvements.

Make the Arduino Play Nice with Your Computer

There are two parts to the Arduino system:

- The Arduino board you hold in your hand (see Figure A-2)

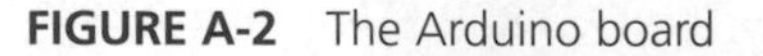

FIGURE A-2 The Arduino board

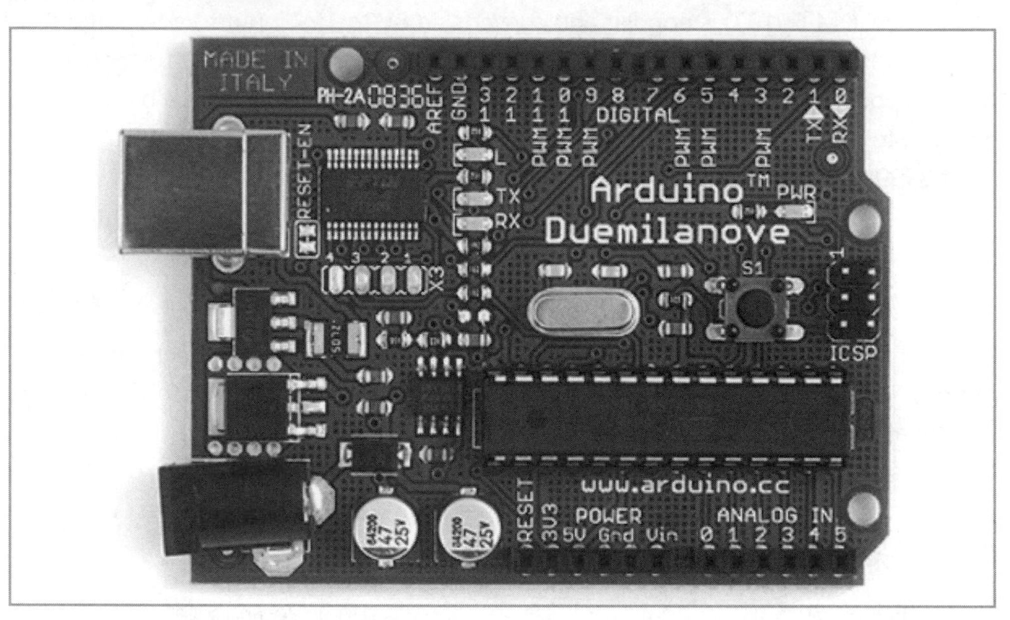

- The Arduino integrated development environment (IDE) you run on your computer

NOTE ***This introduction draws from the online guide (http://arduino.cc/en/Guide/Windows) and*** **Getting Started with Arduino** ***by Massimo Banzi (Sebastopol, CA: Make, 2008).***

Follow these steps to get the system up and running on your computer and write your first program to blink a light:

1. Get an Arduino (SparkFun DEV-00666) and a USB A to B cable (like SparkFun CAB-00512).

2. Go to www.arduino.cc, click the Download the Arduino Software link, and choose the correct download for your operating system (Windows, Mac, or Linux). This tutorial goes through download, install, and first steps with an Arduino Duemilanove on a PC running Windows Vista. (For step-by-step instructions for newer Arduino versions or on other operating systems, go to http://arduino.cc/en/Guide/ HomePage.) Download the file, uncompress it, and save that folder wherever you want, such as in your C:\Program Files folder.

3. Plug the Arduino into the computer with a USB cable. The green power light (marked PWR) on the Arduino board should go on, an orange LED (marked L) next to pin 13 might start flickering, and the Found New Hardware window will pop up. To install the drivers that let your computer talk to the Arduino board, click the recommended option that allows your computer to locate and install the drivers itself. You'll go through this twice, because there are two items that the computer needs to install before it can talk to the Arduino.

4. Once the drivers have been installed successfully, launch the Arduino application by double-clicking the icon in the folder you downloaded in step 2. Check to make sure the correct model of Arduino is selected by selecting Tools | Board from the menu bar and choosing the Arduino Duemilanove, as shown in Figure A-3.

FIGURE A-3 Select the correct board from the Arduino application.

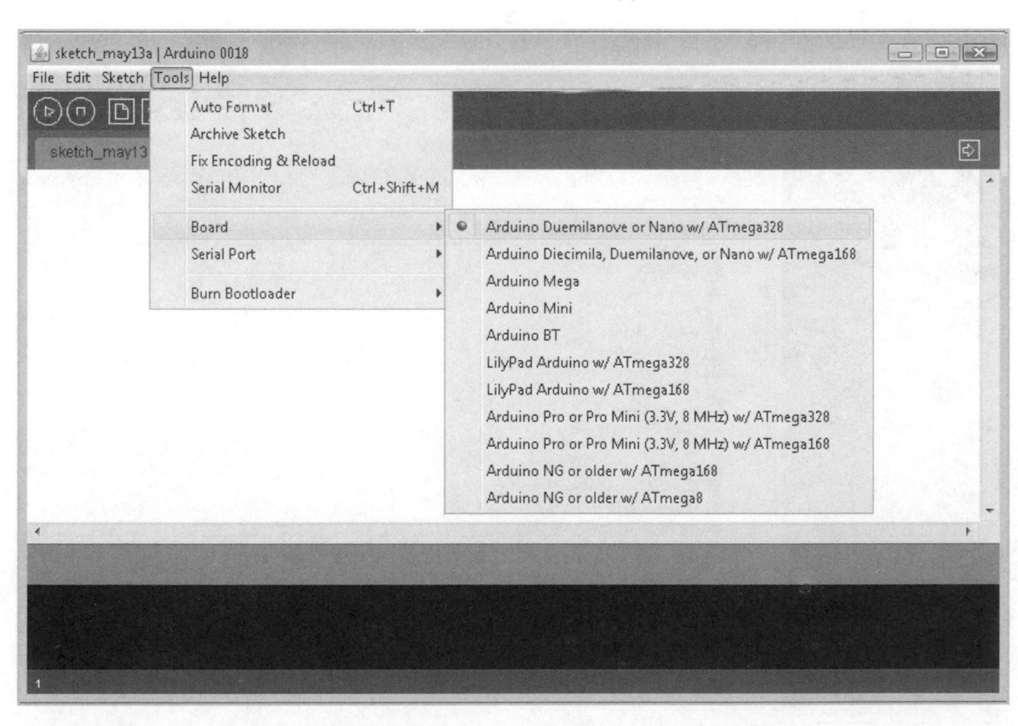

5. Now you need to figure out to which port on your computer your Arduino is hooked up. Open the Device Manager by clicking the Start menu, right-clicking Computer, and choosing Properties. Click Device Manager in the upper-left corner. Expand Ports (COM & LPT). The Arduino will appear as a USB serial port and will have a name like COM3 or COM4, as shown in Figure A-4. Now go back to the Arduino application and choose that port from the Tools | Serial Port menu.

Now everything is set up, and your Arduino and your computer can communicate with each other.

Now Make It Blink

To get the Arduino to do anything for you, you need to give it instructions in a language it understands. These instructions are written in a special code language

FIGURE A-4 Identifying the serial port

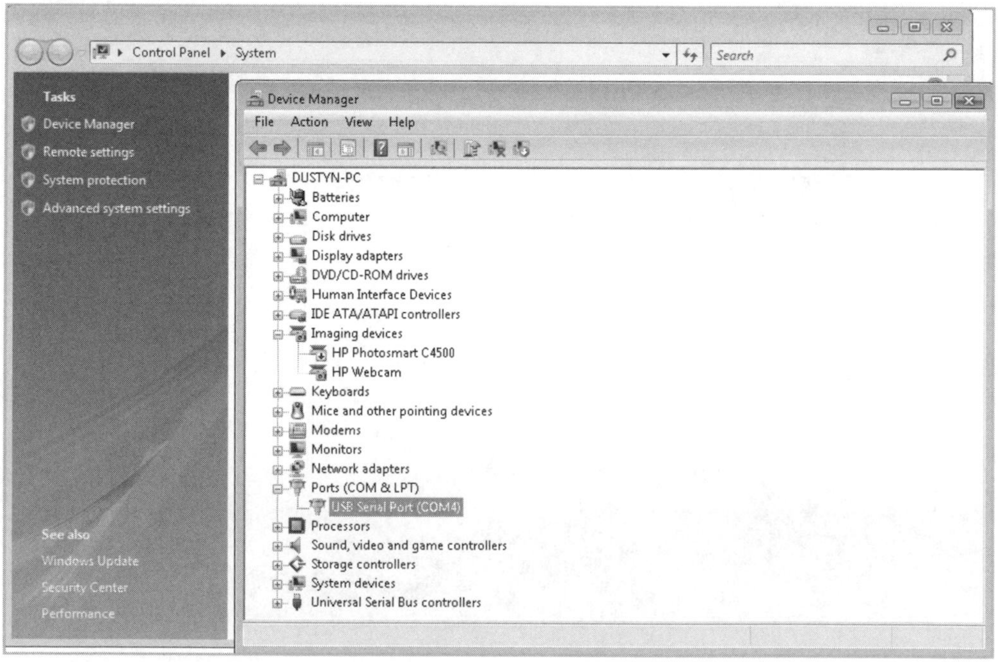

based on C/C++. A set of instructions that you create in the IDE is called a program or a *sketch*. Let's try an example to get familiar with this language.

1. Open the Blink example by navigating to File | Examples | Digital | Blink. This will open in a new window.

2. To get this sketch onto your Arduino board, first click the Verify button (see Figure A-5). This verifies that the code is correct and translates the instructions into a program that the Arduino board can run. It will say "Done Compiling" at the bottom of the Arduino application when this is finished.

FIGURE A-5 Always verify a program before uploading it.

3. Click the Upload button (see Figure A-6). This tells the Arduino to stop whatever it's doing and listen for instructions. In a few seconds, you should see the TX (transfer) and RX (receive) lights on the board flickering. A few seconds later, you should see the tiny orange LED (marked L) next to pin 13 on the board blinking orange—1 second on, 1 second off. At this point, your sketch is stored onboard the Arduino's tiny microcontroller brain. This sketch will live on the Arduino, even if you turn off the board or reset it (until you upload a new sketch).

Congratulations! Now you know how to get the Arduino to do your bidding. If you have any trouble, see the online guide at http://arduino.cc/en/Guide/Troubleshooting.

Now Make It Blink BIG

In the previous example, you probably noticed the words at the top of the sketch are gray and surrounded by a /* at the start and a */ at the end. These symbols are used to give yourself notes about the sketch, but don't actually mean anything to the Arduino. The same applies to anything following a // symbol. The next part of the sketch declares a variable, in this case telling the Arduino which pin the LED is plugged into. This LED is already part of the board, but you can use an external LED on pin 13 as well.

FIGURE A-6 Upload the sketch to your Arduino board.

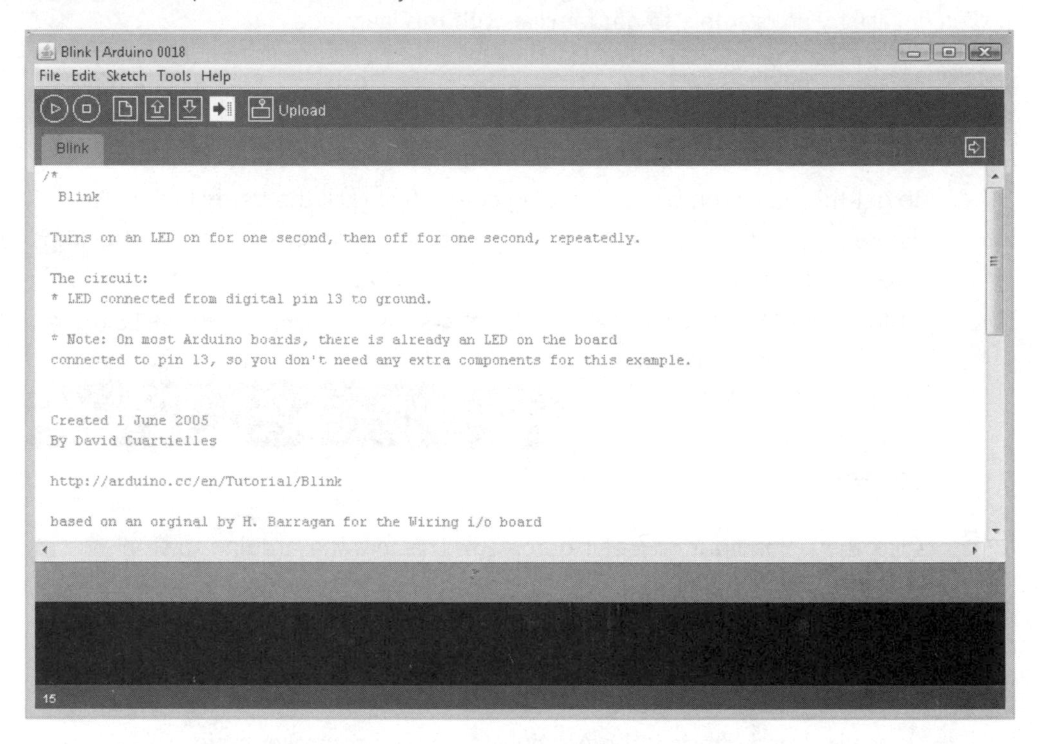

Every Arduino program has a setup method that runs once at the start of the sketch, and a loop method that runs over and over again as long as the Arduino is powered. The setup method here just sets ledPin (pin 13) as an output, so Arduino knows to do something to it instead of listening for something to happen (which would be declared as an INPUT). The loop just says to turn the ledPin HIGH (on) for a delay of 1,000 milliseconds (1,000 milliseconds = 1 second), then LOW (off) for 1 second.

Now if you plug in an LED with the long leg to pin 13 and the short leg directly next to it in ground (marked GND, see Figure A-7), your external LED should blink along with the little one. Don't try this on any other digital pin than 13. The Arduino has a resistor on the board going to pin 13 so the voltage drops to a level that satifies the LED. If you plug the long leg directly into any of the other digital inputs on that row, you'll probably fry it.

FIGURE A-7 Arduino with external LED plugged in

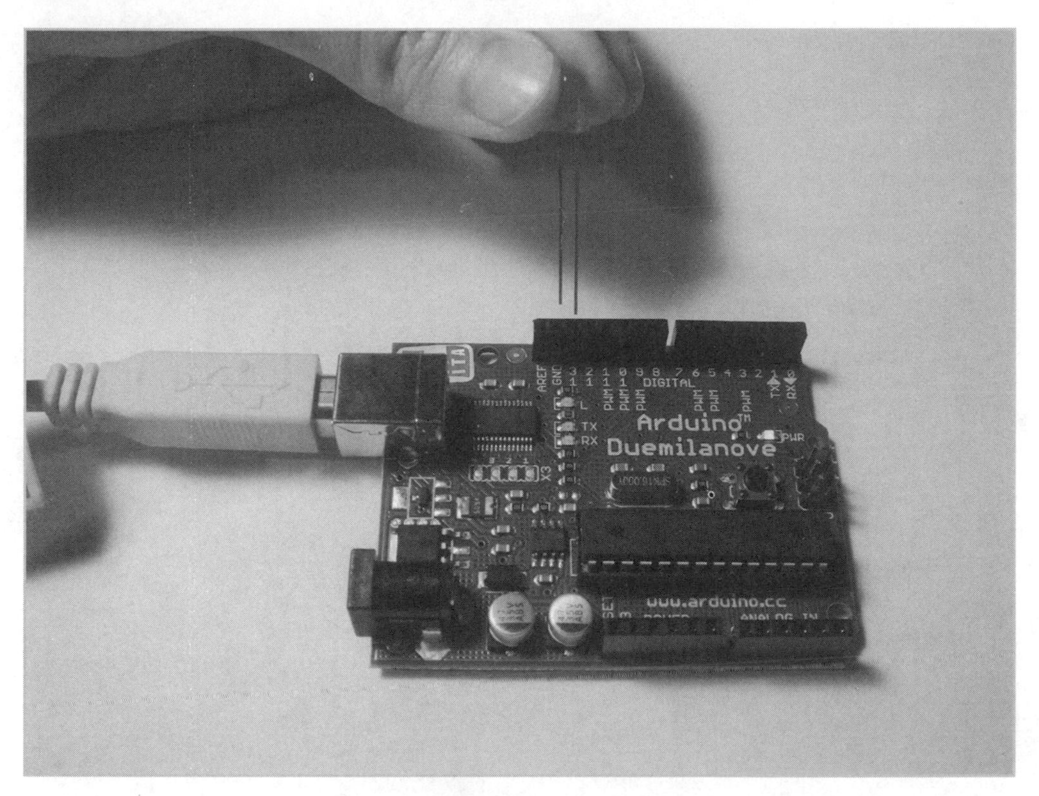

CAUTION ***Do not plug things into (or unplug things from) an Arduino while it is powered. This is bad and could give you some very weird errors, and, worst case, mess up your microcontroller chip (the brain on the Arduino). These are relatively easy to replace, but if you follow this advice, you should never have to do so.***

Now that you know how to affect the outside world with code that lives on your Arduino, you should be ready to tackle the projects in Chapter 6 and beyond. This book does not go into depth on how to write code. Check *Getting Started with Arduino* by Massimo Banzi for an Arduino-specific introduction to programming.

Index

A

D

E

F

G

H

I

J

K

L

M

N

O

P

Q

R

S

T

U

V

W

X

Y

Z